Advanced Functional Metal-Organic Frameworks

Due to the structural flexibility, large surface area, tailorable pore size and functional tenability, metal-organic frameworks (MOFs) can lead to materials with unique properties. This book covers the fundamental aspects of MOFs, their synthesis and modification, including their potential applications in different domains. The major focus is on applications including chemical, biosensors, catalysis, drug delivery, supercapacitors, energy storage, magnetics and their future perspectives.

The volume:

- Covers all aspects related to metal-organic frameworks (MOFs), including characterization, modification, applications and associated challenges
- Illustrates designing and synthetic strategies for MOFs
- Describes MOFs for gas adsorption, separation and purification, and their role in heterogeneous catalysis
- Covers sensing of different types of noxious substances in the aqueous environment
- Includes concepts of molecular magnetism, tunable magnetic properties and future aspects

This book is aimed at graduate students, and researchers in material science, coordination and industrial chemistry, chemical and environmental engineering and clean technologies.

Emerging Materials and Technologies

Series Editor: Boris I. Kharissov

The *Emerging Materials and Technologies* series is devoted to highlighting publications centered on emerging advanced materials and novel technologies. Attention is paid to those newly discovered or applied materials with potential to solve pressing societal problems and improve quality of life, corresponding to environmental protection, medicine, communications, energy, transportation, advanced manufacturing and related areas.

The series takes into account that, under present strong demands for energy, material and cost savings, as well as heavy contamination problems and worldwide pandemic conditions, the area of emerging materials and related scalable technologies is a highly interdisciplinary field, with the need for researchers, professionals and academics across the spectrum of engineering and technological disciplines. The main objective of this book series is to attract more attention to these materials and technologies and invite conversation among the international R&D community.

Emerging Applications of Carbon Nanotubes and Graphene
Edited by Bhanu Pratap Singh and Kiran M. Subhedar

Micro to Quantum Supercapacitor Devices
Fundamentals and Applications
Abha Misra

Application of Numerical Methods in Civil Engineering Problems
M.S.H. Al-Furjan, M. Rabani Bidgoli, Reza Kolahchi, A. Farrokhian, and M.R. Bayati

Advanced Functional Metal-Organic Frameworks
Fundamentals and Applications
Edited by Jay Singh, Nidhi Goel, Ranjana Verma and Ravindra Pratap Singh

Nanoparticles in Diagnosis, Drug Delivery and Nanotherapeutics
Edited by Divya Bajpai Tripathy, Anjali Gupta, Arvind Kumar Jain, Anuradha Mishra, and Kuldeep Singh

For more information about this series, please visit: www.routledge.com/Emerging-Materials-and-Technologies/book-series/CRCEMT

Advanced Functional Metal-Organic Frameworks
Fundamentals and Applications

Edited by
Jay Singh, Nidhi Goel, Ranjana Verma and
Ravindra Pratap Singh

CRC Press
Taylor & Francis Group
Boca Raton London New York

CRC Press is an imprint of the
Taylor & Francis Group, an **Informa** business

Designed cover image: © Shutterstock

First edition published 2023
by CRC Press
6000 Broken Sound Parkway NW, Suite 300, Boca Raton, FL 33487-2742

and by CRC Press
4 Park Square, Milton Park, Abingdon, Oxon, OX14 4RN

CRC Press is an imprint of Taylor & Francis Group, LLC

ISBN: 978-1-032-17164-7 (hbk)
ISBN: 978-1-032-17165-4 (pbk)
ISBN: 978-1-003-25206-1 (ebk)

DOI: 10.1201/9781003252061

Typeset in Times
by MPS Limited, Dehradun

Dedication

Dedicated
to those who care, conserve, and protect but do not destroy the beauty
and unique characteristics of
Kshit (Earth)
Jal (Water)
Pawak (Fire)
Gagan (Sky)
and
Sameera (Air)

Contents

Editors

Dr Jay Singh is currently working as an assistant professor at the Department of Chemistry, Institute of Sciences, Banaras Hindu University, Varanasi, Uttar Pradesh, since 2017. He received his PhD degree in polymer science from Motilal Nehru National Institute of Technology in 2010 and his MSc and BSc from Allahabad University, Uttar Pradesh, India. He had a postdoctoral fellow at National Physical Laboratory, New Delhi, Chonbuk National University, South Korea and Delhi Technological University, Delhi. Dr. Jay has received many prestigious fellowships like CSIR (RA), DST-Young Scientist fellowship, DST-INSPIRE faculty award, etc. He is actively engaged in the development of nanomaterials (CeO_2, NiO, rare-earth metal oxide, Ni, $Nife_2O_4$, Cu_2O, graphene, RGO, etc.), based nanobiocomposite, conducting polymer and self-assembled monolayers based clinically important biosensors for estimation of bioanalaytes such as cholesterol, xanthine, glucose, pathogens and pesticides/toxins using DNA and antibodies. Dr. Jay has published more than 80 international research papers with total citations of more than 3,600 and an h-index being 34. He has completed/running various research projects in different funding agencies. He has many edited/authored books (under pipeline) and has authored more than 20 book chapters of internationally reputed press for publications, namely Elsevier, Springer Nature, IOP, Wiley and CRC. He is actively engaged in fabricating metal oxide-based biosensors for clinical diagnosis, food packaging applications, drug delivery and tissue engineering applications. His research has contributed significantly toward the fundamental understanding of interfacial charge transfer processes and sensing aspects of metal nanoparticles.

Dr Nidhi Goel received her doctorate in chemistry from the Indian Institute of Technology Roorkee, India. After postdoctoral research (Young Scientist) from the Indian Institute of Science Bangalore, Karnataka, India, she started her independent academic career as an assistant professor in the Department of Chemistry, Institute of Science, Banaras Hindu University, Varanasi, India. Her research focuses on developing strategies, based on supramolecular building approaches, for rational construction of functional metal-organic frameworks. Their prospective uses include magnetic, luminescence and sensing applications. She is also working on topics such as drug design, medicinal chemistry and chemical biology. Dr. Goel has published 51 peer-reviewed research papers and book chapters in reputed international journals/books. Moreover, Dr. Goel is editing several books with international publishers. She has been the principal investigator on three research projects from the government of India, and has 15 years of research experience along with four years of teaching experience at the UG and PG level. She is also a regular reviewer of a variety of international reputed journals.

Dr Ranjana Verma received her master's and doctorate in material chemistry from Tezpur University Assam, India, in 2012 and 2017, respectively. Currently, she is working as a postdoctoral fellow under (DSKPDF) UGC funded in the Department of Physics, Banaras Hindu University, Varanasi, India, since 2019. Her research focuses on developing strategies based on synthesis, characterization, properties modification, design, fabrication, nanostructure and photocatalyst, photovoltaic cell and biofuel conversion. Their perspective includes solar photocatalyst and biosensing applications. Dr. Verma has received many prestigious fellowships like DST (NPDF), UGC (DSKPDF), UGC (JRF), CSIR (SRF) awards, etc. In addition, Dr. Verma has published various research papers (15) and has authored more than three book chapters for internationally reputed press, namely Elsevier, Springer Nature and CRC Press.

Dr Ravindra Pratap Singh completed his BSc from Allahabad University India and his MSc and PhD in biochemistry from Lucknow University, India. He is currently working as an assistant professor in the Department of Biotechnology, Indira Gandhi National Tribal University, Amarkantak Madhya Pradesh, India. He has previously worked as a scientist at various esteemed laboratories globally, namely Sogang University, IGR Paris, etc. His work and research interests include biochemistry, biosensors, nanobiotechnology, electrochemistry, material sciences and biosensors applications in biomedical, environmental, agricultural and forensics. He has, to his credit, several reputed national and international honors/awards. Dr. Singh has authored over 60 articles in international peer-reviewed journals, more than 50 book chapters of international repute and has edited nine books. He serves as a reviewer of many reputed international journals and is also a member of many international societies. He is currently also involved in editing various books, which will be published in internationally reputed publication houses, namely IOP Publishing, CRC Press, Elsevier and Springer Nature. Moreover, he is a book series editor of *Emerging Advances in Bionanotechnology*, CRC Press, Taylor and Francis Group. He is also actively involved in guest-editing special issues (SI) for reputed international journals, and one of the latest edited SI by Dr. Singh is "Smart and Intelligent Nanobiosensors: Multidimensional applications" for *Materials Letters*, Elsevier.

Contributors

Sana Ahmed
Japan Advanced Institute of Science
and Technology (JAIST)
1-1Asahidai, Nomi, Ishikawa, Japan

Takashiro Akitsu
Tokyo University of Science
1-3 Kagurazaka, Shinjuku-ku
Tokyo, Japan

Geanne Matos de Andrade
Federal University of Ceará
Rua Cel. Nunes de Melo 1127
Porangabussu, Fortaleza, Ceará, Brazil

Tinku Basu
Amity University
Uttar Pradesh, Noida, India

Stephen Rathinaraj Benjamin
Federal University of Ceará
Rua Cel. Nunes de Melo 1127
Porangabussu, Fortaleza, Ceará, Brazil

Rashi Bhardwaj
Amity University
Uttar Pradesh, Noida, India

Chansi
Amity University
Uttar Pradesh, Noida, India

Nidhi Goel
Banaras Hindu University
Varanasi, Uttar Pradesh, India

Jyoti
Delhi Technological University
Delhi, India

Shabnam Khan
Functional Inorganic Materials Lab
(FIML)
Department of Chemistry
Aligarh Muslim University
Aligarh, Uttar Pradesh, India

Mohammad Yasir Khan
Functional Inorganic Materials Lab
(FIML)
Department of Chemistry
Aligarh Muslim University
Aligarh, Uttar Pradesh, India

Anil Kumar
Delhi Technological University
Delhi, India

Arun Kumar
National Institute of Science Education
and Research
Bhubaneswar, Khurda, Odisha, India

Naresh Kumar
Indian Institute of Technology
Indore, Madhya Pradesh, India

Cansu İlke Kuru
Ege University
İzmir, Turkey
and
Buca Municipality Kızılçullu Science
and Art Center
Izmir, Turkey

Bani Mahanti
Banaras Hindu University
Varanasi, Uttar Pradesh, India

A. Manjceevan
University of Jaffna
Jaffna, Sri Lanka

Sachin Mishra
Kwangwoon University
Nowon-gu, Seoul, South Korea

Shubhranshu Mishra
Banaras Hindu University
Varanasi, Uttar Pradesh, India

Gunjan Nagpure
Indira Gandhi National Tribal
    University
Madhya Pradesh, India

Daisuke Nakane
Tokyo University of Science
1-3 Kagurazaka, Shinjuku-ku, Tokyo,
    Japan

Kamlesh Kumar Nigam
Banaras Hindu University
Varanasi, Uttar Pradesh, India

Reinaldo Barreto Oriá
Federal University of Ceará
1315 Rua Cel. Nunes de Melo
Fortaleza, Ceará,, Brazil

Mrituanjay D Pandey
Banaras Hindu University
Varanasi, Uttar Pradesh, India

Usha Raju
Faculty of DSEU Pusa Campus
Delhi Skill and Entrepreneurship
    University
Delhi, India

Eli José Miranda Ribeiro Júnior
Faculty of Inhumas (FacMais)
Avenida Monte Alegre
n° 100 Residencial – Monte Alegre,
    Inhumas
Goiás-GO, Brazil

Farasha Sama
Department of Chemistry
Indian Institute of Technology Kanpur
Kanpur, Uttar Pradesh, India

Yoshiyuki Sato
Tokyo University of Science
1-3 Kagurazaka, Shinjuku-ku, Tokyo,
    Japan

M. Shahid
Functional Inorganic Materials Lab
    (FIML)
Department of Chemistry
Aligarh Muslim University
Aligarh, Uttar Pradesh, India

Shiva
Banaras Hindu University
Varanasi, Uttar Pradesh, India

Jay Singh
Banaras Hindu University
Varanasi, Uttar Pradesh, India

Ravindra Pratap Singh
Indira Gandhi National Tribal
    University
Madhya Pradesh, India

Arpna Tamrakar
Banaras Hindu University
Varanasi, Uttar Pradesh, India

Fulden Ulucan-Karnak
Ege University
Bornova, Izmir, Turkey

K. Velauthamurty
University of Jaffna
Jaffna, Sri Lanka

Ranjana Verma
Banaras Hindu University
Varanasi, Uttar Pradesh, India

Fahmina Zafar
Inorganic Materials Research
    Laboratory
Department of Chemistry
Jamia Millia Islamia
New Delhi, India

Preface

Due to the structural flexibility, large surface area, tailorable pore size and functional tenability, metal-organic frameworks (MOFs) have become a fascinating class of materials and are continuing to inspire both scientists and engineers. MOFs are the combination of organic and inorganic structure elements that can lead to materials with unique properties, and the choice of initial building units makes it possible to vary some parameters, such as the pore size, density as well as the specific surface area, which opens up new ways to produce materials with tailored physicochemical properties. This ability to fine-tune MOF properties is evident in various areas, both academic and industrial, with potential practical applications including gas storage and separation, heterogeneous catalysis, chemical sensing, biomedical applications, fabrication of membranes or thin films and many more. Hence, these materials play a key role in life every day and their effect on the future of humanity's technology. Therefore, exciting developments, new advancements and foundations, direction for further exploring MOFs for their applications and critical analysis of the research are necessary to provide the field's current status.

The proposed book *Advanced Functional Metal-Organic Frameworks: Fundamentals and Applications* will be an outstanding collection of current research on MOFs and their versatile applications and provide an overview of selected advances that have taken place in the field of MOFs. Moreover, this book will introduce a basic concept of the metal-organic frameworks (synthetic strategies, properties, chemistry and vast applications) to help students and professionals better understand these materials. Besides, this will also help them understand and address day-to-day problems related to environment, human health, national development and security and overcome these challenges by using MOFs. This book will serve as a reference book for professionals, students, scientists, researchers and academicians in this subject area. At last, we hope that readers will find it beneficial to their future research and teaching endeavors.

AIMS AND SCOPE

This book will help to attract a wide range of readers from all the fields as this book will provide the fundamental aspects of MOFs and excellent insight into the synthesis and modification of MOFs as well as their potential applications in different domains. Additionally, this book will also provide a single platform for the scientific community working on different aspects of metal-organic frameworks. This book will offer highlights of new contributions for experienced researchers and provide an introduction to this fascinating research field for novices. Hence, herein this book, we have tried to cover all the facets listed below.

- Overview-what are MOFs?
- Methods for the preparation of MOFs
- Factors affecting the preparation of MOFs
- MOFs for gas adsorption and separation and purification
- MOF-based sensors
- MOFs for catalysis and drug delivery
- MOF-based supercapacitors/energy storage
- Challenges and their prospects

Jay Singh
Nidhi Goel
Ranjana Verma
Ravindra Pratap Singh

Acknowledgments

It gives us immense pleasure to acknowledge Professor Shri Prakash Mani Tripathi; Honorable Vice-Chancellor of Indira Gandhi National Tribal University, Amarkantak, India; Professor Sudhir K Jain, Honorable Vice-Chancellor, Banaras Hindu University, Varanasi, Uttar Pradesh, India; and Institutes of Eminence (IoE), Mistry of Education, India for providing constant assistance in all the possible ways. It is also our great pleasure to acknowledge and express our enormous debt to all the contributors who have provided their quality material to prepare this book. We are grateful to our beloved family members (Dr. Babita Singh, Dr. Naresh Kumar, Miss. Aparajita RB Singh and Miss. Shanvi Singh), who joyfully supported and stood with us in many hours of our absence to finish this book project. We are also thankful to Mr. Kshitij RB Singh of the Department of Chemistry, Institute of Science, Banaras Hindu University, Varanasi, Uttar Pradesh, India for helping us in the finalization of this book project. Thanks are also due to Boris I. Kharissov, Allison Shatkin, Gagandeep Singh and the entire publishing team for their patience and extra care in publishing this book.

Jay Singh
Nidhi Goel
Ranjana Verma
Ravindra Pratap Singh

Acknowledgements

1 Overview: What are Metal-Organic Frameworks?

Gunjan Nagpure, Shiva, Nidhi Goel,
Ranjana Verma, Jay Singh, and
Ravindra Pratap Singh

CONTENTS

1.1 INTRODUCTION

Metal-organic frameworks, or MOFs, emerged as a promising class of crystalline hybrid materials composed of metal clusters or centers and organic linker molecules such as benzene dicarboxylic acid (BDC) that assemble to form a three-dimensional structure (Figure 1.1) [1]. The material is known for its high porosity and enormous internal surface areas, extending up to $6,000m^2/g$. Its components permit various combinations of organic and inorganic blocks, giving rise to millions of different structures [2]. MOFs have received considerable attention due to their topological and fascinating structures with specific properties and applications that represent an eternal desire for many material scientists. The diverse crystalline structure is designed by stitching clusters of metal (additionally known as secondary building units (SBUs)) with organic linkers, mainly nitrogen and carboxylic acid-containing ligands.

In contrast, the single metal nodes known in coordination networks, the polynuclear nature of SBUs can impart thermodynamic stability through solid covalent bonds and architectural/mechanical strength via strong directional bonds that can hold the position of metal centers in MOFs firmly. SBUs serve as stable, rigid, and directional building units in designing the robust crystalline structure of MOFs with prearranged structures and properties [3]. These unique characteristics of SBUs lead

DOI: 10.1201/9781003252061-1

1

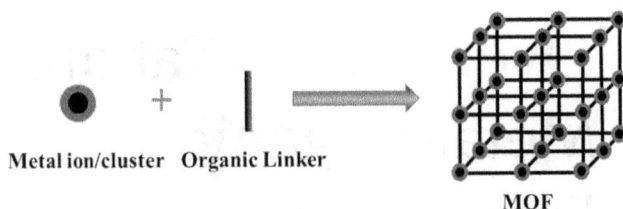

Metal ion/cluster   Organic Linker

MOF

FIGURE 1.1   Scheme for the synthesis of MOF [1] (reproduced with Copyright permission. Coordination Chemistry Reviews 2017).

to the development of MOFs with structural complexity, ultra-high porosity, and mechanical and chemical stability, which also give rise to fantastic framework chemistry [4]. These MOFs' state-of-the-art crystalline porous material has explored its applications in different domains like biomedical, agriculture, environment, sensors, photocatalysis, etc. Similarly, MOFs offer potential applications in sustainable plant growth and development and have gained considerable attention from farmers in the agriculture domain. On the other hand, iron-based MOFs play an essential role as multifunctional materials in agricultural prospects for pesticide delivery and plant nutrient replenishment.

Moreover, these MOFs have also been utilized to detect various environmental pollutants in wastewater treatment, organic dye removal, bioremediation, etc., in an ecological domain. Similarly, MOFs have also been explored as electrochemical sensors used for heavy metal (Pb^{2+}, Cd^{2+}, Hg^{+}) detection in groundwater. Herein, this chapter will discuss the unique synthesis, factors, and properties of MOFs, including the MOF applications.

1.2   SYNTHESIS, FACTORS, AND PROPERTIES OF MOFS

MOFs, a group of porous materials, consist of metal nodes and organic linker molecules. During the last two decades, the synthesis of MOFs has grabbed considerable attention from researchers due to their possibility of obtaining a well-fabricated structure, which has received significant interest in several areas related to porous materials. Moreover, these can be synthesized through conventional, unconventional, and alternative methods. The discussion in detail about the synthesis of MOFs has been illustrated in Chapter 2. Furthermore, the characterization techniques of MOFs are also approachable. Several reports are available in the literature to characterize the MOFs using basic techniques like UV-Visible, FT-IR, and multi-state NMR spectroscopy, including polymerization, high-end morphological (atomic force microscopy), and powder and single-crystal X-ray diffraction techniques [5–8]. With the innovative fabricated design, properties such as high surface area, stability, porosity, conductivity, tunability, and particle morphology can also be customized for specific applications [9]. Porosity and high surface area are considered to be essential characteristics of MOFs. The topology and pore size of the framework can be regulated by selecting appropriate metal nodes and linker molecules. The porosity of MOFs can also play a crucial role in modulating ion transport [10]. Thus, the higher surface area serves to be beneficial for catalytic processes, such as oxygen reduction (ORR) in lithium-oxygen batteries, and represents an opportunity for electrochemical applications. Another characteristic feature of MOFs is facile tunability, which utilizes its performance as precursors and templates in designing functional materials with unique morphologies and chemical compositions [11,12]. According to the last couple of decades' strategy, an exceptional sum of new structures has been reported where; it's been analyzed that MOFs usually experience weak stability such as poor acid/base, water, mechanical, and thermal stability [13].

Nonetheless, MOFs' stability has utilized its potentialities in various practical applications. Thus, MOFs' strength can be improved by increasing coordination bond resistance between SBUs and linker molecules [14]. Additionally, new strategies have been suggested by decorating MOFs' surface or interfaces with hydrophobic and permeable poly dimethyl siloxane (PDMS) thin layers through a vapor deposition technique that can improve the water's stability. Thus, stable MOFs have significantly expanded the applications of MOFs and are widely functional in biomedical, sensing, catalysis, and other domains [15]. Furthermore, various factors such as metal/organic linker ratio, solvents, high temperature, and pH affect the synthesis and properties of MOFs and enhance their vital roles in various fields. Figure 1.2 illustrates the properties of MOFs.

The solvent system plays an essential role in deciding the morphology and MOFs' synthesis. Solvents may coordinate with metal ions and act as a structure-directing agent and as space-filling molecules [16]. Solvents such as diethyl formamide (DEF), dimethylformamide (DMF), dimethyl acetamide (DMA), dimethyl sulphoxide (DMSO), acetone, alcohols, acetonitrile, etc., are used in MOFs' synthesis. These

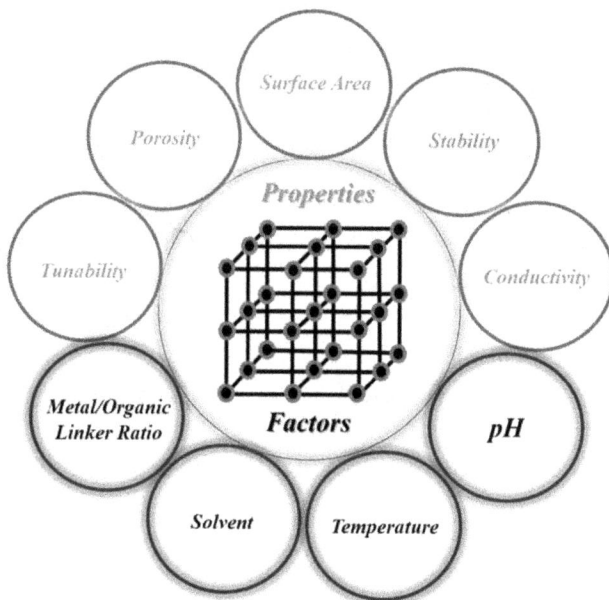

FIGURE 1.2   Schematic overview of MOF properties and factors.

solvents should have a high boiling point as well as polar nature. Thus, because of the solvent's polarity, solubility, and protolysis property of the organic linker, the MOFs' synthesis process is usually affected by the reaction medium. It has been reported that different solvent systems in the same reaction conditions impart MOFs of different morphology [17]. Peng et al. (2022) investigated that coordination solvents mainly affect the basicity generation in MOFs and that strong basicity can be created at low temperatures. In contrast, different solvent systems result in MOF synthesis having different pore sizes [18]. It was further analyzed that MOFs of 4,4'-((5-carboxy-1,3-phenylene bis(oxy)) dibenzoic acid (H_3CPBDA) and cobalt were fabricated using three different solvents viz. DMP, DMF, and DMA of different pore sizes (76.84Å, 72.76Å, 74.37Å) under similar conditions. The alternation in pore size is due to different solvents, which have been associated with the size of the solvent molecule. The variation in pore size of the molecules among three different solvents results in DMP >DMA >DMF. The pore size also gets reduced in the same order [19]. Furthermore, the temperature and pH of the reaction medium have also influenced the MOFs' synthesis. At different pH ranges, different coordination modes are adopted by linker molecules. Hence, Yuan et al. (2013) reported that the interpenetrated network is formed at a higher pH value, and an un-interpenetrated network is formed at a lower pH value. It has also been reported that the color of MOF compounds is also anticipated on the pH of the reaction medium [20]. Moreover, Luo et al. demonstrated via their experimental work, which explored three Co-MOF complexes; $[Co_3(L)_2(BTC)_2].4H_2O$), $[Co_2(L)(HBTC)_2(\mu_2\text{-}H2O) (H_2O)_2].3H_2O$, $[Co_2(L)(BTC) (\mu_2\text{-}OH) (H_2O)_2].2H_2O$ (L = 3,3',5,5'-tetra(1H-imidazole-1-yl)1,1'-biphenyl and BTC = 1,3,5-benzenetricarboxylate) exhibits different color and structure with altered pH value and different adsorption capability. Chu et al. reported that higher dimension compounds are formed at higher pH values. Different pH values affect different MOFs' properties, which are demonstrated in Table 1.1 [21].

Temperature is another critical factor that affects the properties of synthesized MOFs. Thus, due to the high solubility of reactants, high temperature favors higher crystallization, which results in the formation of large crystals of high quality [22]. Dai et al. investigated room temperature has been developed for the green synthesis of robust MOFs, which have received considerable attention from researchers in the sustainable chemistry domain. It has been reported here that a series of highly

TABLE 1.1

Properties of Different MOFs

S. No.	Properties of Co-MOFs	Complex-1	Complex-2	Complex-3
1.	Chemical formula	$[Co_2(L)(HBTC)2(\mu_2\text{-}H_2O) (H_2O)_2].3H_2O$	$[Co_3(L)_2(BTC)_2].4H_2O$	$[Co_2(L)(BTC)(\mu_2\text{-}OH) (H_2O)_2].2H_2O$
2.	pH	5	7	9
3.	Color	Pink	Purple	Brown
4.	Morphology	Monoclinic	Monoclinic	Orthorhombic
5.	Space group	P2/c	C2/c	Pccn

porous M6 oxo-clusters-based MOFs (M=Zr, Hf, Ce) is synthesized at specific room temperature, which includes 8 or 12 connected micro/mesoporous solids associated with different functionalized organic linker molecules. Hence, while maintaining the chemical stability of parent MOFs, the resultant compound shows diverse degrees of defectivity, especially for 12- connected phases [23]. Schoedel et al. demonstrated that the newly developed gel-layer approach facilitates the synthesis of oriented nanoscale films of MOFs on functionalized gold surfaces at specific room temperatures. As a result, organ gels have been used in bulk form in MOF synthesis to acquire single well-shaped crystals for crystallography [24].

Moreover, Bernini et al. reported that two Ho-succinate MOFs are prepared at high temperatures via the hydrothermal method. It was analyzed that the hydrothermal method is thermally more stable than the room temperature [25]. The analysis of various factors that affect MOFs' synthesis and properties has been elaborated in Chapter 3.

1.3 APPLICATIONS OF MOFS

MOFs have utilized their potential applications in various domains such as biomedical, environment, agriculture, nano-sensors, photocatalysis, electrochemical sensors, etc., as depicted in Figure 1.3 [26]. Herein, we are discussing the various applications of MOFs in brief.

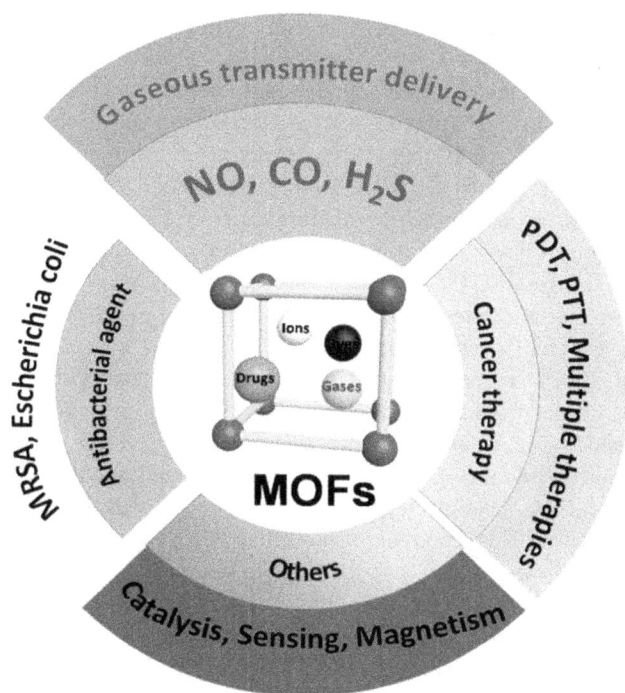

FIGURE 1.3 Schematic representation of MOFs' applications [26] (Copyright permission Nanomaterials 2020 (Creative Commons CC BY 4.0 license).

1.3.1 BIOMEDICAL APPLICATIONS

The exceptional features of MOFs, such as ultra-high surface area, well-defined structure, high porosity, low toxicity, biocompatibility, nanometer pore size, large pore size, and biodegradability, have received considerable attention from researchers in the chemical science domain. Due to this, MOFs possess great potential in biomedical applications, including antibacterial properties, bioimaging, biosensing, drug delivery, and biocatalysis. Due to extensive surface area (1,000 to 10,000 m^2/g), high porosity, and magnetic properties, MOFs can be utilized as drug carriers to target specific body sites and release drugs in a controlled manner. As a result, several types of research have been done with MOFs for drug delivery systems. Zeng et al. (2022) reported that MOFs have gained substantial attention from researchers due to their high porosity and promising drug delivery applications. They designed redox-responsive, pH-sensitive folic acid functionalized MOFs as drug carriers improved Bufalin (Buf) antitumor activity against breast cancer. Buf is an active ingredient of the traditional Chinese medicine Chansu and is found effective for treating breast cancer, but due to high toxicity and poor solubility and high side effects, its use is limited. As a result, it was demonstrated that, as compared to free Buf, the constructed FA-MOF/Buf nanoparticles were found to be more effective in treating breast cancer with improved water solubility and stability, enhanced anticancer activity, enhanced cytotoxicity, reduced side effects with developed drug delivery system [27] demonstrated, due to the unique properties of MOFs, such as high drug loading capacity, tunable porous structure, outstanding biocompatibility, and structural diversity, MOFs have been explored as drug nanocarriers in the biomedical domain. Here, it was noted that by incorporating high-Z metal ions (radiosensitizers for radiotherapy) or porphyrin-based ligands (photosensitizers for photodynamic therapy) into the scaffolds of MOFs, then MOFs themselves can act as an anti-therapeutic age and can be used as drug delivery vehicles in the biomedical domain [28]. Alves et al. (2021) reported MOFs associated with high photosensitizers (PSs) emerged as inorganic building units (metal ions/clusters), which play an effective role as a co-adjuvant in photodynamic therapy (PDT). Whereas PS-based MOFs effectively decrease antioxidant species and alleviate hypoxia in solid tumors to yield ROS, they act as a contrasting agent for imaging-guided chemotherapy for cancer treatment [29]. Hamedi et al. (2021) reported that cyclodextrin-based metal-organic frameworks (CD-MOFs) serve as promising nano-carriers for drug delivery due to their tunable properties, including high porosity and high specific surface area, easy fabrication, and novel chemical structure. Along with drug delivery applications, it also represents eco-friendly and biocompatible features that lead to biomedical applications [30]. Whereas Zeng et al. (2021) reported that polyethylene glycol (PEG)–based nanoscale MOFs (n MOFs) show promising applications in targeted cancer imaging and drug delivery. As a result, in this study, Doxorubicin (DOX) served as a therapeutic drug agent and a fluorescent label which was then transported into UiO-66-PEG associated with sufficient drug loading efficiency [31]. Chen et al. (2020) reported that as per synthesis strategies, porphyrin-based MOFs such as porphyrin@MOFs, composite porphyrinic MOFs, and porphyrinic MOFs have been widely used and explored for various biomedical applications due to their excellent electrochemical and photophysical properties [32]. Some typical examples of porphyrin@MOFs such as DTPP(Zn)-l$_2$ doped UiO-66 and TCPP doped UiO-66, displayed as effective nano-sized photosensitizers and show photodynamic therapy efficacy for *in-vitro* diagnosis and treatment of liver cancer [33,34]. Bahrani et al. (2019) investigated that Zn-based MOFs as drug carriers/nanoencapsulator show promising biomedical applications due to their non-toxic and biodegradable nature. Zn-MOFs serve as magnetic resonance imaging (MRI) agents without using ionizing radiations or radioactive nuclides released as effective nano-carriers for cancer imaging and therapy [35]. Moreover, MOF-based biosensors serve as promising candidates in biomedical applications. According to analysis, quantitative and qualitative sense are two MOF-based biosensors used to detect and quantify the presence of compounds in prominent research areas. MOFs possess various advantageous sensing applications and gained attention in the field, such as gas storage and separation, catalysis, ion exchange, and many more, due to their well-defined properties such as large surface areas and well-defined high porosity [36]. Quantitative sensing involves the interaction of an analyte with MOF, whereas lanthanide-based MOFs enhance the sensing ability and are used for visual observation of sensing events. On the other hand, qualitative sensing involves using MOF luminescence to visualize and locate the cellular region of interest using optical microscopy [37]. Qualitative sensing encompasses using MOFs as contrast agents for MRI or confocal microscopy, which serves as a useful technique for biomedical diagnosis and treatment. Thus, due to ideal properties such as tunable synthesis and flexibility, MOFs have shown potential biosensing applications that allow smart designs of frameworks as sensors that target optical or MR imaging [38,39].

1.3.2 AGRICULTURAL APPLICATIONS

As the population, globalization, and industrialization increase rapidly, the demand for food and food products grows. The agricultural domain is considered the most important sector as it helps increase the country's economy. But the use of chemical pesticides, fertilizers, insecticides, and composites pollutes the quality of agricultural land and water quality that affects the lifestyle of human beings, plants, and animals in certain ways [40]. Therefore, recent projects have investigated that MOF-based platforms are considered the most effective in offering potential sustainable plant growth and development applications. Shan et al. (2020) also demonstrated potential iron-based MOF (Fe-MOFS) applications as multifunctional materials in agricultural prospects for plant nutrient replenishment and pesticide delivery because iron is an essential micronutrient for crop growth and development. Currently, researchers have explored the feasibility of MOFs as a suitable platform for novel applications in plant protection. For example, the healthy function of 1,3,5-benzenetricarboxylate (Fe-MIL-100) as an iron micronutrient enhanced wheat growth and was prepared as the fungicide carrier azoxystrobin. As a result, azoxystrobin-mediated Fe-MOFs (AZOX@Fe-MIL-100) exhibit effective fungicidal activities against two pathogenic fungi-tomato late blight (*Phytophthora infestans*) and wheat head scab

(*Fusarium graminearum*) [40]. Hence, due to rapid population growth, the demand for food has increased, which exerted unprecedented pressure on the global agrochemical industry. As a group of novel photocatalytic materials, MOFs possess ultrahigh tunability and porosity properties to remove toxic pollutants from the environment efficiently. Moreover, Wen et al. (2021) studied the potential efficacy of MOF as a photocatalyst for the degradation of agricultural pollutants such as antibiotics and pesticides from water [41]. Abdelhameed et al. (2019) reported that porous iron-based MOFs modified with an ethylenediaminetetraacetic dianhydride (EDTA) component were adopted as an important strategy in promoting the growth and quality parameters of plants. The potential effect of Fe-sources such as ferrous-ethylenediaminetetraacetic acid (Fe-EDTA), Fe-MOF, and Fe-MOF-EDTA is reported as the most efficient in accelerating growth rate, protein, and chlorophyll content, and inflammatory enzyme activity of the plant *Phaseolus vulgaris* [42]. Zhang et al. reported that when treated with ethylene, MOF enhances the ripening process in climacteric fruits like bananas and avocados by diminishing tissue firmness and facilitating the color change, indicating the sign of fruit ripening [43]. Vlasova et al. investigated that aluminum, titanium, and zinc-based MOFs act as sorbents in purifying and enhancing the psychochemical chemical properties such as odor and taste of unrefined vegetable oils like sunflower linseed and olive oils [44]. Anstoetz et al. investigated novel applications of porous iron-based oxalate-phosphate-amine-MOF (OPA-MOF) to promote seedling growth, grain yields, and nutrient uptake of wheat (*Triticum aestivum L.*) plants as well as soil dynamics in incubated soil. Further, it has also been reported that OPA-MOF-type fertilizers can be used as novel enhanced N-fertilizers, which serve to be more useful in alkaline calcareous soil and helps in maintaining the pH balance of soil [45]. As per the increase in population growth, the demand of food is also substantially increasing in the agriculture sector. Food wastage is also a major concern for sustainable health and agriculture with food production. Various classical food preservation techniques such as pasteurization, fermentation, freeze-drying, and microwaves are available to reduce food waste. But, as per analysis, these techniques exhibit major shortcomings, such as altering food's taste and color. Moreover, Sultana et al. (2022) reported that MOFs served as antimicrobial, oxygen scavengers, ethylene scavengers, and moisture absorbers and developed as an active agent followed by major applications in extending food shelf life and inactive packaging of food materials [46]. Further studies have reported that the overuse of different organophosphate (OP) groups of pesticides (malathion, chlorpyrifos, parathion, ethion, and methyl parathion) in soil has led to their accumulation of vegetables and food materials, causing major health-related issues. Hence, reliable sensors have to be developed for detecting neurotoxic organophosphates (OP). Chansi et al. (2020) reported MOF directed electrosynthesis of anisotropic gold nanorods (aAuNR), neurotransmitter acetylcholinesterase (AChE), and cysteamine (Cys) functionalization in fabricating a novel electrochemical bio probe aAuNR/AChE/Cys/MOF that offers specificity, good stability, and anti-interference properties in sensing and quantifying residues of organophosphate pesticides in vegetable samples (*Solanum melongena, Momordica charantia Linn, Abelmoschus esculentus, Capsicum annum*), etc. [47].

As a result, due to versatile properties such as high surface area, tailorable pore size, and exposed active sites, MOFs have been explored as electrochemical sensors that are used for the detection of heavy metal ions like Pb^{2+}, Cd^{2+}, Cu^{2+}, and Hg^{2+} in groundwater including phenolic compounds, antibiotics, and pesticides [48,49]. Liu et al. (2022) demonstrated the potential efficiency of dye-encapsulated azo-terephthalate MOF-derived nano-sensors in designing an antiviral agent, dufulina-nalyzed by acid-phosphate (ACP) enzyme for pesticide sensing in food safety and environmental assessment [50].

1.3.3 ENVIRONMENTAL APPLICATIONS

Over the last few years, the field of MOFs has witnessed a remarkable upgrade in a wide array of environmental applications [51]. It has been noted that MOFs serve as robust photocatalysts in disintegrating and removing recalcitrant organic pollutants from the environment [52]. Musarurwa et al. (2022) investigated advanced appli-cations of chitosan-based MOFs such as Uio, MIL, HKUST, and ZIF as adsorbents in the remediation of various environmental pollutants [53]. Zhang et al. (2022) reported MILs series of MOFs have attracted a lot of attention as effective en-vironmental adsorbents with large specific surface areas and abundant adsorption sites in treating contaminated water containing heavy metals (cadmium, mercury, lead, chromium, and arsenic), etc. [54]. Kalimuthu et al. (2022) investigated that MOFs with high crystallinity derived from waste polyethylene terephthalate (PET) bottles were considered promising adsorbents for the quick removal of arsenate from contaminated water [55]. Pan et al. (2022) reported HKUST-1 derived porous carbon has received considerable attention as adsorbents, synthesized by the hydrothermal method under an oxygen-free atmosphere for quick removal of tet-racycline (TC) from water [56]. Guo et al. (2021) studied that MOFs have been extensively used as electrochemical sensors in agricultural systems and the food industry. Here, 2D and 3D copper-based MOFs (Cu-MOFs) are reported as most effective in the electrochemical detection of hydrogen peroxide (H_2O_2) in milk samples [57]. Li et al. (2021) noted that poly(ethyleneimine) (PEI) tailored MIL-101(Cr) (MILP) with a low-cost strategy was employed to extract rhenium (ReO_4^-) and uranium (Uo_2^{2+}) from wastewater [58]. Tang et al. (2020) studied the potential efficiency of bimetallic (Fe and Cu) MOF material ($Fe_xCu_{1-x}(BDC)$) in Fenton-like degradation of sulfamethoxazole (SMX) antibiotics in the presence of hydrogen peroxide and plays a valuable role in the remediation of environmental pollution [59]. Wang et al. (2015) reported that MOFs containing transition metals as nodes had gained considerable interest due to their significant morphological and structural variability. MOFs containing Cu, Zn, Co, Cd, and Fe centers have been revealed as photocatalysts for the degradation of organic pollutants under UV/visible sources [60]. MOF-based nano-sensors show their highly relevant applications in the en-vironmental domain. MOF-based nano-sensors help to detect various toxic inter-mediates, pollutants, and heavy metals such as mercury (Hg), copper (Cu), lead (Pb), chromium (Cd), arsenic (As), and cadmium (Cd) from the soil, water, and air. Zhang et al. (2022) studied the significance of MOFs-based fluorescence sensor for visually detecting both fluoride F⁻and Al_3^+ from drinking water, which causes severe health disorders such as fluorosis. A ratiometric fluorescence sensor was fabricated by integrating in situ rhodamine B (RhB) into zirconium-based MOF, UiO-66-NH₂ [61]. Lei at al. (2021) reported practical application of rapid, sensitive, and accurate MOF-based ratiometric luminescent sensors in water-sensing pollutants like antibiotics and D2O [62].

1.3.4 Miscellaneous Applications

Apart from the above applications, MOFs show applications in catalysis, gas storage, nanoparticle precursors, electrochemistry, luminescence, and technology sensors. The advanced properties such as tunable porosity, high surface area, and diversity in metal ions and organic linkers assemble MOFs highly fascinating as heterogeneous catalysts. Hence, in some MOFs, coordinated metal ions and organic linkers could directly be used as catalytic centers for catalytic reactions. Because of different catalytic sites, MOFs are unique heterogeneous catalytic systems; for example, without any mutual interference, acid and base areas can be accurately installed within one MOF. Liu et al. (2016) reported that MIL-101 MOFs are fabricated by post-synthetic modification and dually functionalized with sulfo and amino groups for one-pot deacetalization Knoevenagel condensation. Here, it was proved that being the ammonium group rather than sulfo one, MOFs takes the zwitterionic form with the catalytic acid site [63]. Du et al. (2021) reported MOFs exhibit excellent photocatalysis and adsorption efficiency due to their adaptability, high porosity, structural tailorability, and good crystal form. MOFs with good photocatalysis and photodegradation properties have explored their applications in dye degradation and in removing toxic antibiotics chemicals from groundwater and surface water [64]. Kaur et al. (2016) investigated a novel hybrid structure of cadmium telluride quantum dots (CdTe QDs), associated with emporium-MOF (Eu-MOF), have been reported as photocatalysts. Further study has revealed that QD/Eu-MOF synthesizes the novel nanocomposite material carrying photocatalytic properties. Hence, results have shown that QD/Eu-MOF is a promising catalyst in the rapid degradation of Rhodamine 6G dye [65].

Moreover, MOFs have shown their practical applications in gas storage and separation. Different methods are known for storing gases functionally, but these methods face many disadvantages. These methods tend to be highly expensive for practical purposes; hence, simpler, safer, and cheaper storage methods have to be introduced. Several studies have demonstrated that MOF plays an essential role as a promising adsorbent compared to other porous materials due to its tuneable pore structure, high surface area, high separation selectively, and increased storage capacity. Zhai et al. (2022) reported that ultrastable (Sc_3O), organic frameworks regulated by a micropore combine strategy show prominent efficiency as adsorbents for acetylene storage and purification [66]. Song et al. (2015) investigated that MIL-68 MOF with different pore sizes and high stability exhibits promising applications in enhanced CO_2 gas storage [67]. Furthermore, MOFs have explored their potential in sensing applications along with gas storage and separation applications. Hitabatuma et al. (2022) investigated that MOFs represent abundant functional groups, high porosity, and tunable physical and chemical properties that demonstrate promising sensing applications.

1.4 CONCLUSION AND PROSPECTS

Today, MOFs emerged as the most promising novel crystalline material, consisting of a new class of coordination bonds between metal clusters (metal-oxalate clusters and metal-carboxylate clusters) and organic linkers (oxygen or nitrogen donors), forming a three-dimensional structure. The MOFs are fabricated by stitching metal/metal clusters (also known as SBU) along organic linkers, including nitrogen-containing ligands and carboxylic acid. The unique properties like stability, pore size, conductivity, reversibility, high surface area, and particle morphology make MOFs different from porous materials. This chapter discusses synthesis properties and applications of MOFs in various domains. The synthesis of MOFs has received considerable attention from researchers due to its ability to obtain varieties of interesting pleasing structures that make MOF a unique porous material compared to other porous materials. Different synthesis techniques have been introduced for MOFs that offer unique, cost-effective, and eco-friendly methods that facilitate the synthesis of MOFs.

Moreover, due to exceptional features such as biocompatibility, biodegradability, and ultra-high surface area, MOFs have utilized their novel applications in various domains like biomedical, agricultural, and environmental. MOFs can be used as drug carriers to target specific body sites and release drugs in that affected area. MOFs also explored their potentialities as photocatalysts in drug delivery, diagnosis, imaging, and treating cancer. The agricultural sector is considered as one of the largest sectors that offer potential applications in sustainable growth and development of plants. It has been reported that oxalate-phosphate amine MOF-type fertilizers have been used as N-type fertilizers, maintaining in the fertility of soil and nourishing plants.

Moreover, different MOFs are highly used in developing organic chemical-free nano-pesticides and nano-insecticides. Moreover, MOFs have also been used as nano-sensors to detect nutrient levels and pollutants in the soil. Although as population's growth, health, safety, and regulations are entirely dependent on agricultural-derived products, the agriculture sector has demanded a lot of attention of researchers to develop nano-fertilizers for further use in the future.

Similarly, MOFs also show versatile utilities in the environmental domain by helping in detecting inorganic and organic pollutants; remediation of heavy metals; and removing heavy metals from the soil, water, and air. MOFs also play a crucial role in treating wastewater that contaminates freshwater. Nowadays, MOFs possess absorption and photocatalysis properties that carry the unique ability to degrade methylene blue dye from water. The advantage of MOFs in various domains also includes several drawbacks that need to be resolved in the future. MOFs show suitable prospects to combat these drawbacks and hold a promising future.

REFERENCES

[1] Ren, Jianwei, Mpho Ledwaba, Nicholas M. Musyoka, Henrietta W. Langmi, Mkhulu Mathe, Shijun Liao, and Wan Pang. 2017. "Structural Defects in Metal–Organic Frameworks (MOFs): Formation, Detection and Control Towards Practices of Interests." *Coordination Chemistry Reviews* 349: 169–197. 10.1016/j.ccr.2017.08.017.

[2] Safaei, Mohadeseh, Mohammad Mehdi Foroughi, Nasser Ebrahimpoor, Shohreh Jahani, Ali Omidi, and Mehrdad Khatami. 2019. "A Review on Metal-Organic Frameworks: Synthesis and Applications." *TrAC Trends in Analytical Chemistry* 118 (September): 401–425. doi: 10.1016/j.trac.2019.06.007.

[3] Ha, Junsu, Jae Hwa Lee, and Hoi Ri Moon. 2020. "Alterations to Secondary Building Units of Metal–Organic Frameworks for the Development of New Functions." *Inorganic Chemistry Frontiers* 7 (1): 12–27. doi: 10.1039/C9QI01119F.

[4] Schoedel, Alexander. 2020. "Secondary Building Units of MOFs." In *Metal-Organic Frameworks for Biomedical Applications*, 11–44. Elsevier. doi: 10.1016/B978-0-12-816984-1.00003-2.

[5] Lucier, B.E.G., Shoushun Chen, and Yining Huang. 2018. "Characterization of Metal-Organic Frameworks: Unlocking the Potential of Solid-State NMR." *Accounts of Chemical Research* 51 (2): 319–330. 10.1021/acs.accounts.7b00357.

[6] Howarth, Ashlee J., Aaron W. Peters, Nicolaas A. Vermeulen, Timothy C. Wang, Joseph T. Hupp, and Omar K. Farha. 2017. "Best Practices for the Synthesis, Activation, and Characterization of Metal–organic Frameworks." *Chemistry of Materials* 29 (1): 26–39. 10.1021/acs.chemmater.6b02626.

[7] Li, Minyuan, and Mircea Dincă. 2011. "Reductive Electrosynthesis of Crystalline Metal–Organic Frameworks." *Journal of the American Chemical Society* 133 (33): 12926–12929. doi: 10.1021/ja2041546.

[8] Al-Kutubi, Hanan, Jorge Gascon, Ernst J.R. Sudhölter, and Liza Rassaei. 2015. "Electrosynthesis of Metal-Organic Frameworks: Challenges and Opportunities." *ChemElectroChem* 2 (4): 462–474. doi: 10.1002/celc.201402429.

[9] Wu, Hao Bin, and Xiong Wen (David) Lou. 2017. "Metal-Organic Frameworks and Their Derived Materials for Electrochemical Energy Storage and Conversion: Promises and Challenges." *Science Advances* 3 (12): eaap925. doi: 10.1126/sciadv.aap9252.

[10] Furukawa, Hiroyasu, Kyle E. Cordova, Michael O'Keeffe, and Omar M. Yaghi. 2013. "The Chemistry and Applications of Metal-Organic Frameworks." *Science* 341 (6149): 1230444. doi: 10.1126/science.1230444.

[11] Yuan, Shuai, Liang Feng, Kecheng Wang, Jiandong Pang, Matheiu Bosch, Christina Lollar, and Yujia Sun, et al. 2018. "Stable Metal–Organic Frameworks: Design, Synthesis, and Applications." *Advanced Materials* 30 (37): 1704303. doi: 10.1002/adma.201704303.

[12] Lu, Weigang, Zhangwen Wei, Zhi-Yuan Gu, Tian-Fu Liu, Jinhee Park, Jihye Park, Jian Tian, et al. 2014. "Tuning the Structure and Function of Metal–Organic Frameworks via Linker Design." *Chemical Society Reviews* 43 (16): 5561–5593. doi: 10.1039/C4CS00003J.

[13] Burtch, Nicholas C., Himanshu Jasuja, and Krista S. Walton. 2014. "Water Stability and Adsorption in Metal–Organic Frameworks." *Chemical Reviews* 114 (20): 10575–10612. doi: 10.1021/cr5002589.

[14] Bosch, Mathieu, Muwei Zhang, and Hong-Cai Zhou. 2014. "Increasing the Stability of Metal-Organic Frameworks." *Advances in Chemistry* 2014 (September): 1–8. doi: 10.1155/2014/182327.

[15] Zhang, Wang, Yingli Hu, Jin Ge, Hai-Long Jiang, and Shu-Hong Yu. 2014. "A Facile and General Coating Approach to Moisture/Water-Resistant Metal–Organic Frameworks with Intact Porosity." *Journal of the American Chemical Society* 136 (49): 16978–16981. doi:10.1021/ja509960n.

[16] Yakovenko, Andrey A., Zhangwen Wei, Mario Wriedt, Jian-Rong Li, Gregory J. Halder, and Hong-Cai Zhou. 2014. "Study of Guest Molecules in Metal–Organic Frameworks by Powder X-Ray Diffraction: Analysis of Difference Envelope Density." *Crystal Growth & Design* 14 (11): 5397–5407. doi:10.1021/cg500525g.

[17] Soni, Sanju, Parmendra Kumar Bajpai, and Charu Arora. 2018. "A Review on Metal-Organic Framework: Synthesis, Properties and Application." *Characterization and Application of Nanomaterials* 2 (2). doi:10.24294/can.v2i2.551.

[18] Peng, Song-Song, Guo-Song Zhang, Xiang-Bin Shao, Chen Gu, Xiao-Qin Liu, and Lin-Bing Sun. 2022. "Generation of Strong Basicity in Metal–Organic Frameworks: How Do Coordination Solvents Matter?" *ACS Applied Materials & Interfaces* 14: 8058–8065. February, acsami.1c24299. doi:10.1021/acsami.1c24299.

[19] Huang, Wen-Huan, Guo-Ping Yang, Jun Chen, Xin Chen, Cui-Ping Zhang, Yao-Yu Wang, and Qi-Zhen Shi. 2013. "Solvent Influence on Sizes of Channels in Three New Co(II) Complexes, Exhibiting an Active Replaceable Coordinated Site." *Crystal Growth & Design* 13 (1): 66–73. doi:10.1021/cg301146u.

[20] Yuan, Fei, Juan Xie, Huai-Ming Hu, Chun-Mei Yuan, Bing Xu, Meng-Lin Yang, Fa-Xin Dong, and Gang-Lin Xue. 2013. "Effect of PH/Metal Ion on the Structure of Metal–Organic Frameworks Based on Novel Bifunctionalized Ligand 4′-Carboxy-4,2′:6′,4″-Terpyridine." *CrystEngComm* 15 (7): 1460. doi:10.1039/c2ce26171e.

[21] Chu, Qian, Guang-Xiang Liu, Taka-aki Okamura, Yong-Qing Huang, Wei-Yin Sun, and Norikazu Ueyama. 2008. "Structure Modulation of Metal–Organic Frameworks via Reaction PH: Self-Assembly of a New Carboxylate Containing Ligand N-(3-Carboxyphenyl)Iminodiacetic Acid with Cadmium(II) and Cobalt(II) Salts." *Polyhedron* 27 (2): 812–820. doi:10.1016/j.poly.2007.11.029.

[22] Seetharaj, R., P.V. Vandana, P. Arya, and S. Mathew. 2019. "Dependence of Solvents, PH, Molar Ratio and Temperature in Tuning Metal Organic Framework Architecture." *Arabian Journal of Chemistry* 12 (3): 295–315. doi:10.1016/j.arabjc.2016.01.003.

[23] Dai, Shan, Farid Nouar, Sanjun Zhang, Antoine Tissot, and Christian Serre. 2021. "One-Step Room-Temperature Synthesis of Metal(IV) Carboxylate Metal—Organic Frameworks." *Angewandte Chemie International Edition* 60 (8): 4282–4288. doi:10.1002/anie.202014184.

[24] Schoedel, Alexander, Camilla Scherb, and Thomas Bein. 2010. "Oriented Nanoscale Films of Metal-Organic Frameworks By Room-Temperature Gel-Layer Synthesis." *Angewandte Chemie International Edition* 49 (40): 7225–7228. doi:10.1002/anie.201001684.

[25] Bernini, Maria C., Elena V. Brusau, Griselda E. Narda, Gustavo E. Echeverria, C. Gustavo Pozzi, Graciela Punte, and Christian W. Lehmann. 2007. "The Effect of Hydrothermal and Non-Hydrothermal Synthesis on the Formation of Holmium(III) Succinate Hydrate Frameworks." *European Journal of Inorganic Chemistry* 2007 (5): 684–693. doi:10.1002/ejic.200600860.

[26] Zhang, Mengdan, Ruirui Qiao, and Jinming Hu. 2020. "Engineering Metal–Organic Frameworks (MOFs) for Controlled Delivery of Physiological Gaseous Transmitters." *Nanomaterials* 10(6): 1134. 10.3390/nano10061134.

[27] Zeng, Hairong, Chao Xia, Bei Zhao, Mengmeng Zhu, HaoYue Zhang, Die Zhang, Xin Rui, Huili Li, and Yi Yuan. 2022. "Folic Acid–Functionalized Metal-Organic Framework Nanoparticles as Drug Carriers Improved Bufalin Antitumor Activity Against Breast Cancer." *Frontiers in Pharmacology* 12 (January): 747992. doi:10.3389/fphar.2021.747992.

[28] Wei, Qin, Yihan Wu, Fangfang Liu, Jiao Cao, and Jinliang Liu. 2022. "Advances in Antitumor Nanomedicine Based on Functional Metal–Organic Frameworks beyond Drug Carriers." *Journal of Materials Chemistry B* 10 (5): 676–699. doi:10.1039/D1 TB02518J.

[29] Alves, Samara Rodrigues, Italo Rodrigo Calori, and Antonio Claudio Tedesco. 2021. "Photosensitizer-Based Metal-Organic Frameworks for Highly Effective Photodynamic Therapy." *Materials Science and Engineering: C* 131 (December): 112514. doi:10.1016/j.msec.2021.112514.

[30] Hamedi, Asma, Anastasia Anceschi, Alessia Patrucco, and Mahdi Hasanzadeh. 2021. "A γ-Cyclodextrin-Based Metal–Organic Framework (γ-CD-MOF): A Review of Recent Advances for Drug Delivery Application." *Journal of Drug Targeting* 30(December):1–13. doi:10.1080/1061186X.2021.2012683.

[31] Zeng, Yawen, Jinling Xiao, Yiyang Cong, Jia Liu, Yiming He, Brian D. Ross, Haixing Xu, Yihua Yin, Hao Hong, and Wenjin Xu. 2021. "PEGylated Nanoscale Metal–Organic Frameworks for Targeted Cancer Imaging and Drug Delivery." *Bioconjugate Chemistry* 32 (10): 2195–2204. doi:10.1021/acs.bioconjchem.1c00368.

[32] Chen, Jiajie, Yufang Zhu, and Stefan Kaskel. 2021. "Porphyrin-Based Metal–Organic Frameworks for Biomedical Applications." *Angewandte Chemie International Edition* 60 (10): 5010–5035. doi:10.1002/anie.201909880.

[33] Zhou, Le-Le, Qun Guan, Yan-An Li, Yang Zhou, Yu-Bin Xin, and Yu-Bin Dong. 2018. "One-Pot Synthetic Approach toward Porphyrinatozinc and Heavy-Atom Involved Zr-NMOF and Its Application in Photodynamic Therapy." *Inorganic Chemistry* 57 (6): 3169–3176. doi:10.1021/acs.inorgchem.7b03204.

[34] Park, Jihye, Qin Jiang, Dawei Feng, and Hong-Cai Zhou. 2016. "Controlled Generation of Singlet Oxygen in Living Cells with Tunable Ratios of the Photochromic Switch in Metal–Organic Frameworks." *Angewandte Chemie* 128 (25): 7304–7309. doi:10.1002/ange.201602417.

[35] Bahrani, Sonia, Seyyed Alireza Hashemi, Seyyed Mojtaba Mousavi, and Rouhollah Azhdari. 2019. "Zinc-Based Metal–Organic Frameworks as Non-toxic and Biodegradable Platforms for Biomedical Applications: Review Study." *Drug Metabolism Reviews* 51 (3): 356–377. doi:10.1080/03602532.2019.1632887.

[36] Miller, Sophie E., Michelle H. Teplensky, Peyman Z. Moghadam, and David Fairen-Jimenez. 2016. "Metal-Organic Frameworks as Biosensors for Luminescence-Based Detection and Imaging." *Interface Focus* 6 (4): 20160027. doi:10.1098/rsfs.2016.0027.

[37] Kreno, Lauren E., Kirsty Leong, Omar K. Farha, Mark Allendorf, Richard P. Van Duyne, and Joseph T. Hupp. 2012. "Metal–Organic Framework Materials as Chemical Sensors." *Chemical Reviews* 112 (2): 1105–1125. doi:10.1021/cr200324t.

[38] Rocca, Joseph Della, Demin Liu, and Wenbin Lin. 2011. "Nanoscale Metal–Organic Frameworks for Biomedical Imaging and Drug Delivery." *Accounts of Chemical Research* 44 (10): 957–968. doi:10.1021/ar200028a.

[39] Horcajada, Patricia, Tamim Chalati, Christian Serre, Brigitte Gillet, Catherine Sebrie, Tarek Baati, Jarrod F. Eubank, et al. 2010. "Porous Metal–Organic-Framework Nanoscale Carriers as a Potential Platform for Drug Delivery and Imaging." *Nature Materials* 9 (2): 172–178. doi:10.1038/nmat2608.

[40] Shan, Yongpan, Lidong Cao, Bilal Muhammad, Bo Xu, Pengyue Zhao, Chong Cao, and Qiliang Huang. 2020. "Iron-Based Porous Metal–Organic Frameworks with Crop Nutritional Function as Carriers for Controlled Fungicide Release." *Journal of Colloid and Interface Science* 566 (April): 383–393. doi:10.1016/j.jcis.2020.01.112.

[41] Wen, Yinghao, Mingbao Feng, Peng Zhang, Hong-Chai Zhou, Virender K. Sharma, and Xingmao Ma. 2021. "Metal Organic Frameworks (MOFs) as Photocatalysts for the Degradation of Agricultural Pollutants in Water." *ACS ES&T Engineering* 1 (5): 804–826. doi:10.1021/acsestengg.1c00051.

[42] Abdelhameed, Reda M., Reda E. Abdelhameed, and Hedaya A. Kamel. 2019. "Iron-Based Metal-Organic-Frameworks as Fertilizers for Hydroponically Grown Phaseolus Vulgaris." *Materials Letters* 237 (February): 72–79. doi:10.1016/j.matlet.2018.11.072.

[43] Zhang, Boce, Yaguang Luo, Kelsey Kanyuck, Gary Bauchan, Joseph Mowery, and Peter Zavalij. 2016. "Development of Metal–Organic Framework for Gaseous Plant Hormone Encapsulation To Manage Ripening of Climacteric Produce." *Journal of Agricultural and Food Chemistry* 64 (25): 5164–5170. doi:10.1021/acs.jafc.6b02072.

[44] Vlasova, E.A., S.A. Yakimov, E.V. Naidenko, E.V. Kudrik, and S.V. Makarov. 2016. "Application of Metal–Organic Frameworks for Purification of Vegetable Oils." *Food Chemistry* 190 (January): 103–109. doi:10.1016/j.foodchem.2015.05.078.

[45] Anstoetz, Manuela, Terry J. Rose, Malcolm W. Clark, Lachlan H. Yee, Carolyn A. Raymond, and Tony Vancov. 2015. "Novel Applications for Oxalate-Phosphate-Amine Metal-Organic-Frameworks (OPA-MOFs): Can an Iron-Based OPA-MOF Be Used as Slow-Release Fertilizer?" Edited by Shuijin Hu. *PLOS ONE* 10 (12): e0144169. doi:10.1371/journal.pone.0144169.

[46] Sultana, Afreen, Ajay Kathuria, and Kirtiraj K. Gaikwad. 2022. "Metal–Organic Frameworks for Active Food Packaging. A Review." *Environmental Chemistry Letters* 20: 1479–1495. doi:10.1007/s10311-022-01387-z.

[47] Chansi, Pragadeeshwara Rao R., Irani Mukherjee, Tinku Basu, and Lalit M. Bharadwaj. 2020. "Metal Organic Framework Steered Electrosynthesis of Anisotropic Gold Nanorods for Specific Sensing of Organophosphate Pesticides in Vegetables Collected from the Field." *Nanoscale* 12 (42): 21719–21733. doi:10.1039/D0NR04480F.

[48] Kajal, Navdeep, Vishavjeet Singh, Ritu Gupta, and Sanjeev Gautam. 2022. "Metal Organic Frameworks for Electrochemical Sensor Applications: A Review." *Environmental Research* 204 (March): 112320. doi:10.1016/j.envres.2021.112320.

[49] Singh, Kshitij R.B., Vanya Nayak, Jay Singh, Charles Oluwaseun Adetunji, and Ravindra Pratap Singh. 2021. "Introduction: Potentialities of Bionanomaterials towards the Environmental and Agricultural Domain." In *Bionanomaterials for Environmental and Agricultural Applications*. IOP Publishing. doi:10.1088/978-0-7503-3863-9ch1.

[50] Liu, Jintong, Lin Yao Ye, Yan Yang Mo, and Hong Yang. 2022. "Highly Sensitive Fluorescent Quantification of Acid Phosphatase Activity and Its Inhibitor Pesticide Dufulin by a Functional Metal–Organic Framework Nanosensor for Environment Assessment and Food Safety." *Food Chemistry* 370 (February): 131034. doi:10.1016/j.foodchem.2021.131034.

[51] Ghosh, Sujit K., and Susumu Kitagawa. 2019. "Introduction." In *Metal-Organic Frameworks (MOFs) for Environmental Applications*, 1–4. Elsevier. doi:10.1016/B978-0-12-814633-0.00001-6.

[52] Wen, Yinghao, Peng Zhang, Virender K. Sharma, Xingmao Ma, and Hong-Cai Zhou. 2021. "Metal-Organic Frameworks for Environmental Applications." *Cell Reports Physical Science* 2 (2): 100348. doi:10.1016/j.xcrp.2021.100348.

[53] Musarurwa, Herbert, and Nikita Tawanda Tavengwa. 2022. "Advances in the Application of Chitosan-Based Metal Organic Frameworks as Adsorbents for Environmental Remediation." *Carbohydrate Polymers* 283 (May): 119153. doi:10.1016/j.carbpol.2022.119153.

[54] Zhang, Hanshuo, Xin Hu, Tianxiao Li, Yuxuan Zhang, Hongxia Xu, Yuanyuan Sun, Xueyuan Gu, Cheng Gu, Jun Luo, and Bin Gao. 2022. "MIL Series of Metal Organic Frameworks (MOFs) as Novel Adsorbents for Heavy Metals in Water: A Review." *Journal of Hazardous Materials* 429 (May): 128271. doi:10.1016/j.jhazmat.2022.128271.

[55] Kalimuthu, Pandi, Youjin Kim, Muthu Prabhu Subbaiah, Daewhan Kim, Byong-Hun Jeon, and Jinho Jung. 2022. "Comparative Evaluation of Fe-, Zr-, and La-Based Metal-Organic Frameworks Derived from Recycled PET Plastic Bottles for Arsenate Removal." *Chemosphere* 294 (May): 133672. doi:10.1016/j.chemosphere.2022.133672.

[56] Pan, Junyao, Xueting Bai, Yiyao Li, Binhao Yang, Peiyu Yang, Fei Yu, and Jie Ma. 2022. "HKUST-1 Derived Carbon Adsorbents for Tetracycline Removal with Excellent Adsorption Performance." *Environmental Research* 205 (April): 112425. doi:10.1016/j.envres.2021.112425.

[57] Guo, Xiangjian, Chuyan Lin, Minjun Zhang, Xuewei Duan, Xiangru Dong, Duanping Sun, Jianbin Pan, and Tianhui You. 2021. "2D/3D Copper-Based Metal-Organic Frameworks for Electrochemical Detection of Hydrogen Peroxide." *Frontiers in Chemistry* 9 (October): 743637. doi:10.3389/fchem.2021.743637.

[58] Li, Jie, Yan Zhang, Yi Zhou, Fei Fang, and Xuede Li. 2021. "Tailored Metal-Organic Frameworks Facilitate the Simultaneously High-Efficient Sorption of UO2+ and ReO4- in Water." *Science of The Total Environment* 799 (December): 149468. doi:10.1016/j.scitotenv.2021.149468.

[59] Tang, Juntao, and Jianlong Wang. 2020. "Iron-Copper Bimetallic Metal-Organic Frameworks for Efficient Fenton-like Degradation of Sulfamethoxazole under Mild Conditions." *Chemosphere* 241 (February): 125002. doi:10.1016/j.chemosphere.2019.125002.

[60] Wang, Sa-Sa, and Guo-Yu Yang. 2015. "Recent Advances in Polyoxometalate-Catalyzed Reactions." *Chemical Reviews* 115 (11): 4893–4962. doi:10.1021/cr500390v.

[61] Zhang, Min, Rui Liang, Ke Li, Ting Chen, Shuangjun Li, Yongming Zhang, Dieqing Zhang, and Xiaofeng Chen. 2022. "Dual-Emitting Metal–Organic Frameworks for Ratiometric Fluorescence Detection of Fluoride and Al^{3+} in Sequence." *Spectrochimica Acta Part A: Molecular and Biomolecular Spectroscopy* 271 (April): 120896. doi:10.1016/j.saa.2022.120896.

[62] Lei, Mingyuan, Fayuan Ge, Xiangjing Gao, Zhiqiang Shi, and Hegen Zheng. 2021. "A Water-Stable Tb-MOF As a Rapid, Accurate, and Highly Sensitive Ratiometric Luminescent Sensor for the Discriminative Sensing of Antibiotics and D2O in H2O." *Inorganic Chemistry* 60 (14): 10513–10521. doi:10.1021/acs.inorgchem.1c01145.

[63] Liu, Hui, Fu-Gui Xi, Wei Sun, Ning-Ning Yang, and En-Qing Gao. 2016. "Amino- and Sulfo-Bifunctionalized Metal–Organic Frameworks: One-Pot Tandem Catalysis and the Catalytic Sites." *Inorganic Chemistry* 55 (12): 5753–5755. doi:10.1021/acs.inorgchem.6b01057.

[64] Du, Chunyan, Zhuo Zhang, Guanlong Yu, Haipeng Wu, Hong Chen, Lu Zhou, Yin Zhang, et al. 2021. "A Review of Metal Organic Framework (MOFs)-Based Materials for Antibiotics Removal via Adsorption and Photocatalysis." *Chemosphere* 272 (June): 129501. doi:10.1016/j.chemosphere.2020.129501.

[65] Kaur, Rajnish, Kowsalya Vellingiri, Ki-Hyun Kim, A.K. Paul, and Akash Deep. 2016. "Efficient Photocatalytic Degradation of Rhodamine 6G with a Quantum Dot-Metal Organic Framework Nanocomposite." *Chemosphere* 154 (July): 620–627. doi:10.1016/j.chemosphere.2016.04.024.

[66] Lv, Hong-Juan, Jian-Wei Zhang, Yu-Cheng Jiang, Shu-Ni Li, Man-Cheng Hu, and Quan-Guo Zhai. 2022. "Micropore Regulation in Ultrastable [Sc3O]-Organic Frameworks for Acetylene Storage and Purification." *Inorganic Chemistry* 61:3553–3562. doi:10.1021/acs.inorgchem.1c03562.

[67] Hu, Yue, Bin Lin, Peng He, Youyong Li, Yining Huang, and Yang Song. 2015. "Probing the Structural Stability of and Enhanced CO_2 Storage in MOF MIL-68(In) under High Pressures by FTIR Spectroscopy." *Chemistry - A European Journal* 21 (51): 18739–18748. doi:10.1002/chem.201502980.

2 Methods for the Preparation of Metal-Organic Frameworks

Nidhi Goel, Shiva, Shubhranshu Mishra, and Naresh Kumar

CONTENTS

2.1 INTRODUCTION

Metal-organic frameworks (MOFs) are the inorganic-organic hybrid compounds with a unique structure and composition, constructed chemically by utilizing various metal clusters or ions and organic linkers [1,2]. Customarily, alkaline earth metals, transition elements, or lanthanides are used as metal ions/clusters because of their low electronegativity and high electro-positivity. Additionally, a wide variety of coordination numbers, oxidation states, and geometries are shown by them and thus, they display synthetic and structural diversity. On the other side, bivalent/trivalent carboxylic acids or *N*-containing compounds as organic linkers are used for the construction of robust MOFs with flexible or dynamic frameworks [3]. The organic linkers with one or more donor nitrogen or oxygen atoms such as carboxylates, pyridyl, cyano compounds, polyamines, crown ethers, etc. are used for

DOI: 10.1201/9781003252061-2

17

the purpose of bridging between metal nodes in MOFs [4,5]. Therefore, by the use of pertinent metal clusters or ions and organic linkers, one can assemble a distinct framework.

In last two decades, synthesis of MOFs has fascinated incredible attention and the reason behind it is the possibility of obtaining large variety of structures with high porosity. Due to their structural flexibility, enormous surface area, tailorable pore size, and thermal stability, MOFs show the extensive applications in a large number of fields including sensing, separation, gas adsorption, as well as storage, catalysis, drug delivery, and many more [6–8]. There are several methods that are developed for the preparation of MOFs [9,10]. Though several factors, such as various interactions, functionality of multitopic organic linkers, and environment play key roles in designing MOFs with desired physical and chemical properties. Herein, we have described various conventional and unconventional as well as alternative synthetic methods, which will provide a vast knowledge about the formation of these well-defined frameworks.

2.2 CONVENTIONAL METHODS

Conventional synthesis is a term used for those reactions in which conventional electric heating is used. It is a dynamic tool to accelerate the invention of novel MOFs. There are various methods in this category, which are as follows (Figure 2.1) [11].

2.2.1 HYDRO-OR SOLVOTHERMAL METHODS

According to the definition provided by Rabenau, solvothermal reactions are those reactions that occur on the temperature above the boiling point of the solvent under autogenous pressure and the process that is carried out below room temperature at an ambient pressure range are known as non-solvothermal reactions. In this type of method, a mixture of metal salt and organic linker in solvent are kept in Teflon-lined vessel and then heated at a temperature in the range of 50–260°C [11,12]. Generally, polar solvents are used in this method, such as acetonitrile, acetone, diethyl formamide, dimethyl formamide, ethyl alcohol, methyl alcohol, water, or their mixtures. It is worth mentioning that if the solvent used in this process is water, then this is known as a hydrothermal method. This method is also known as a direct precipitation reaction [13]. It is noteworthy here that the reaction temperature

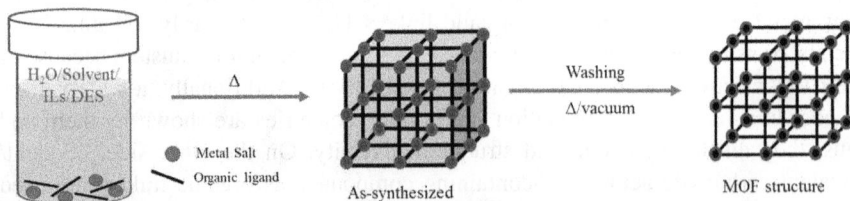

FIGURE 2.1 Hydro (solvo) thermal and Ionothermal methods for the synthesis of MOFs (Reproduced with permission (Creative Commons CC BY 4.0 license) from Magu et. al., 2019 [11]).

influences the morphology of the product as well as an increase in temperature can facilitate bond formation if kinetically inert ions are used. Horcajada et al. designed and synthesized a Fe-MOF, through the hydrothermal or solvothermal method, and explored their ability as a transport vehicle to deliver anticancer drug oridonin [14]. Goel et al. synthesized $\{[Tb(L_1)(L_2)_{0.5}(NO_3)(DMF)].DMF\}_n$ under solvothermal condition with 1,10-phenanthroline and 3,3',5,5'-azobenzenetetracarboxylic acid. The synthesized MOF showed high selective sensing for volatile organic compound (acetone) and powerful explosive (2,4,6-trinitrophenol) through a fluorescence quenching mechanism [15]. Li et al. developed a porous Cu-MOF, $[Cu(tdc)(H_2O)]_n.n$ (DMA), by using $CuCl_2$ as well as thiophene-2,5-dicarboxylic acid through solvothermal reaction, which exhibited high uptake capacity for light hydrocarbons [16]. Sensitive identification of Cu^{2+} and UO_2^{2+} has been achieved by a luminescent Eu-MOF $(Eu_2(MTBC)(OH)_2(DMF)3(H_2O)_4]\cdot2DMF\cdot7H_2O)$ designed and synthesized by Wang and co-researchers under a solvothermal environment [17]. Manos et al. investigated a Zr-MOF by reacting $ZrCl_4$ and 2-alkyl-amino-terephthalic acid in dimethyl formamide, which displayed high sorption capability for the toxic $SeCN^-$ [18]. Kamal et. al. synthesized nickel-based MOF-74, and it showed high CO_2 uptake (5.80 mmol/g), which was quite higher than the elemental synthesis [19]. Nelyubina and co-workers synthesized the known MOF-5 $(Zn_4O(BDC)_3)$, (BDC = terephthalate anion) by the solvothermal method in autoclaves prepared by 3D printing from polypropylene [20]. Jaroniec and co-workers synthesized MOF-520(Al), MOF-5(Zn), ZIF-8(ZnO–Zn), MOF-74(Mg) by the solvothermal technique, with high surface areas and large micropore volumes that were the main factors for the C_6H_6 adsorption [21]. Olawale and colleagues reported a $[Ni(II)(Tpy)(Pydc)]\cdot2H_2O$ MOF was synthesized by the use of nickel nitrate, terpyridine, and pyridine dicarboxylate through the solvothermal method [22].

2.2.2 IONOTHERMAL METHOD

This method uses the ionic liquids (ILs) and can be considered as a sub-class of the solvothermal method. Like dissolve like is the basis of using ILs as solvents [11]. These ILs are found appropriate to dissolve the necessary inorganic part of the synthesized MOFs. These are eco-friendly than organic solvents because of their low vapors, non-flammability, and high thermal stability. The cations as well as anions of ionic liquids produce the diversity in the reaction environments and employed the variety of frameworks [23]. Literature confirm that a wide number of MOFs have been fabricated by using ILs (especially obtained from 1-alkyl-3-methylimidazolium). The use of deep eutectic solvents (DESs), which are obtained by mixing two or more compounds, is also permitted. The DESs show the lower melting point compared to both of its constituents, exhibit the similar solvent properties like ILs, and are used as structure directing agents [24,25]. Ahn et al. prepared HKUST-1 through a conventional ionothermal synthesis route in an oven [26]. Chou et al. synthesized a novel coordination polymer, $[RMI]_2[Co_2(BTC)_2(H_2O)_2]$ (RMI = 1-alkyl-3-methylimidazolium, alkyl) by ionothermal reaction of cobalt salts and 1,3,5-benzenetricarboxylic acid), and discussed their physicochemical properties [27]. Rujiwatra et al. developed two zinc coordination polymers, $EMIm_2[Zn_3(C_6H_4(COO)_2)_3Cl_2]$ and $BMIm_2[Zn_3(C_6H_4(COO)_2)_3Cl_2]$, where the ionic liquids 1-ethyl-3-methylimidazolium chloride (EMIm-Cl) and 1-butyl-3-methylimidazolium chloride (BMIm-Cl) act as solvents [28]. Jiao et al. synthesized a series of Eu-(H)BDC compounds, [RMI][Eu$_2$(BDC)$_3$Cl] under ionothermal conditions, and checked their selectivity towards aniline [29]. Uk Kwon et al. prepared [RMI]$_2$[Ni$_3$(BTC)$_2$(OAc)$_2$] and [RMI]$_2$[Ni$_3$(HBTC)$_4$(H$_2$O)$_2$] by using Ni(OAc)$_2$ and 1,3,5-benzenetricarboxylic acid in various ILs, [RMI]X (R = ethyl, n-propyl, n-butyl; X = Cl, Br, I), and demonstrated their relative thermodynamic stability [30]. Chen et al. investigated a new coordination polymer, [(EEIM)NaCu(phth)$_2$]$_n$ (phth = terephthalate), through the mixture of ILs such as 1-ethyl-3-methylimidazolium tetrafluoroborate and chiral 1-ethyl-3-methylimidazolium-l-lactate, which act as solvents, and also evaluate the role of these ILs in ionothermal synthesis [31]. A drawback of this method is that the guest (solvent) molecules remain included in the framework and the process of its removal without the decomposition of framework can be a tough task. Ahmad et. al. investigated a Zn-MOF by using zinc nitrate hexahydrate and benzene-1,3,5-tricarboxylic acid or benzene-1,4-dicarboxylate in C1$_6$PyBr ionic liquid at room temperature. In this process, ILs accelerated the formation of MOFs and shortened the reaction time (6 hours) [32].

2.3   UNCONVENTIONAL METHODS

Mechano-chemical method of synthesis is considered as unconventional method because of its various features [33]. Primarily, this method involves the splintering of intramolecular bonds through mechanical efforts accompanied by the chemical transformation (Figure 2.2) [11,13]. The method is performed at room temperature by the means of ball mill or by grinding manually, and is considered to be nature-friendly because of its ability to ensue under solvent free conditions [34,35]. Out of 12 principles of green chemistry, this method follows 1 principle i.e., less hazardous chemical synthesis. Moreover, herein the high quantitative yield is also obtained within 10–60 minutes. Thus, the method is proven an advantageous technique.

James and colleagues prepared MOFs by using this method for the first time in 2006. The group prepared a microporous MOF (Cu(INA)$_2$) by grinding the isonicotinic acid (INA) along with copper acetate in the absence of heat for 10 minutes. Milner et al. synthesized the isomeric frameworks Mg$_2$(m-dobdc) (m-dobdc^{4-} = 2,4-dioxidobenzene-1,5-dicarboxylate) and Mg$_2$(dobpdc) (dobpdc^{4-} = 4,4'-dioxidobiphenyl-3,3'-dicarboxylate) under solvent-free conditions. The obtained

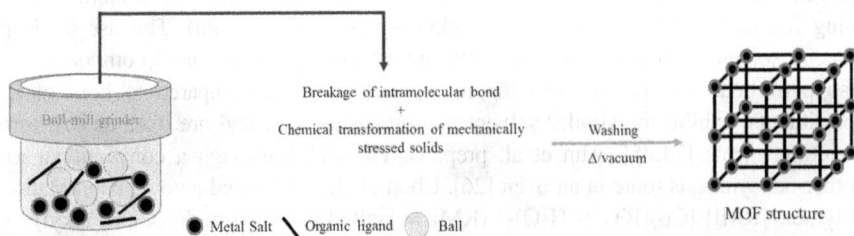

FIGURE 2.2   Mechanochemical method for the synthesis of MOFs (Reproduced with permission (Creative Commons CC BY 4.0 license) from Magu et. al., 2019 [11]).

MOFs displayed high crystallinities, surface areas and permanent porosity [36]. In addition, it was found that some researchers use a minute amount of solvent, and this reaction condition may lead to hasten the mechano-chemical reactions by enhancing the reactants' mobility. This mechano-chemical process is termed liquid-assisted grinding (LAG), where liquid displays the structure directing properties. HKUST-1 $(Cu_3(BTC)_2)$ has been synthesized mechanochemically [37]. In this report, copper acetate was found to be a preferred starting material, and the ratio of cupric ion to that of the Benzene tri-carboxylic acid was found to be 3:2. The synthesis through LAG leads to the superior crystalline products bearing improved absorption and adsorption properties. The product synthesized by less solvent condition, also displayed the BET surface area augmented to 1,364 m^2/kg instead of 1,084 m^2/kg through LAG after the addition of methanol (100 micromililitre). Recently, both ions and liquids are used to synthesize the MOFs, and this method is known as ion and liquid assisted grinding (ILAG) [38]. Friscic et al., synthesized Imidazolate-based MOFs by ILAG method [39]. Milner et al. reported M_2(dobdc) MOFs (M = Mg, Mn, Co, Ni, Cu, Zn; dobdc^{4-} = 2,5-dioxidobenzene-1, 4-dicarboxylate) by using this method at room temperature within 5 minutes only [37]. Wang et al. synthesized two pillared-layer MOFs via the mechanochemical method at room temperature. The mechanochemically synthesized MOFs displayed high purity and porosity as compared to traditional solvothermal method [40].

2.4 ALTERNATIVE METHODS

In addition to the above-mentioned routes i.e., conventional and unconventional method, alternative modes have also been endeavored for the preparation of MOFs. It is well known that chemical reactions require some form of energy as input. In conventional solvothermal methods, supply of energy is furnished by electric heating, whereas in unconventional mechano-chemical method, it is supplied through mechanically. Apart from these, in alternative methods, the energy supply can also be done in other ways like electric potential, EM radiation, ultrasound waves, etc., and it has been found that the crystallization rate, particle size, and morphologies differ from those synthesized by conventional and unconventional methods. Therefore, diversity in the topology of MOFs has been shown that cannot be observed otherwise. In the biomedical field, nanocrystalline and non-toxic MOFs with high loading capacities are also dreamed of, and microwave assisted as well as sonochemical methods are envisioned for this purpose. Additionally, for commercial use, the establishment of inexpensive as well as rapid routes is important for large-scale production and energy-efficient processes. A few of them are discussed below.

2.4.1 MICROWAVE-ASSISTED METHOD

In this method, energy supply is required to carry out the reaction, and this is obtained by the microwave radiation (MW). Microwave ovens regulate the temperature and pressure during the course of the reaction, which can be monitored easily, and hence a much-precise control of reaction conditions is made possible.

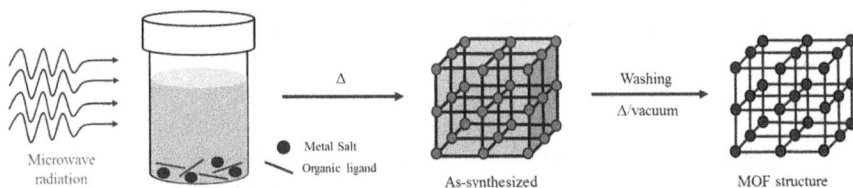

FIGURE 2.3 Microwave assisted method for the synthesis of MOFs (Reproduced with permission (Creative Commons CC BY 4.0 license) from Magu et. al., 2019 [11]).

Numerous MOFs have been constructed by the use of this method. The goal of MOF synthesis through the MW-assisted method is to accelerate the crystallization and improvement in the purity of the product [41–44]. It is also used to synthesize the polymorphs as well as nanoscale products. The basis of microwave-assisted synthesis is the interaction of mobile electric charge with electromagnetic waves. The charge is found on the polar solvent molecules or ions. In this process, the substrate mixture formed in the appropriate solvent system is moved to the Teflon vessel prior to place in the microwave unit followed by heating for the adequate time at particular temperature, as shown in Figure 2.3 [11]. The polar molecules present in the solution tend to align themselves in the EM field; thus, the orientation of the molecules is changed permanently. Moreover, when suitable frequency is applied, molecular collisions take place due to which the kinetic energy and the temperature of the system are increased.

Because of this type of direct interaction between radiation and polar molecules, this method demonstrates a very dynamic method of heating. Customarily, it has been observed that MW-assisted synthesis is faster than conventional electric heating for the synthesis of MOFs. Cr-MIL-100 was the first MOF synthesized by MW-assisted heating. It took 4 hours under the microwave irradiation at 220°C for the synthesis of MOF with 44% yield as compared to conventional electric heating [45,46]. IRMOF-1 has also been synthesized by the MW-assisted route, which produced the crystals of higher quality and showed better adsorption property for CO_2 [47]. Also, the synthesis of HKUST-1 through this route revealed that this method was better to rapidly synthesize crystals of HKUST-1 in the 10–20 micrometer range with high yield as compared to other classic methods [48–50]. Kustov et al. synthesized CPO-27-Mg under microwave irradiation that exhibited an adequately high drug loading for aspirin and paracetamol [51]. The Hf-UiO-66 and Zr-UiO-66 MOFs, prepared in less time and high yield by using the microwave irradiation, showed very small size [52]. Devarayapalli et al. reported the rapid synthesis procedure of Ni-MOF through the microwave method, and the synthesized MOF showed high photocatalytic efficiency [53].

2.4.2 ELECTROCHEMICAL METHOD

Generally, MOF powders on an industrial scale are synthesized by this method. Researchers at BASF synthesized MOFs through the electrochemical method for the first time [54]. They synthesized few MOFs by using Mg, Co, Cu, and Zn as

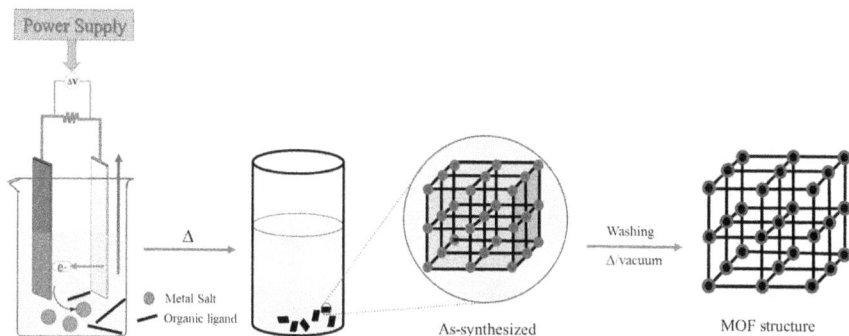

FIGURE 2.4 Electrochemical method for the synthesis of MOFs (Reproduced with permission (Creative Commons CC BY 4.0 license) from Sillanpää et al., 2021 [55]).

cathodes, with H_3BTC and H_2BDC as organic linkers. Herein, instead of metal salts, the metal ions (provided continuously via anodic dissolution) react with the organic linkers dissolved in the reaction medium containing the conducting salt. Metal deposition on the cathode can be averted by using polar protic solvents but hydrogen gas is evolved in this process. MOFs containing ionic liquids have also been synthesized via this route (Figure 2.4) [55].

The solvothermal method was used to synthesize the HKUST-1 by using pure ethanol or a mixture of ethanol/water as solvent system under ambient pressure in electrochemical synthesis [49,55]. The obtained dark blue powder with 1,820 m^2/g surface area showed the inferior quality during the sorption studies. This was probably due to the inclusion of organic linkers as well as conducting salts in the pores during the crystallization process. A $Cu_3(HHTP)_2$ MOF has been fabricated with superior crystalline quality having the value of electrical conductivity around 0.087 S cm^{-1} [56]. Easun and co-researchers synthesized an amine-based Mn-MOF with 93% yield for the first time by using an electrochemical method. The fabricated MOF showed a high CO_2 (92.4 wt%) and H_2 (12.3 wt%) uptake [57]. Bloch and colleague also applied this method for the synthesis of Ti^{III}-MIL-101, which exhibited the superior quality as compared to the conventional routes [58].

2.4.3 SONOCHEMICAL METHOD

In this technique, the reaction mixture is kept under the high energy ultrasound radiation of 20 kHz to 10 MHz (Figure 2.5) [55]. This method is considered to be rapid as well as eco-friendly for the synthesis of MOFs. The primary intention of using this method was to develop a distinct, fast, energy-efficient and room-temperature feasible procedure. Though the wavelength of used ultrasound radiation is plenty higher than the molecular, there is no chemical reaction that occurs due to the direct interface of molecules and ultrasound radiation. But under the low-pressure conditions, these radiations interact with the liquids that resulted in the formation of small bubbles/cavities due to a pressure drop below the vapor pressure of used solvent(s). Due to the diffusion of solute vapor into the volume of a bubble, an alternating pressure is maintained and under the influence of this alternating

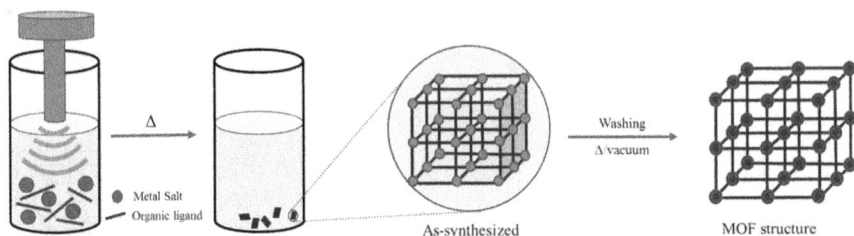

FIGURE 2.5 Sonochemical method for the synthesis of MOFs (Reproduced with permission (Creative Commons CC BY 4.0 license) from Sillanpää et al., 2021 [55]).

pressure, the bubble grows and attains its maximum size, which becomes highly unstable and debacles [59]. In this process, high energy is released with extremely high pressure (~1,000 bar) and high temperature (~5,000 K).

This synthetic approach was used to fabricate the MOFs in 2008, but still it is needing more exploration. Qiu et al. synthesized a $Zn_3(BTC)_2$ MOF by using this method [60]. Compared to solvothermal synthesis, the resulted MOF was achieved under mild conditions and lesser time. The HKUST-1 MOF has been synthesized in just 5 minutes by using an ultrasonic bath [61]. The method was also used by Ahn and colleagues for the fabrication of small size (5–25 μm) crystals of a MOF within less time of 30 minutes as compared to the conventional solvothermal synthetic procedure, which took 24 hours [62]. The same synthetic strategy was again used by Ahn and co-researchers for the preparation of a MOF in 40 minutes using 1-methyl-2-pyrrolidone as a solvent. The crystals of this MOF also showed the small size of 5-20 mm along with high CO_2 adsorption capacity value of 1315 mg g^{-1} [63]. Morsali and co-workers investigated a 3-D Zn-MOF via the sonochemical method, and resulted in the formation of uniform nanoplates by increasing the irradiation time and decreasing the concentration [64]. The research team of Joo also succeeded in the preparation of a Mg- MOF by using this method [65].

2.4.4 MICROEMULSION METHOD

The microemulsion method has been extensively used for the construction of nanoparticles [66]. Recently, this method has also been used to synthesize MOFs. Li et al. first time reported a long rod Zn-MOF by using $Zn(NO_3)_2$ and 1,3,5-benzenetricarboxylic acid in ionic liquid microemulsions. This method can be easily carried out at room temperature and no energy input is required [67]. Moreover, the morphology of the synthesized MOFs could be controlled through this method at room temperature. The ionic liquid-containing microemulsion system was used by Sun and colleagues for the synthesis of nanoscale zeolitic imidazolate frameworks (NZIFs) with superior thermal stabilities [68]. Huang et al. prepared a Co-MOF-74 with a controllable morphology and size. Bagherzadeh and co-workers synthesized Fe-MIL-88A nanorods showing well-defined morphology [69]. Han and co-workers used the 1,3,5-benzenetricarboxylic acid and lanthanum (III) nitrate for the preparation of a La-MOFs via the microemulsion technique [70].

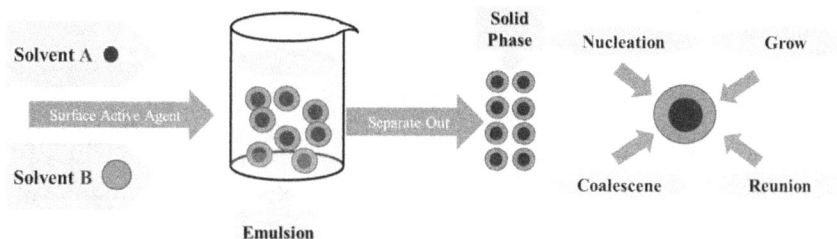

FIGURE 2.6 Microemulsion method for the synthesis of MOFs (Reproduced with permission (Creative Commons CC BY 4.0 license) from Zhu et. al., 2018 [66]).

In this method, water microemulsions consist of water droplets of nanometer size which are stagnant by a surfactant on the organic phase. Additionally, the kinetics of nucleation as well as crystal growth was also controlled by the micelles of microemulsions (Figure 2.6) [66]. The prime utility of this method is that dimensions of the nanoscale materials can be guarded but the high cost and environmental pollutant surfactants are the major disadvantage of this method [71].

2.4.5 DRY-GEL CONVERSION METHOD

The method is extensively used to synthesize the zeolites and zeolite membranes. In this method, when the amorphous alumino-silicate gel comes in contact with volatile amines and water vapors, it is converted into the crystalline zeolite (Figure 2.7) [55, 72–74]. This comprises of two separate methods including (i) the vapor phase transport method and (ii) steam-assisted crystallization method [75]. The complete transformation of gel to crystalline materials with large yield is noticeable benefits of using this method. The method was firstly used by Li and colleagues for the synthesis of UiO-66 MOF with strong catalytic activity and reusability during the esterification. They also did not use any kind of amide solvents as well as hydrochloric acid during the preparation of discussed MOF [76]. Motkuri et al. prepared Ni-MOF-74 produced through the dry-gel conversion method, which exhibited a very high yield with improved sorption characteristics [77]. Recently, Dong et al. and Jhung et al. used this method to synthesize ZIF-8 and Fe-MIL-100, respectively.

2.4.6 MICROFLUIDIC METHOD

It is necessary for a process to be continuous and rapid as well as feasible to be employed commercially or industrially, and this synthetic approach is regarded as a suitable candidate for this purpose [11,78]. Lately, HKUST-1 MOF has been prepared by using this method. In this method, two types of substrate mixtures are required. Therefore, hydrated copper acetate and polyvinyl alcohol were added in water at room temperature and hence the first aqueous phase was formed. After that, an organic linker solution was formed by adding H_3BTC in 1-octanol and heated to 60°C so that it gets dissolved completely. After the preparation of both mixtures,

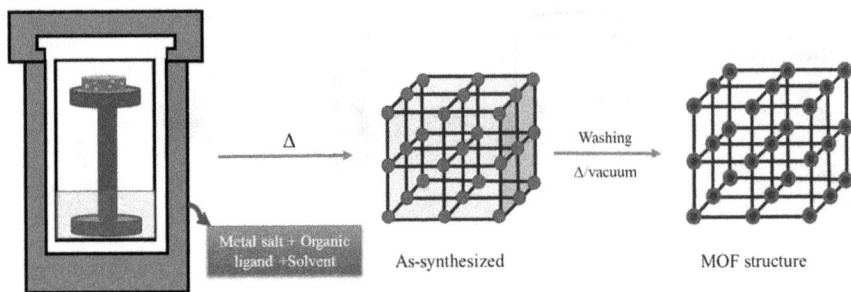

FIGURE 2.7 Dry-gel conversion method for the synthesis of MOFs (Reproduced with permission (Creative Commons CC BY 4.0 license) from Sillanpää et al., 2021 [55]).

they were supplied to a T-junction with the help of syringe pump. In the T-junction, the aqueous solution droplets are continuously formed in the organic phase. HKUST-1 capsule shells with capsule walls of approximately 4 mm thickness were formed at a liquid-liquid interface. From the outside, a macro-porous structure is possessed by them, while densely packed crystallites are present inside (Figure 2.8) [11,79,80].

Schneider et al. developed ZIF-8 particles (from ca. 300 to 900 nm) within 10 min by using an environmentally friendly microfluidic process. [81]. Yao et al. synthesized zirconium-MOF by a microfluidic method [82]. Compared with the other traditional methods, microfluidic synthesis was more efficient.

2.4.7 SLOW EVAPORATION AND DIFFUSION METHOD

Both methods are executed at room temperature and do not require any energy supply to perform the reaction. In a slow evaporation method, the solution of reagents is mixed together and left for slow evaporation to favor nucleation and crystal growth and when critical concentration is reached, crystals are formed (Figure 2.9) [83].

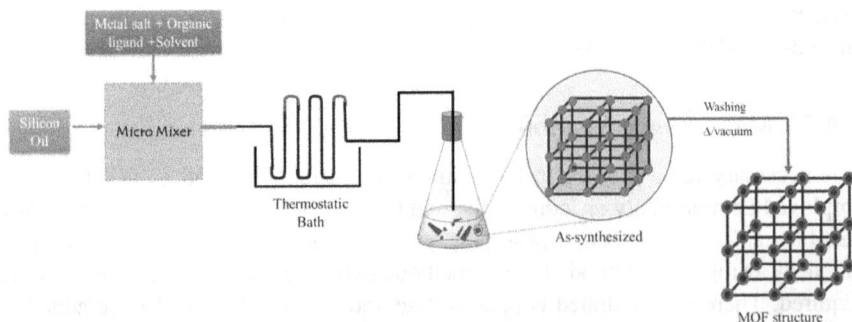

FIGURE 2.8 Microfluidics method for the synthesis of MOFs (Reproduced with permission (Creative Commons CC BY 4.0 license) from Magu et al., 2019 [11]).

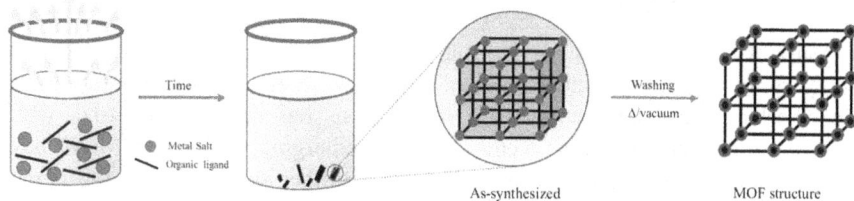

As-synthesized   MOF structure

FIGURE 2.9 Slow evaporation method for the synthesis of MOFs (Reproduced with permission from Arya et al., 2020 [83] Order Number: 5292940104585).

To accelerate this process, mixtures of low boiling point solvents are used [84,85]. Xu et al. synthesized three MOFs, [Cu(2,3-pydc)(bpp)]·2.5H$_2$O, [Zn(2,3-pydc)(bpp)]·2.5H$_2$O, and [Cd(2,3-pydc)(bpp)(H$_2$O)]·3H$_2$O at room temperature, and reported their emissions properties [86]. Watkins and co-workers also prepared another 1D supramolecular MOF using 2,6-pyridinedicarboxylic acid and cobalt(II) by slow evaporation, also investigated its catalytic properties [87]. In the diffusion method, solutions of reactants are kept one above another and between them a layer of solvent is placed (Figure 2.10).

The gels have been used occasionally as diffusion as well as crystallization media, and the crystallization occurs in the interface between the layers. Yaghi and colleagues reported Zn$_4$O(BDC)$_3$.(DMF)$_8$(C$_6$H$_5$Cl) (MOF-5) by this diffusion technique. The resulted MOF was found to exhibit the superior pore volume and apparent surface area as compared to most of the porous crystalline zeolites [88]. Hmadeh et al. synthesized MOF-199 in a hydrogel medium using a reaction-diffusion framework [89]. Goel synthesized two Cu-MOFs by the diffusion method, and demonstrated their magnetic properties [90]. Nidhi et al. also prepared four new mixed ligand-based Cd-MOFs through this method [91]. The synthesized MOFs also exhibited an attractive luminescent property, and showed selective as well as sensitive detection of 4-nitrophenol. A diffusion technique was also carried out for the preparation of MOF-70 [92].

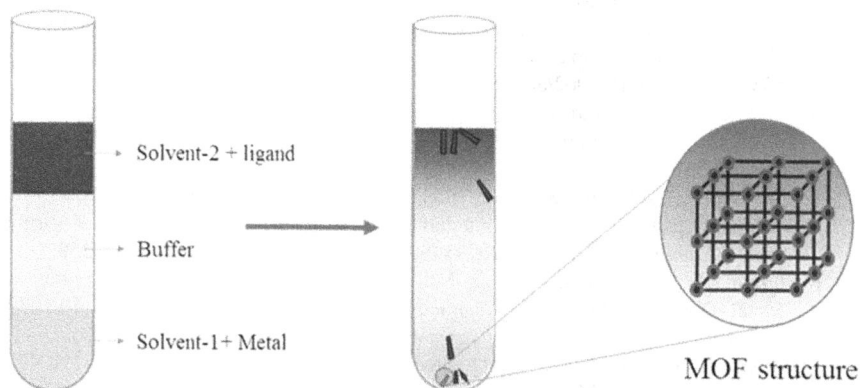

Solvent-2 + ligand

Buffer

Solvent-1+ Metal

MOF structure

FIGURE 2.10 Diffusion method for the synthesis of MOFs.

2.5 CONCLUSIONS

Although the field of metal organic frameworks has grown rapidly, still there are plenty of things left to understand in this field. In the preparation of desired framework, the conventional method of synthesis has been widely used, while unconventional and alternative methods are just emerging. Only a few MOFs have been reproduced by these methods. Thus, the further optimization of synthetic conditions and organized studies are needed to perform. The increasing awareness to design the novel approaches regarding alteration of morphology as well as the size of MOFs in order to mold its properties is the primary goal of scientists in this century. This chapter has summarized various synthetic strategies used to fabricate the hybrid MOFs with superior chemical properties and wide applications. To boost the wide variety of applications exhibited by MOFs along with their mechanistic studies is a significant challenge and needs more exploration for further work.

ACKNOWLEDGMENT

N.G gratefully acknowledges the financial support from UGC, New Delhi (Letter No. F.30-431/2018(BSR), M-14-55) and BHU (Letter No. R/Dev/D/IoE/Seed Grant/2020-23). Authors also thank BHU and IIT Indore for providing the research work facilities.

REFERENCES

[1] Zhou, H., Long, J. R., & Yaghi, O. M. 2012. "Introduction to Metal-Organic Frameworks." *Chemical Reviews*. 112 (2), 673–674. doi:10.1021/cr300014x

[2] Zhou, H. J., & Kitagawa, S. 2014. "Metal Organic Frameworks (MOFs)." *Chemical Society Reviews*. 43 (16), 5415–5418. doi:10.1039/C4CS90059F

[3] Tan, J.-C., & Civaller, B. 2015. "Metal–Organic Frameworks and Hybrid Materials: From Fundamentals to Applications." *CrystEngComm*. 17 (2), 197–198. doi:10.1039/c4ce90162b

[4] Hu, Y. H., & Zhang, L. 2010. "Hydrogen Storage in Metal–Organic Frameworks." *Advanced Materials*. 22 (20), E117–E130. doi:10.1002/adma.200902096

[5] Yuan, D., Zhao, D., Sun, D., & Zhou, H.-C. 2010. "An Isoreticular Series of Metal-Organic Frameworks with Dendritic Hexacarboxylate Ligands and Exceptionally High Gas-Uptake Capacity." *Angewandte Chemie International Edition*. 49 (31), 5357–5361. doi:10.1002/anie.201001009

[6] Pettinari, C., Marchetti, F., Mosca, N., Tosi, G., & Drozdov, A. 2017. "Application of Metal–Organic Frameworks." *Polymer International*. 66 (6), 731–744. doi:10.1002/pi.5315

[7] Walker, G. C., Konda, S. S. M., Maji, T. K., & Kirk, S. S. 2020. "Preface to the "Metal–Organic Frameworks: Fundamental Study and Applications" Joint Virtual Issue." *Langmuir*. 36 (49), 14901–14903. doi:10.1021/acs.langmuir.0c03350

[8] Czaja, A. U., Trukhan, N., & Muller, U. 2009. "Industrial Applications of Metal–Organic Frameworks." *Chemical Society Reviews*. 38 (5), 1284. doi:10.1039/B804680H

[9] Stock, N. & Biswas, S. 2012. "Synthesis of Metal-Organic Frameworks (MOFs): Routes to Various MOF Topologies, Morphologies, and Composites." *Chemical Reviews*. 112 (2), 933–969. doi:10.1021/cr200304e

[10] Rasheed, T., Rizwan, K., Bilal, M., & Iqbal, H. M. N. 2020. "Metal-Organic Framework-Based Engenered Materials-Fundamentals and applications." *Molecules.* 25 (7), 1598. doi-10.3390/molecules25071598

[11] Anumah, A., Louis, H., Zafar, S-U., Hamzat, A. T., Amusan, O. O., Pigweh, A. I., Akakuru, O. U., Adeleye, A. T., & Thomas O. M. 2019. "Metal-Organic Frameworks (MOFs): Recent Advances in Synthetic Methodologies and Some Applications." *Chemical Methodologies.* 3 (3), 283–305. doi:10.22034/CHEMM.2 018.139807.1067

[12] Safaei, M., Foroughi, M. M., Ebrahimpoor, N., Jahani S., Omidi, A., & Khatami, M. 2019. "A Review on Metal-Organic Frameworks: Synthesis and Applications." *Trends in Analytical Chemistry.* 118, 401–425. doi:10.1016/j.trac.2019.06.007

[13] Soni, S., Bajpai, P. K., & Arora. C. 2020. "A Review on Metal-Organic Framework: Synthesis, Properties and Application." *Characterization and Application of Nanomaterials.* 3 (2), 87–106. doi:10.24294/CAN.V2I2.551

[14] Horcajada, P., Chalati, T., Serre, C., Gillet, B., Sebrie, C., Baati, T., Eubank, J. F., Heurtaux, D., Clayette, P., Kreuz, C., Chang, J.-S., Hwang. Y. K., Marsaud,V., Bories, P.-N., Cynober, L., Gil, S., Ferey, G., Couvreur, P., & Gref, R. 2009. "Porous Metal–Organic-Framework Nanoscale Carriers as a Potential Platform for Drug Delivery and Imaging." *Nature Materials.* 9 (2), 172–178. doi:10.1038/nmat2608

[15] Goel, N., & Kumar, N. 2018. "A Dual-Functional Luminescent Tb(III) Metal–Organic Framework for the Selective Sensing of Acetone and TNP in Water." *RSC Advances.* 8 (20), 10746–10755. doi:10.1039/C7RA13494K

[16] Li, W.-W., Guo, Y., & Zhang, W. H. 2017. "A Porous Cu(II) Metal-Organic Framework: Synthesis, Crystal Structure and Gas Adsorption Properties." *Journal of Molecular Structure.* 1143, 20–22. doi:10.1016/j.molstruc.2017.04.068

[17] Liu, W., Wang, Y., Song, L., Silver, M. A., Xie, J., Zhang, L., Chen, L., Diwu, J., Chai, Z., & Wang S. 2019. "Efficient and Selective Sensing of Cu2+ and UO22+ by a Europium Metal-Organic Framework." *Talanta.* 196, 515–522. doi:10.1016/ j.talanta.2018.12.088

[18] Pournara, A. D., Rapti, S., Valmas, A., Margiolaki, I., Andreou, E., Armatas, G. S., Tsipis, A. C., Plakatouras, J. C., Giokas, D. L., & Manos M. J. 2021. "Alkylamino-Terephthalate Ligands Stabilize 8-Connected Zr4+ MOFs with Highly Efficient Sorption for Toxic Se Species." *Journal of Material Chemistry A.* 9 (6), 3379–3387. doi:10.1039/d0ta11653j

[19] Kamal, K., Bustam, M. A., Ismail, M., Grekov, D., Shariff, A. M., & Pre P. 2020. "Optimization of Washing Processes in Solvothermal Synthesis of Nickel-Based MOF-74." *Materials.* 13 (12), 2741. doi:10.3390/ma13122741

[20] Denisov, G. L., Primakov, P. V., Korlyukov, A. A., Novikov, V. V., & Nelyubina, Y. V. 2019. "Solvothermal Synthesis of the Metal-Organic Framework MOF-5 in Autoclaves Prepared by 3D Printing." *Russian Journal of Coordination Chemistry.* 45 (12), 836–842. doi:10.1134/S1070328419120030

[21] Gwardiak, S., Szczęśniak, B., Choma, J., & Mietek J. 2018. "Benzene Adsorption on Synthesized and Commercial Metal–Organic Frameworks." *Journal of Porous Materials.* 26 (3), 775–783. doi:10.1007/s10934-018-0678-0

[22] Olawale, M. D., Obaleye, J. O., & Oladele, E. O. 2020. "Solvothermal Synthesis and Characterization of Novel [Ni(II)(Tpy)(Pydc)].2H2O Metal-Organic Framework as an Adsorbent for the Uptake of Caffeine Drug from Aqueous Solution." *New Journal of Chemistry.* 44 (43), 18780–18791. doi:10.1039/D0NJ04316H

[23] Oh, H. C., Jung, S., Ko, I. J., & Choi E. Y. 2018. "Ionothermal Synthesis of Metal-Organic Framework." In *Recent Advancements in the Metallurgical Engineering and Electrodeposition,* edited by Uday Al-Naib, Dhanasekaran, Vikraman , and K. Karuppasamy . London: IntechOpen. doi:10.5772/intechopen.79156

[24] Lee, Y. R., Kim, J., & Ahn W. S. 2013. "Synthesis of Metal-Organic Frameworks: A Mini Review." *Korean Journal of Chemical Engineering.* 30 (9), 1667–1680. doi: 10.1007/s11814-013-0140-6

[25] Lin, Z., Wragg, D. S., Warren, J. E., & Morris R. E. 2007. "Anion Control in the Ionothermal Synthesis of Coordination Polymers." *Communications.* 129 (34), 10334–10335. doi: 10.1021/ja0737671

[26] Kim, S. H., Yang, S. T., Kim, J. & Ahn W. S. 2011. "Sonochemical Synthesis of Cu3(BTC)2 in a Deep Eutectic Mixture of Choline Chloride/dimethylurea." *Bulletin of the Korean Chemical Society.* 32 (8), 2783–2786. doi: 10.5012/bkcs.2011.32.8.2783

[27] Ko, I. J., Oh, H. C., Cha, Y. J., Han, C. H., & Choi E. Y. 2017. "Ionothermal Synthesis of a Novel 3D Cobalt Coordination Polymer with a Uniquely Reported Framework: [BMI]$_2$[Co$_2$(BTC)$_2$(H$_2$O)$_2$]." *Advances in Materials Science and Engineering.* 2017, 1–6. doi: 10.1155/2017/3237247

[28] Tapala, W., Prior, T. J., & Rujiwatra A. 2014. "Two-Dimensional Anionic Zinc Benzenedicarboxylates: Ionothermal Syntheses, Structures, Properties and Structural Transformation." *Polyhedron.* 68, 241–248. doi: 10.1016/j.poly.2013.10.036

[29] Feng, H. J., Xu, L., Liu, B., & Jiao H. 2016. "Europium Metal-Organic Frameworks as Recyclable and Selective Turn-off Fluorescence Sensors for Aniline Detection." *Dalton Transactions.* 45 (43), 17392–17400. doi: 10.1039/c6dt03358j

[30] Xu, L., Yan, S., Choi, E. Y., Lee, J. Y., & Kwon, Y. U. 2009. "Product Control by Halide Ions of Ionic Liquids in the Ionothermal Syntheses of Ni–(H)BTC Metal–Organic Frameworks." *Chemical Communications.* 23, 3431. doi: 10.1039/b902223f

[31] Xiahou, Z. J., Wang, Y. L., Liu, Q. Y., Wei, J. J., & Chen, L. L. 2013. "Ionothermal Synthesis of a 3D Heterometallic Coordination Polymer Based on the Rod Shaped Copper(II)–Sodium(I)-Carboxylate Secondary Building Units with a PCU Topology." *Inorganic Chemistry Communications.* 38, 62–64. doi: 10.1016/j.inoche.2013.09.076

[32] Ahmad, H., Yusoh, N. A., Jumbri, K., Basyaruddin, M., & Rahman A. 2020. "Ionothermal Synthesis of Zn-Based Metal Organic Frameworks in Pyridinium Ionic Liquid." *Malaysian Journal of Analytical Sciences.* 24(2), 159–164. http://pkukmweb.ukm.my/mjas/

[33] Klimakow, M., Klobes, P., Thünemann, A. F., Rademann, K., & Emmerling, F. 2010. "Mechanochemical Synthesis of Metal–Organic Frameworks: A Fast and Facile Approach toward Quantitative Yields and High Specific Surface Areas." *Chemistry of Materials.* 22 (18), 5216–5221. doi: 10.1021/cm1012119

[34] Głowniak, S., Szczęśniak, B., Choma, J., & Jaroniec, M. 2021. "Mechanochemistry: Toward Green Synthesis of Metal–Organic Frameworks." *Materials Today.* 46, 109–124. doi: 10.1016/j.mattod.2021.01.008

[35] Miao, Y.-R., & Suslick, K. S. 2018. "Mechanochemical Reactions of Metal-Organic Frameworks." *Supramolecular Chemistry, Advances in Inorganic Chemistry.* 71, 403–434. doi: 10.1016/bs.adioch.2017.11.001

[36] Wang, Z., Li, Z., Ng, M., & Milner, P. J. 2020. "Rapid Mechanochemical Synthesis of Metal-Organic Frameworks using Exogenous Organic Base." *Dalton Transactions.* 49 (45), 16238–16244. doi: 10.1039/D0DT01240H

[37] Yang, H., Orefuwa, S., & Goudy, A. 2011. "Study of Mechanochemical Synthesis in the Formation of the Metal–Organic Framework Cu$_3$(BTC)$_2$ for Hydrogen Storage." *Microporous and Mesoporous Materials.* 143 (1), 37–45. doi: 10.1016/j.micromeso.2011.02.003

[38] Tao, C.-A., & Wang, J.-F. 2020. "Synthesis of Metal Organic Frameworks by Ball-Milling." *Crystals*. 11 (1), 15. doi:10.3390/cryst11010015

[39] Beldon, P. J., Fábián, L., Stein, R. S., Thirumurugan, A., Cheetham, A. K., & Friščić, T. 2010. "Rapid Room-Temperature Synthesis of Zeolitic Imidazolate Frameworks by Using Mechanochemistry." *Communications*. 122 (50), 9834–9837. doi:10.1002/ange.201005547

[40] Gao, T., Tang, H.-J., Zhang, S.-Y., Cao, J.-W., Wu, Y.-N., Chen, J., & Chen K.-J. 2021. "Mechanochemical Synthesis of Three-Component Metal-Organic Frameworks for Large Scale Production." *Journal of Solid-State Chemistry*. 303, 122547. doi:10.1016/j.jssc.2021.122547

[41] Klinowski, J., Paz, F. A. A., Silvab, P., & Rocha J. 2011. "Microwave-Assisted Synthesis of Metal–Organic Frameworks." *Dalton Transactions*. 40 (2), 321–330. doi:10.1039/c0dt00708k

[42] Blanita, G., Ardelean, O., Lupu, D., Borodi, G., Mihet, M., Coros, M., Vlassa, M., Ioan, M., Coldea, I., & Popeneciu, G. 2011. "Microwave Assisted Synthesis of MOF-5 at Atmospheric Pressure." *Revue Roumaine de Chimie*. 56(6), 583–588.

[43] Khan, N. A., & Sung H. J. 2015. "Synthesis of Metal-Organic Frameworks (MOFs) with Microwave or Ultrasound: Rapid Reaction, Phase-Selectivity, and Size Reduction." *Coordination Chemistry Reviews*. 285, 11–23. doi:10.1016/j.ccr.2014.10.008

[44] Ge, J., Wu, Z., Huang, X., & Ding, M. 2019. "An Effective Microwave-Assisted Synthesis of MOF235 with Excellent Adsorption of Acid Chrome Blue K." *Journal of Nanomaterials*. 2019, 1–8. doi:10.1155/2019/4035075

[45] Jhung, S. H., Lee, J.-H, Yoon, J. W., Serre, C., Férey, G., & Chang, J.-S. 2007. "Microwave Synthesis of Chromium Terephthalate MIL-101 and Its Benzene Sorption Ability." *Advanced Materials*. 19 (1), 121–124. doi:10.1002/adma.200601604

[46] RUI, Zhengqiu, LI, Quanguo, CUI, Qun, WANG, Haiyan, CHEN, Haijun, & YAO, Huqing (2014). Adsorption Refrigeration Performance of Shaped MIL-101-Water Working Pair. *Chinese Journal of Chemical Engineering*, 22(5), 570–575 10.1016/s1004-9541(14)60076-8

[47] Lu, C.-M., Liu, J., Xiao, K., & Harris, A. T. 2010. "Microwave Enhanced Synthesis of MOF-5 and its CO_2 Capture Ability at Moderate Temperatures Across Multiple Capture and Release Cycles." *Chemical Engineering Journal*. 156 (2) 465–470. doi:10.1016/j.cej.2009.10.067

[48] Seo, Y.-K., Hundal, G., Jang, I. T., Hwang, Y. K., Jun, C.-H., & Chang J.-S. 2009. "Microwave Synthesis of Hybrid Inorganic–Organic Materials Including Porous $Cu_3(BTC)_2$ from Cu(II)-Trimesate Mixture." *Microporous and Mesoporus Materials*. 119 (1–3), 331–337. doi:10.1016/j.micromeso.2008.10.035

[49] Schlesinger, M., Schulze, S., Hietschold, M., & Mehring M. 2010. "Evaluation of Synthetic Methods for Microporous Metal–Organic Frameworks Exemplified by the Competitive Formation of $[Cu_2(btc)_3(H_2O)_3]$ and $[Cu_2(btc)(OH)(H_2O)]$." *Microporous and Mesoporous Materials*. 132 (1–2), 121–127. doi:10.1016/j.micromeso.2010.02.008

[50] Khan, N. A., Haque, E., & Jhung, S. H. 2010. "Rapid Syntheses of a Metal–Organic Framework Material Cu3(BTC)2(H2O)3 under Microwave: A Quantitative Analysis of Accelerated Syntheses." *Physical Chemistry Chemical Physics*. 12 (11), 2625. doi:10.1039/b921558a

[51] Kudelin, A. I., Papathanasiou, K., Isaeva, V., Caro, J., Salmi, T., & Kustov L. M. 2021. "Microwave-Assisted Synthesis, Characterization and Modeling of CPO-27-Mg Metal-Organic Framework for Drug Delivery." *Molecules*. 26 (2), 426. doi:10.3390/molecules26020426

[52] Dang, Y. T., Hoang, H. T., Dong, H. C., Bui, K. B. T., Nguyen, L. H. T., Phan, T. B., Kawazoe, Y., & Doan, T. L. H. 2020. "Microwave-Assisted Synthesis of Nano Hf- and Zr-based Metal-Organic Frameworks for Enhancement of Curcumin Adsorption." *Microporous and Mesoporous Materials.* 298, 110064. doi: 10.1016/j.micromeso.2020.110064

[53] Devarayapalli, K. C., Vattikuti, S. V. P., TVM, S., Yoo, K. S., Nagajyothi, P. C., & Shim, J. 2019. "Facile Synthesis of Ni-MOF using Microwave Irradiation Method and Application in the Photocatalytic Degradation." *Materials Research Express.* 6 (11), 1150h3. doi: 10.1088/2053-1591/ab5261

[54] Mueller, U., Puetter, H., Hesse, M., Schubert, M., Wessel, H., Huff, J., & Guzmann, M. 2011. Method for Electrochemical Production of a Crystalline Porous Metal Organic Skeleton Material, United States BASF Aktiengesellschaft(Ludwigshafen, DE), Patent: 7968739. https://www.freepatentsonline.com/7910732.html

[55] Joseph, J., Iftekhar, S., Srivastava, V., Fallah, Z., Zaree, E. N., & Sillanpää, M. 2021. "Iron-based Metal-organic Framework: Synthesis, Structure and Current Technologies for Water Reclamation with Deep Insight into Framework Integrity." *Chemosphere.* 284, 131171. 10.1016/j.chemosphere.2021.131171

[56] Liu, Y., Wei, Y., Liu, M., Bai, Y., Wang, X., Shang, S., Chen, J., & Liu, Y. 2020. "Electrochemical Synthesis of Large Area Two-Dimensional Metal–Organic Framework Films on Copper Anodes." *Angewandte Chemie International Edition.* 60 (6), 2887–2891. doi: 10.1002/anie.202012971

[57] Asghar, A., Iqbal, N., Noor, T., Kariuki, B. M., Kidwell, L., & Easun, T. L. 2021. "Efficient Electrochemical Synthesis of a Manganese-based Metal–Organic Framework for H_2 and CO_2 Uptake." *Green Chemistry.* 23 (3), 1220–1227. doi: 10.1039/d0gc03292a

[58] Antonio, A. M., Rosenthal, J., & Bloch, E. D. 2019. "Electrochemically Mediated Syntheses of Titanium(III)-Based Metal–Organic Frameworks." *Journal of the American Chemical Society.* 141 (29), 11383–11387. doi: 10.1021/jacs.9b05035

[59] Vaitsis, C., Sourkounib, G., & Argirusis, C. 2020. "Sonochemical Synthesis of MOFs." *Metal-Organic Frameworks for Biomedical Applications.* 223–244. doi: 10.1016/B978-0-12-816984-1.00013-5

[60] Qiu, L.-G., Li, Z.-Q., Wu, Y., Wang, W., Xu, T., & Jiang, X. 2008. "Facile Synthesis of Nanocrystals of a Microporous Metal–Organic Framework by an Ultrasonic Method and Selective Sensing of Organoamines." *Chemical Communications.* 31, 3642. doi: 10.1039/B804126A

[61] Li, Z.-Q., Qiu, L.-G., Xu, T., Wu, Y., Wang, W., Wu, Z.-Y., & Jiang, X. 2009. "Ultrasonic Synthesis of the Microporous Metal–Organic Framework $Cu_3(BTC)_2$ at Ambient Temperature and Pressure: An Efficient and Environmentally Friendly Method." *Materials Letter.* 63 (1), 78–80. doi: 10.1016/j.matlet.2008.09.010

[62] Son, W.-J., Kim, J., Kim, J., & Ahn W.-S. 2008. "Sonochemical Synthesis of MOF-5." *Chemical Communications.* 47, 6336. doi: 10.1039/b814740j

[63] Jung, D.-W., Yang, D.-A., Kim, J., Kim, J., & Ahn W. S. 2010. "Facile Synthesis of MOF-177 by a Sonochemical Method using 1-methyl-2-Pyrrolidinone as a Solvent." *Dalton Transactions.* 39 (11), 2883. doi: 10.1039/b925088c

[64] Abdollahi, N., Masoomi, M. Y., Morsali, A., Junk, P. C., & Wang, J. 2018. "Sonochemical Synthesis and Structural Characterization of a New Zn(II) Nanoplate Metal–Organic Framework with Removal Efficiency of Sudan Red and Congo Red." *Ultrasonics Sonochemistry.* 45, 50–56. doi: 10.1016/j.ultsonch.2018.03.001

[65] Tahmasian, A., Morsali, A., & Joo, S. W. 2013. "Sonochemical Syntheses of a One-Dimensional Mg(II) Metal-Organic Framework: A New Precursor for Preparation of MgO One-Dimensional Nanostructure." *Journal of Nanomaterials.* 2013, 1–7. doi: 10.1155/2013/313456

[66] Zhan, H., Bian, Y., Yuan, Q., Ren, B., Hursthouse, A., & Zhu, G. 2018. "Preparation and Potential Applications of Super Paramagnetic Nano-Fe$_3$O$_4$." *Processes*. 6 (4), 33. doi:10.3390/pr6040033

[67] Ye, R., Ni, M., Xu, Y., Chen, H., & Li, S. 2018. "Synthesis of Zn-Based Metal–Organic Frameworks in Ionic Liquid Microemulsions at Room Temperature." *RSC Advances*. 8 (46), 26237–26242. doi:10.1039/c8ra04573a

[68] Zheng, W., Hao, X., Zhao, L., & Sun, W. 2017. "Controllable Preparation of Nanoscale Metal–Organic Frameworks by Ionic Liquid Microemulsions." *Industrial & Engineering Chemistry Research*. 56 (20), 5899–5905. doi:10.1021/acs.iecr.7b00694

[69] Bagherzadeh, E., Zebarjad, S. M., Hosseini, H. R. M., & Khodaei, A. 2022. "Interplay Between Morphology and Band Gap Energy in Fe-MIL-88A Prepared via a High Temperature Surfactant-Assisted Solvothermal Method." *Materials Chemistry and Physics*. 277, 125536. doi:10.1016/j.matchemphys.2021.125536

[70] Shang, W., Kang, X., Ning, H., Zhang, J., Zhang, X., Wu, Z., Mo, G., Xing, X., & Han, B. 2013. "Shape and Size Controlled Synthesis of MOF Nanocrystals with the Assistance of Ionic Liquid Mircoemulsions." *Langmuir*. 29 (43), 13168–13174. doi:10.1021/la402882a

[71] Raptopoulou, C. P. 2021. "Metal-Organic Frameworks: Synthetic Methods and Potential Applications." *Materials* 14 (2), 310. doi:10.3390/ma14020310

[72] Xu, W., Dong, J., Li, J., Li, J., & Wu, F. 1990. "A Novel Method for the Preparation of Zeolite ZSM-5." *Journal of the Chemical Society Chemical Communications*. 10, 755. doi:10.1039/C39900000755

[73] Koekkoek, A. J. J., Degirmenci, V., & Hensen, E. J. M. 2011. "Dry Gel Conversion of Organosilane Templated Mesoporous Silica: From Amorphous to Crystalline Catalysts for Benzene Oxidation." *Journal of Materials Chemistry*. 21 (25), 9279. doi:10.1039/c1jm10779h

[74] Luo, Y., Tan, B., Liang, X., Wang, S., Gao, X., Zhang, Z., & Fang, Y. 2020. "Dry Gel Conversion Synthesis and Performance of Glass-Fiber MIL-100(Fe) Composite Desiccant Material for Dehumidification." *Microporous and Mesoporous Materials*. 297, 110034. doi:10.1016/j.micromeso.2020.110034

[75] Matsukata, M., Ogura, M., Osaki, T., Rao, P. R. H. P., Nomura, M., & Kikuchi, E. 1999. "Conversion of Dry Gel to Microporous Crystals in Gas Phase." *Topics in Catalysis*. 9(1-2), 77–92. doi:10.1023/a:1019106421183

[76] Lu, N., Zhou, F., Jia, H., Wang, H., Fan, B., & Li, R. 2017. "Dry-Gel Conversion Synthesis of Zr-Based Metal–Organic Frameworks." *Industrial & Engineering Chemistry Research*. 56 (48), 14155–14163. doi:10.1021/acs.iecr.7b04010

[77] Das, A. K., Vemuri, R. S., Kutnyakov, I., McGrail, B. P., & Motkuri, R. K. 2016. "An Efficient Synthesis Strategy for Metal-Organic Frameworks: Dry-Gel Synthesis of MOF-74 Framework with High Yield and Improved Performance." *Scientific Reports*. 6 (1), 28050. doi:10.1038/srep28050

[78] Schoenecker, P. M., Belancik, G. A., Grabicka, B. E., & Walton, K. S. 2012. "Kinetics Study and Crystallization Process Design for Scale-up of UiO-66-NH$_2$ Synthesis." *American Institute of Chemical Engineers Journal* 59 (4), 1255–1262. doi:10.1002/aic.13901

[79] Ameloot, R., Vermoortele, F., Vanhove, W., Roeffaers, M. B. J., Sels, B. F., & Vos, D. E. D. 2011. "Interfacial Synthesis of Hollow Metal–Organic Framework Capsules Demonstrating Selective Permeability." *Nature Chemistry*. 3 (5), 382–387. doi:10.1038/nchem.1026

[80] Witters, D., Vergauwe, N., Ameloot, R., Vermeir, S., Vos, D. E. D., Puers, R., Sels, B., & Lammertyn, J. 2012. "Digital Microfluidic High-Throughput Printing of Single Metal-Organic Framework Crystals." *Advanced Materials*. 24 (10), 1316–1320. doi:10.1002/adma.201104922

[81] Kolmykov, O., Commenge, J.-M., Alem, H., Girot, E., Mozet, K., Medjahdid, G., & Schneider, R. 2017. "Microfluidic Reactors for the Size-Controlled Synthesis of ZIF-8 Crystals in Aqueous Phase." *Materials and Design*. 122, 31–41. doi:10.1016/j.matdes.2017.03.002

[82] Bingchen, S., Cong, L., Yingya, L., Anjie, W., Yao, W., Jian, Z., & Weixu, L. 2019. "Microfluidic Synthesis of UiO-66 Metal-organic Frameworks Modified with Different Functional Groups." *Chemical Journal of Chinese Universities*. 40(7), 1365–1373. doi:10.7503/cjcu20180831

[83] Arya, S., Mahajan, P., Gupta, R., Srivastava, R., Tailor, K. N., Satapathi, S., Sumathi, R. R., Datt, R., & Gupta, V. 2020. "A Comprehensive Review on Synthesis and Applications of Single Crystal Perovskite Halides". *Progress in Solid State Chemistry*. 60, 100286. 10.1016/j.progsolidstchem.2020.100286

[84] Halper, S. R., Do, L., Stork, J. R., & Cohen, S. M. 2006. "Topological Control in Heterometallic Metal–Organic Frameworks by Anion Templating and Metalloligand Design." *Journal of the American Chemical Society*. 128 (47), 15255–15268. doi:10.1021/ja0645483

[85] Du, M., Li, C.-P., & Zhao, X.-J. 2006. "Metal-Controlled Assembly of Coordination Polymers with the Flexible Building Block 4-Pyridylacetic Acid (Hpya)." *Crystal Growth & Design*. 6 (1), 335–341. doi:10.1021/cg0502542

[86] Wang, G.-H., Li, Z.-G., Jia, H.-Q., Hu, N.-H., & Xu, J.-W. 2008. "Metal–Organic Frameworks based on the Pyridine-2,3-Dicarboxylate and a Flexible Bispyridyl Ligand: Syntheses, Structures, and Photoluminescence." *CrystEngComm*. 11 (2), 292–297. doi:10.1039/B809557D

[87] Murinzi, T. W., Hosten, E., & Watkins, G. M. 2017. "Synthesis and Characterization of a Cobalt-2,6-Pyridinedicarboxylate MOF with Potential Application in Electrochemical Sensing." *Polyhedron*. 137, 188–196. doi:10.1016/j.poly.2017.08.030

[88] Yaghi, O. M., Li, H., Eddaoudi, M., & O'Keeffe, M. 1999. "Design and Synthesis of an Exceptionally Stable and Highly Porous Metal-Organic Framework." *Nature*. 402 (6759), 276–279. doi:10.1038/46248

[89] Al-Ghoul, M., Issa, R., & Hmadeh, M. 2017. "Synthesis, Size and Structural Evolution of Metal–Organic Framework-199 via a Reaction–Diffusion Process at Room Temperature." *CrystEngComm*. 19 (4), 608–612. doi:10.1039/C6CE02436J

[90] Goel, N. 2016. "Study of New Cu(II) Metal-Organic Frameworks: Syntheses, Structural, Gas Sorption and Magnetic Properties." *Inorganica Chimica Acta*. 450, 330–336. doi:10.1016/j.ica.2016.06.009

[91] Goel, N., & Kumar, N. 2019. "Study of Four New Cd(II) Metal-Organic Frameworks: Syntheses, Structures, and Highly Selective Sensing for 4-Nitrophenol." *Inorganica Chimica Acta*. 503, 119352. doi:10.1016/j.ica.2019.119352

[92] Rosi, N. L., Kim, J., Eddaoudi, M., Chen, B., O'Keeffe, M., & Yaghi, O. M. 2005. "Rod Packings and Metal–Organic Frameworks Constructed from Rod-Shaped Secondary Building Units." *Journal of the American Chemical Society*. 127 (5), 1504–1518. doi:10.1021/ja045123o

3 Properties and Factors Affecting the Preparation of Metal-Organic Frameworks

Chansi, Rashi Bhardwaj, and Tinku Basu

CONTENTS

DOI: 10.1201/9781003252061-3

3.1   INTRODUCTION

In the last two decades, the chemistry of porous substances has revolutionized through booming progress portrayed over the canvas of metal-organic frameworks (MOFs). Aided by their stunning structures and attributes, MOFs have an immense command over the chemical community. The thrust is towards the designing and synthesis of novel MOFs employing facile approaches with minimum reactants to achieve spectacular properties and maximum surface area through an inter-penetrating interlinking resulting finely tuned porous network [1]. The varied connection of metal and organic linkers tend to achieve a fascinating variety of architectures and topologies, with enduring porosities, easily altered surface traits, and thermal resistance, along with their innumerable potential applications [2–6].

Aided by their wide scope in application towards advanced materials, precise tuning of synthetic methods, availability of ligands with various coordination nodes for metal ion/s, and the influences of numerous factors have become prominent in this arena [7]. Crystal engineering of MOF surfaces has been achieved through the analysis of their 3D structure and optimization of their growth methodologies [8]. The alignment of the reactants in coordination polymers greatly occurs through their interaction in a solid state leading to the synthesis of small building blocks later, by self-assembly processes, MOFs growth is based on the same interactions. In the tryst to understand the synthesis of MOF, a lot of experimentation has been performed by chemists to decipher the laws for the synthesis of tailored MOF structures [9].

From the structural viewpoint, the three-dimensional framework of MOF can be projected through the assortment of reaction components. The physical attributes of the connecting linker, such as their firmness or flexibility, length, size, weight, and linear or nonlinear geometry, execute a vital role in the production of individual macromolecular constructions [10]. Hence, the prudent choice of inorganic nodes and organic ligands make MOFs a good matrix for target-specific applications [11]. Besides coordinating functional groups (e.g., carboxylates, amines), various other functional groups such as amino(-NH$_2$), halo(-X), hydroxyl(-OH), ether(-O-), thiol (-SH), etc. could be introduced into the ligands. Additionally, multiple functional groups could be mounted at a specific position on the ligand by the implementation of positional regioisomers, which assist to explicate the individual impacts that each substituent levies on the organic linkers [12]. Therefore, methodical structural and functional features can be created into a pristine framework. Flexible MOFs have inherent abilities to display different structural alterations or dynamic performances toward different external stimuli, due to their breathing behavior that can be attained by the aromatic exchange within the organic ligand [13]. The bridging length and non-coordinative side groups of organic linkers can perform a significant role in the framework for their flexibility and diversity. Recently, specific organic inserts that serve as growth inhibitors of structures are directly synchronized with the inorganic nodes to design the final 2D hybrid material [14]. These groups tend to modulate the physio-chemical behavior of MOF, enhancing their reactivity and uptake capabilities.

The aim is to design the extended solids through molecular building networks to monitor the basic architecture of the building blocks towards the development of MOF that occurs through the rigid entities termed as *secondary building units (SBU)*. The design of MOF structures through SBU imparts structural rigidity and predicts the topology of the framework. Besides all these, predetermined control of structure, properties, and function can be achieved by several synthetic conditions that play a definite role to form crystals of MOFs [15].

Hence, the present chapter investigates the intricate relationship between various factors and properties affecting the synthesis of MOF. A brief overview of types of organic ligands along with their assembling strategies (positional isomeric and aromatic substitutional effect) controlling the resulting MOF topology and their role in the synthesis of MOF has been discussed. Topological modulation arising due to the regiospecific isomerism of carboxylate, pyridine, and mixed ligands is discussed in detail. The role of the spacer/modulator in guiding the dimensionality in the structural framework has been illustrated. Further, the function of secondary building units (SBUs) in producing a predictable and rigid framework is reviewed [16,17]. The excellent modulation of size and morphology can be manipulated by the solvent composition compelled by the bonding of organic ligands and the solvent molecules [18]. The possible distribution of the linker molecule promoting the self-assembly of the MOF controlling its microstructure is relied on a judicious selection of the solvent [19]. Temperature plays a vital role in altering the morphology, dimensionality, and pore volume, employing them for varied applications. The pH of the medium has a wider potential in regulating these properties as well as monitoring the coloration of synthesized crystals. Inspired by their discerning effect on the structural characteristics of linkers and in the assembling modes of ligands, these parameters have been elaborately discussed.

3.2 ASSEMBLING STRATEGIES OF ORGANIC LIGANDS

MOF design is relied upon over two main interacting components of secondary building units, (SBUs); the metallic centers along with bridging organic ligands. The properties and functionality may perhaps evolve either from individual secondary building units or the nature and type of connections between them. Further, the synthesis of each unit is governed by the types of ligands and their assembling strategies.

3.2.1 Types of Organic Ligands

A large number of organic ligands are categorized based on coordinating groups such as carboxylate ligands, porphyrin ligands, and nitrogen-containing heterocyclic ligands. Heterocyclic linkers (polypyridyl-ligands, pyridyl-ligands) improvise functional attributes to synthesized MOF structures.

3.2.1.1 Carboxylate Ligands

Amongst various organic ligands, carboxylate ligands have been used as multidentate linkers by coordinating with metal nodes as hydrogen donors and acceptors to design supramolecular assemblies; hence, a variety of MOF metal carboxylates have been created. MOF structures designed utilizing aryl dicarboxylates have been synthesized by π–π assembling, along with hydrogen-bonded interaction to synthesize captivating crystal structures ranging from a 1D linear structure to a 3D framework [20–22]. Aromatic heterocyclic carboxylic acids [23] possessing multiple proton donor and acceptor electrons along with the presence of nitrogen and oxygen in carboxylates, associate with metal ions to form accurate complex structures with exceptional network topology [24]. Ditopic carboxylate linkers have been explored since the start of the MOF research, partly due to their ready availability and perhaps to a certain extent due to easily perceivable structures formed by combining different SBUs. Interestingly, the twisted angle of linear dicarboxylates has a profound influence on determining both the dimensionality as well as the topology of the final architecture.

3.2.1.2 Porphyrin-Based Ligands

Porphyrin-based ligands are significantly important in MOF synthesis, particularly in light-harvesting, sensing, and catalysis. Employing porphyrins into MOF platforms annihilates the routes for dimerization and reduces the solubility and chemical stability concerns so often associated with porphyrin structural chemistry. Thus, the presentation of porphyrins into MOFs as linkers for constructing photosensitive building units and post-synthetic metalation of metal-free porphyrins lead to alteration of catalytic property of porphyrins-based MOFs.

3.2.1.3 Heterocyclic Ligands

Compared to carboxylate and other homocyclic ligands, reviewed previously the coordination of heterocyclic organic ligands (polypyridyl) provides a superior mechanism. The immediacy amongst nearby pyridyl groups surrounded by the linker has been appropriate for the development of metal linking sites that could accept multiple diatomic metal cations. Monophosphonic and biphosphonic acids have been investigated to produce MOFs of the pillared-layered structures devoid of any certain void spaces. A limited porosity has been observed due to the occupancy of free phosphonic acid molecules in between the phosphonate layers. Chelation of metal by linkers utilizing three coplanar chelators is termed as *pincer ligand*. In organometallic chemistry, their important role has been proven for catalytic applications. An aim to design novel MOF structures has led to the involvement of mixed organic ligands in their constructions.

Presently a variety of mixed ligand MOFs possessing greater tenability and functionality have been developed. Recent studies have proven that linkers used in MOF synthesis are not behaving as inert structural units rather they yield high-end benefits if their functional properties are utilized. This has led to a noteworthy cultural exit to create more and more MOF with intricately functionalized organic ligand groups. Further, their assembling pattern determines the topology and functionality of the framework, and hence discussed in detail in subsequent sections.

3.2.2 Positional Isomeric Effect

The organic ligand possessing a similar coordination number but located at different positions displays a positional isomeric effect for structural assembly in MOF synthesis. The positional isomeric effect of different types of ligands is discussed.

3.2.2.1 Effect of Carboxylate Ligand

Isophthalic acid and terephthalic are the simplest examples of having binding tendencies at different angles, finally forming different structural assemblies while synthesizing MOF. Jubaraj B. Baruah et al. [25] have studied the ten different positional isomers of naphthalene dicarboxylates and utilized them in the synthesis of various structural MOFs. They act as unrestricted scaffolds to alter the directionalities and structures of a MOF.

Chen et al. [26] have demonstrated the synthesis of five isomeric ligands using a central naphthyl core attached to two peripheral isophthalate moieties in various ways to synthesize varied copper-based MOF. The synthesized MOF displays different topologies (ssa and nbo) and also displays varied methane gas absorption behavior. Two novel 3D pillar-layered MOFs [27] driven by positional isomerism of aromatic multicarboxylate anion have been synthesized with unique (3,5)-linked binodal gra topology and (4,4,5,9)-linked 4-nodal topology, respectively. They display varied photoluminescence behavior.

Zhao et al. [28] have demonstrated the positional isomeric effect of zwitter ionic ligand (1-carboxylatomethylpyridinium-3-carboxylate and 4-carboxylate analog for the synthesis of 2D and 3D topologies of Pb-based MOFs.

3.2.2.2 Effect of Mixed Ligand

A large number of studies have been devoted to positional isomeric effect displayed by polycarboxylate and pyridyl mixed linkers for the construction of different MOFs [29–36].

The systematic positional effect of ligands has been studied by Fan et al. [37] in the synthesis of Cu-based MOF using 5-(pyridine-3-yl) isophthalate linkers with positional substitution of methyl groups yielding four different MOF. It has been detected that methyl substitution at the *para* position of N atom in pyridine results in framework **ZJNU-27** with *eea*-type distinct topology, whereas when methyl groups are attached to different positions, *rtl*-type topology is reported, as seen in Figure 3.1.

Jin-long Zhu et al. [38] have recently reported the Mn(II)-organic frameworks (MOFs) based on the multi-functional 5-iodoisophthalic acid (H_2iip) and a pair of positional isomers N,N'-bis(3-pyridylmethyl)oxalamide (3bpmo) and N,N'-bis(4-pyridylmethyl)oxalamide (4bpmo), $\{[Mn(iip)(3bpmo)0.5 \cdot (H_2O)] \cdot 0.5(3bpmo)\}_n$, and $\{[Mn(iip)(4bpmo) \cdot (H_2O)] \cdot (H_2O)\}_n$ for studying their effect on conduction of proton and additionally in luminescent-sensing. 2D structures are pillared by the

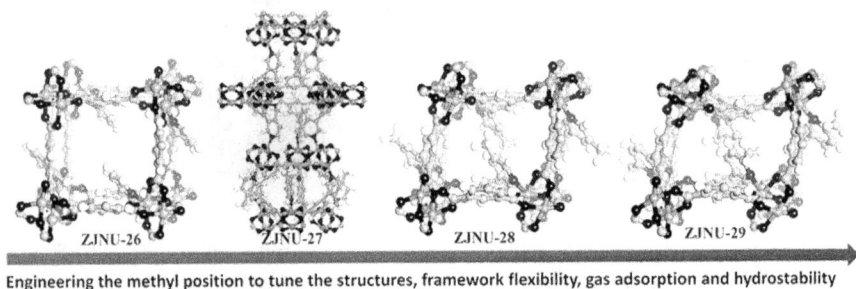

Engineering the methyl position to tune the structures, framework flexibility, gas adsorption and hydrostability

FIGURE 3.1 Engineering topology of MOFs through tuning methyl position (Reproduced with permission [37].

lattice 3bpmo particle via 1D hydrogen-bonded tape, directing the formation of the 3D supramolecular framework. Both isomers display variation in redox characteristics as displayed in electrochemical studies.

Crystal engineering aspect of positional isomerism [39] has been demonstrated by attaching two pyridylmethyl groups (PMD) at either side of nitrogen in PMD cores leading to synthesis of isomeric organic complexes, 2-PMPMD and 4-PMPMD (2/4-PMPMD = N, N'-bis(2/4-pyridylmethyl)-pyromellitic diimide), which are amassed into various supramolecular systems by lone pair-π connections. The structures acquired possess exceptional photochromic performances with completely distinct photo responsive rate and coloration contrast, which are assigned to the separate interfacial networks of electron donors/acceptors.

Zhou et al. [40] have described the structural and functional transformation of Cu (II)-tricarboxylate (Figure 3.2) frameworks by utilizing pairs of unsymmetrical biaryl tricarboxylate ligands altered with a methyl group and a pyridinic-N atom at separate positions. It has been noted in diffraction experiments that three different kinds of topological shapes are attributed to the steric influence of the methyl group and the chelating effect related to the pyridinic-N atom.

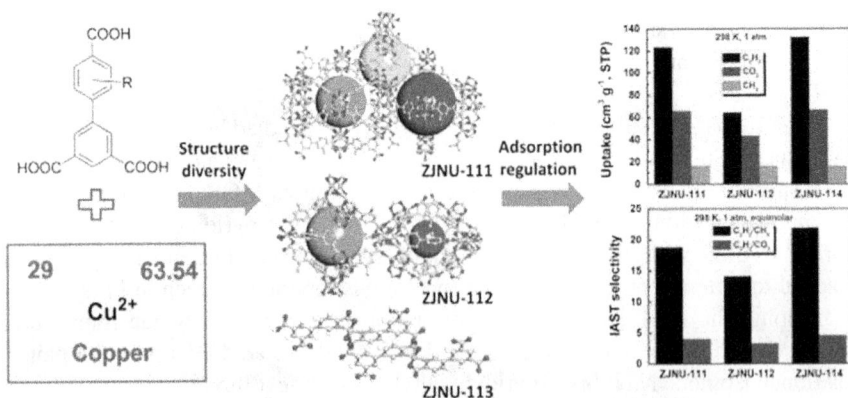

FIGURE 3.2 Modulation of Cu MOF using bi-aryl tricarboxylate ligand using substituent effect (Reproduced with permission Lin et al. 2021).

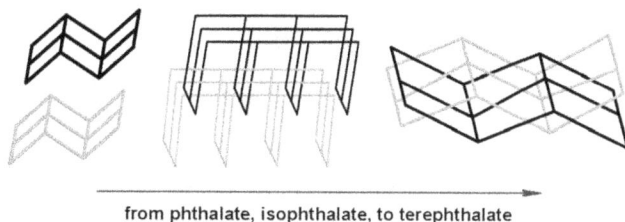

from phthalate, isophthalate, to terephthalate

FIGURE 3.3   Varied structural framework synthesized using benzene dicarboxylate isomers (Reproduced with permission Du, Jiang, and Zhao et al. 2007).

Du et al. have synthesized a series of nine mixed metal MOF [41] using bent dipyridyl linker 4-amino-3,5-bis(4-pyridyl)-1,2,4-triazole (bpt) and three benzene dicarboxylate isomers (pa = phthalate, ip = isophthalate, and tp = terephthalate), respectively, with different metal ions such as Co^{2+}, Ni^{2+}, Cu^{2+}, Zn^{2+}, and Cd^{2+}. Figure 3.3 displays multidimensional coordination modes connecting the metal node to align a 1-D chain or ribbon arrays, which is extended via the exo-bidentate bpt linkers establishing a variety of coordination framework, with ornamented (3,6) topology, 2-D bilayer (twofold interpenetration), 3-D poly threaded architecture (1-D + 2-D), and two-fold interpenetrating porous lattice of (4,4) layers.

3.2.3   Aromatic Substituent Eeffects

Aromatic substitution in ligands has a key role in affecting the synthesis and behavior of MOFs. They perform twin modalities; firstly, they provide interconnecting functionalities for extension of the coordination network through aromatic stacking and hydrogen bonding. Secondly, they might serve as inert substituents enhancing the steric hindrance affecting the electronic binding capabilities of ligands, finally affecting the framework of synthesized MOF. The effect of aromatic substitution in an organic ligand during MOF synthesis is discussed in detail.

3.2.3.1   Effect of Substitution on Carboxylate Ligands

An array of zinc (II)-based DMOFs (DMOF = dabco MOF, dabco = 1,4-diazabicyclo[2.2.2]octane) has been presented using regioisomerism and substitution of BDC organic ligands [42] to study their differential breathing behavior [42]. NH_2–OMe combination has showcased noteworthy variations in elasticity (breathing behavior) by merely varying the positions of functional groups (from 2,3 to 2,5), as displayed in Figure 3.4. The density of n electrons over the benzene ring is a crucial factor for flexibility deviations in the regioisomeric MOF system.

Two positional isomeric ligands [43] and one of their derivatives with an attached nitro-group (−3,4′,5-bpca (H₃L₁),3,3′,5-bpca(H₃L₂) and 3-nitro-5,3′,5′-bpca (H₃L₃) along with ancillary ligands have been manipulated for attaining structural modularity.

Intermolecular interactions between an organic ligand [44] and a substituent polar side groups to functionalize benzenes units have been studied. The strongest categories of intermolecular binding are found to be: (i) amide lone pair donating atoms (N, O)

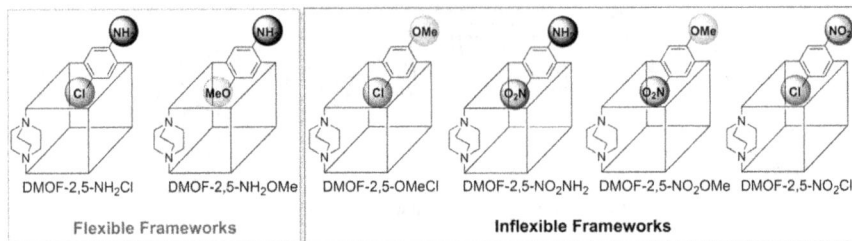

FIGURE 3.4 Effect of functional group position on the flexibility of MOF (Reproduced with permission Hahm et al. 2016).

present in side groups and also C of CO_2 and (ii) hydrogen bonding connections amongst acidic protons (of COOH and SO_3H groups) and CO_2 oxygen (enhancement of 3–4 kJ mol^{-1}). Both interactions have a large effect on splitting the CO_2 molecule.

Imidazole dicarboxylate ligands display various coordination modes. The introduction of a massive aromatic group at the 2-position of the imidazole ring indicates strong coordination ability [45]. This ability has prompted for the introduction of methyl or hydroxyl groups to alter the phenyl unit for preparing two organic ligands, 2-(p-hydroxylphenyl)-1H-imidazole-4,5-dicarboxylic acid (p-OHPhH$_3$IDC), and 2-(p-methylphenyl)-1H-imidazole-4,5-dicarboxylic acid (p-MePhH$_3$IDC) for the synthesis of 2D to 3D structures.

3.2.3.2 Effect of Substitution on Mixed Ligands

Mixed ligand approaches have been labored to achieve symbiotic functional properties of ligands. Further, the aromatic substitution of these ligands improvises their behavior manifold. Six cobalt-based MOF [46] have been effectively produced by a solvothermal method using a mixed-ligand methodology to study their adsorption properties for CO_2 and C_2 hydrocarbons, revealing incorporation of amino and bromo entities can strengthen the adsorption behavior. The influence of mixed organic linker over bandgap variation of polymorphic MOFs, UiO-66, and MIL-140A having different organic substitution (NO_2, NH_2) have been studied [47], as shown in Figure 3.5.

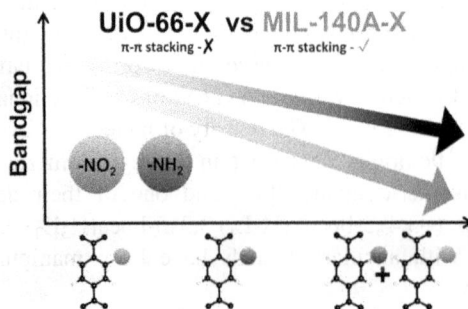

FIGURE 3.5 Effect of aromatic substitution using mixed ligand MOF (Reproduced with permission Cedeno et al. 2021).

Substituent engineering [48] of microenvironment and size of pores using substituent alteration has been utilized for enhancing the C_2H_6/C_2H_4 separation by synthesis of four isoreticular MOF using 5-(pyridin-3-yl)isophthalate ligands with a variety of substitution. The introduction of a nonpolar methyl group has improved separation efficiency compared to its parent compound measured through the absorption isotherms. 1,2,4-triazoles [49] have been produced by the reaction of a suitable aliphatic, aromatic, or heterocyclic primary amine with diformhydrazide.

Co-crystallizations of 2,3,5,6-tetramethyl pyrazine (TMP) [50] with different substituted aromatic compounds have been executed, resulting in the formation of three co-crystals (TMP: DNS = 1:1; TMP: CSA = 1:2; TMP: DCA = 1:2). The Hirshfeld surface investigation has unveiled that N–H⋯O/O–H⋯N strong hydrogen bonds and other intermolecular connections such as H⋯H, C–H⋯π, π⋯π, and lone-pair⋯π have contributed in a collaborative way to stabilize the supramolecular assemblies. Additionally, thermodynamic findings have revealed that halogen bonding exerted a positive effect on the stability of co-crystal structures.

3.2.4   SPACER EFFECT

Conjugated organic linkers help in the design of rigid organic ligands [51,52]. Synthesis of MOFs with rigid organic ligands leads to the formation of an open network with cavities. Hence, modulator ligands (spacers) have been introduced for modification in structures. On increasing the spacer length, the interpenetration within the structure tends to decrease the space available [53]. However, a new strategy implies with the use of flexible dynamic spacers that expand and contract on the effect of external stimuli. The spacer effect alters the porosity, dimensionality, and stability of synthesized frameworks and have been discussed.

3.2.4.1   Effect on Dimensionality

Yaghi et al. [44] have reported the distribution of triethylamine into zinc salt and dicarboxylate ligand dispersed in a solvent (DMF and chlorobenzene) and named it MOF-5. Further, they reported the series of H_2BDC derivatives and aligned these aromatic carboxylate ligands to form a series of MOF (IRMOF) using base prototypes as MOF-5. The synthesized MOF possesses 2D to 3D structures. Equated to single linker-based MOFs, mixed linkers have achieved a novel degree of cogent strategy and structure, involving the synergistic organization of distinct linkers to metals and the subsequent formation of network. The mixed ligands strategy can be well adopted by utilizing structural motifs serving as rod spacers [54]. The incorporation of malonic acid imparted stability to the structure; otherwise, it might lead to interpenetration in the structure.

Pyridyltetrazoles (pytz) are captivating linkers, exploited for the production of MOFs, owing to their potential to link multiple metal nodes and significant dissimilarities in tethering lengths. For tuning the dimensionality in MOF multivariate pytz, 4,4′-bipy, phen, and 4-pytz have been used as secondary inserts due to their (i) comparable molecular dimensions; (ii) alike synthesis methods; (iii) rare coordination positions; (iv) rigid/chelate coordination capability for tuning the dimensionality.

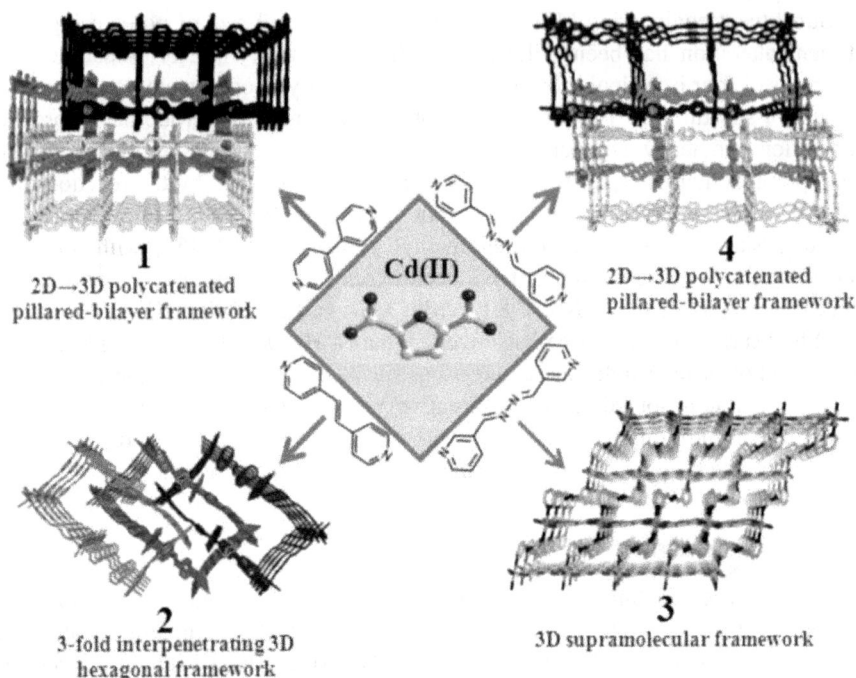

FIGURE 3.6 Different topologies of MOFs formed using auxiliary ligand-assisted structural variation (Reproduced with permission Singh and Nagaraja 2015).

Four new metal-organic frameworks (MOFs) designed using a Cd(II) ion have been efficaciously produced using a mixed ligand route (Figure 3.6).

The resulting structure is varied from a 2D → 3D polycatenated pillared-bilayer structure with **SP** 2-periodic net topology, a threefold interpenetrating 3D hexagonal framework structure with **sqc**-net topology. A photoluminescence study divulges emissions from compounds owing to ligand-based charge transfer (n → π* and π → π*) transitions. The luminescent emission of **2** can be reduced by adding a trace amount of 2,4,6-trinitrophenol (TNP), indicating its prospective application for extremely selective and sensitive recognition of TNP [55]. A glutamic acid-derived linker has been utilized for the cohort of chiral MOF constructions vacillating from 1D to 2D through a complex hierarchy of coordination and hydrogen bonds that comprise the encapsulation of solvent molecules [56]. Ionic MOF [57] exhibits interesting functions owing to the presence of counter ions within their channels.

3.2.4.2 Effect on Stability

Organic spacers can serve as linkers for tuning the binding intensity and directionality and engage in the main role of the construction of MOFs with a novel framework and enhanced properties [59]. Base organic ligands (pyridine, imidazole, and, triazole) have been modified by incorporating a different spacer to synthesize angular rigid organic ligands to design and synthesize well-defined MOF structures.

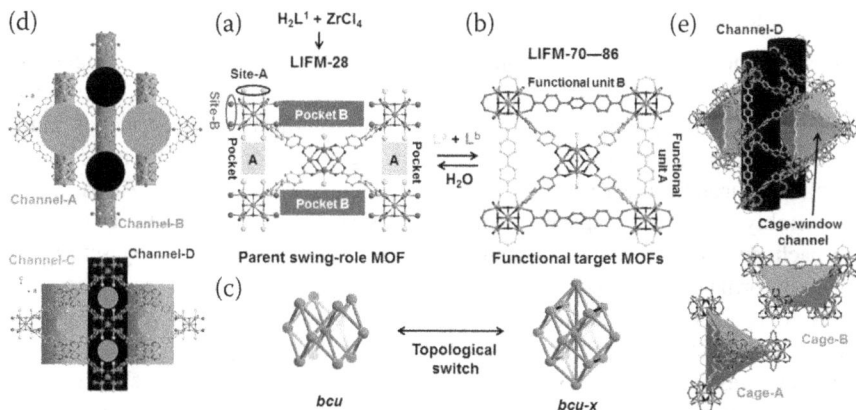

FIGURE 3.7 (a) Procedure to synthesize parent LIFM-28 for designing multifunctional MOFs (b) Alteration of a series of functional MOFs (LIFM-70–86) mediated through spacer incorporation. (c) inter-conversion of Frame work architecture topology. (d) Different channels incorporated within the parent MOF. (e) Functionalized MOF with straight and cage-window channels (Reproduced with permission C.-X. Chen, Wei, et al. 2017).

Recently robust Zr-MOF (LIFM-28) [58,59] containing expendable coordination nodes for an additional spacer installation has been synthesized (Figure 3.7).

A unique methodology of reversible installation/uninstallation is opted to use a different spacer for a different task within a single-parent MOF. LIFM-70–86 MOF with isomorphous series but altered functional entities have been synthesized. The pivotal feature of parent MOF for multirole switching ascends through the de-functionalization dynamics.

Yang et al. have reported the reaction of zinc(II) chloride and BTAH with four distinct aromatic/aliphatic acids under comparable hydrothermal conditions to produce four novel mixed-ligand metal-organic frameworks (MOFs), $[Zn_5(\mu_3\text{-}OH)_2(BTA)_2(tp)_3]_n$ (self-penetrating MOF with pentanuclear zinc clusters), $[Zn(BTA)(chdc)_{0.5}]_n$ (3-D infinite network), $[Zn(BTA)(ap)_{0.5}]_n$, and $[Zn(BTA)(gt)_{0.5}]_n$ (racemic coordination framework with a one-dimensional Zn-BTA helix chain motif) (BTAH = benzotriazole, H_2tp = terephthalic acid, H_2chdc = 1,4-cyclohexanedicarboxylic acid, H_2ap = adipic acid, and H_2gt = glutaric acid. The complex has compositional stability at elevated temperatures and also display strong solid-state luminescence emissions (E.-C. Yang et al. 2007). Recently, the two-linker system, with two spacers has been employed to provide variability in the structure. Constructing MOF with two ligands exhibits a controlled assembly with one ligand (main spacer) and metal center as fixed, the other ligand (secondary spacer) serves a structure-directing function to obtain various dimensional frameworks. A mixed ligands MOF with rod-spacer $[Mn_3(HCOO)_2(D\text{-}cam)_2(DMF)_2]_n$, [60] cam=camphorate has been utilized for coating capillary column used in gas chromatographic separation. The framework exhibits specific polarized spots in the three mixed organic ligands of $HCOO^-$, D-cam, and DMF, such as the rich polarized C=O, C-O, N-C=O groups while it holds high thermal stability.

3.2.4.3 Effect on Porosity

A post-synthetic variable-spacer installation (PVSI) approach [61,62] aligned on flexible proto-Zr_6-MOF LIFM-28, introduces structural evolution between a narrow-pore (np) and a large pore (lp) phase. Synthesis of a series of multi-functional MOFs is created via pinpointed sparing of the adjunct water molecules to Zr_6 cluster through linear spacers with varied groups of substituents. As assembly/disassembly of diverse spacers using the PVSI approach bestows a facetious approach not only to precisely control the breathing performance by setting the elastic deformation and also by fine-tuning the surface of the pores to enhance gas sorption capacity and separation ability, increasing the BET effective surface area, and to provide stability to the overall framework.

Elena Ilyes et al. [63] have reported a neutral 3D metal-organic framework, $[Cu_2(mand)_2(hmt)] \cdot H_2O$, using binuclear Cu_2O_2 alkoxo-bridged nodes, created by the dual deprotonated mandelic acid. A connection between the nodes has been achieved by using hexamethylenetetramine (hmt) spacers, which act as connective bridging ligands and by carboxylate groups.

3.3 SYNTHESIS OF MOF BASED ON DIFFERENT SBUs

The previous section states that MOF design and synthesis greatly rely on the linkage of metal and organic ligands and the synthesis parameters under controlled reaction conditions. However, a common approach is to extend the ligands to achieve the desired porosity within the MOF. No doubt, it has a major limitation as they form interpenetrated structures with low porosity. However, binding of metal ions with multitopic organic linkers has led to the synthesis of many positively charged framework structures albeit while exchanging the guest and removal of solvents from the pores, again results in collapsing of the framework. Moreover, the structures designed by these methodologies are not predictable as the precise monitoring of the components is difficult to achieve.

To combat these issues, Yaghi and his group [64] have proposed the concept of reticular chemistry, where the metal nodes have been replaced by secondary building units (SBUs) to synthesize materials on demand. Introducing SBU tactics has led to the extension of precision chemistry from molecular complexes into two-dimensional and three-dimensional frameworks, for designing coherent constructs using functional building sites. By designing polynuclear cluster points, also termed SBUs, thermodynamic stability can be imparted via strong covalent bonds and mechanical/architectural stability can be achieved via strong directional bonds by locking the central metal in MOF. Hence, the choice of an appropriate SBU has a special significance in modular synthesis of MOF to design robust, predictable MOF structures with superior physical properties and upgraded practical applications.

Amongst varied types of SBUs, metal oxygen-based tetranuclear and dinuclear $[Cu_2(COO)_4]$ SBUs have been broadly explored and produced via self-assembly.

In a pursuit to synthesize a new MOF, using $[NR_4]^+_{(2+b-a)}[WS_{4-n}O_n Cu_aX_b]_{(2+b-a)}$, ($n = 0$–2, X = Cl⁻, Br⁻, I⁻, CN⁻, SCN⁻; R = Me, Et, Pr, has been synthesized to form a hetero-thiometalic (W-Cu-S) structure, yet it is not very popular. In the subsequent section, both metal-oxygen, and W-Cu-S clusters have been discussed in detail.

3.3.1 METAL OXYGEN-BASED MOF

Multifaceted carboxylate functionalities are well known for chelation of metal ions as well as locking their sites into M-O-C clusters. These SBUs are quite rigid as the metal has been locked at a particular position by the oxygen linkage of the carboxylates; thus, the strategy is a replacement of a single metal at the vertex by the SBU and further joined with an organic linker (rigid) to impart structural stability. The group of vertices present at the nodes of the framework has been termed decoration, leading to the synthesis of open structure, avoiding the issue of interpenetration. M-O-C clusters have been reported for the synthesis of four net tetrahedral nodes and six net octahedral nodes, respectively. It is worth mentioning that the size of the rings (pores) and void spaces can be increased by decoration, augmentation, or combinations. For instance, a cubic structure with four rings can be stretched by octahedra carboxylate carbon atoms for producing a five-connected structure with eight-ring pores as well as a void space, highly prominent in comparison to an original structure.

Accounts of MOF research indicate that discrete multidentate carboxylic acid–based linkers integrated to basic zinc acetate (Td) motifs as well as paddlewheel copper acetate (D4h) are trained modules with symmetric structures for the polymerization to occur. Theoretically, every monocarboxylate linker in the structural complex can be substituted with a bi, tri, or a multicarboxylate functionalities to polymerize the SBU into an expanded network [65]. Figure 3.8 illustrates different metal oxide node structures and different MOFs obtained using BDC as a linker.

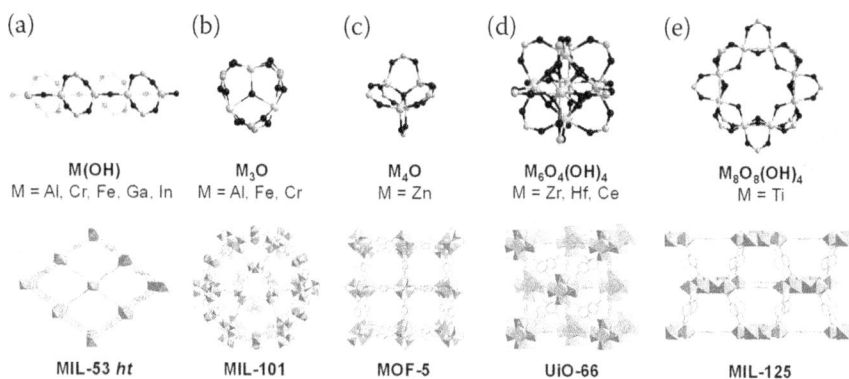

FIGURE 3.8 Demonstration node structures of metal oxide and crystal configuration of five different MOFs that include terephthalic acid as linkers: (a) MIL-53 ht, (b) MIL-101, (c) MOF-5, (d) UiO-66, and (e) MIL-125. Color code: metal atoms and metal–oxygen are denoted by polyhedra (light blue), oxygen (red), carbon (gray) (Reproduced with permission D. Yang et al. 2020).

Metal oxide groups in MOFs are shielded to limit their accumulation by linkers and by ionization. However, it tends to accumulate when these protective environments are removed. When highly charged metal ions are present in MOFs, they enhance their stability. Expanding MOF chemistry offers a pool of chances to enhance their reactivities.

3.3.2   W-Cu-S Clusters

In the quest for the synthesis of novel SBUs, interesting new SBUs have been observed by altering the array of Cu^+ ions wrapped along with the core $[WS_{4-n}O_n]^{2-}$ (n = 0–2) and by modifying the synthesis to demonstrate different geometries and connections. Distinct from other M-O-C crystals, these SBUs hold varied core structures. For example, $[WS_4Cu_6]^{4+}$, a SBU is tetrahedral, whereas in $[WS_4Cu_4]^{2+}$ a SBU is square planar. Principally elementary W–S–Cu architecture can have varied charged entities by the amalgamation of various anions, including halide or –CN ions, etc., eventually permitting the synthesis of neutral and charged MOF. The coordination units, copper ions, and the accessible spaces within the molecule describe the stability of W–Cu–S cluster-based MOFs. These MOFs have been structured with varying nodes (trinuclear to heptanuclear), reported with an exceptional twofold interpenetrated 3D network.

Liang and H. G. Zheng et al. have documented thio-metalate W–Cu–S cluster. The synthesized MOF exhibits an interpenetrating structure within its cationic and anionic framework simultaneously. A fourfold interpenetrating diamond like structure with large 11.124 Å × 11.129 Å channels have been obtained, with disordered assembly of solvent water molecules. Through the delocalization of π-electron cloud from heavy metal atoms containing a core of cluster along with organic linker strengthening the cluster carcass, an array of non-linear optical (NLO) properties has emerged.

Solvatochromism is a phenomenon of a huge shifting in the absorption spectra as an effect of variation in solvent identity. Such a noticeable alteration in a material's color is one of the easiest and perhaps the most powerful tool for transducing a sensing signal. Recently, a polymer $\{[(WS_4Cu_3)2(m\text{-}bip)_6](CuCl_3)(H_2O)_8\}n$ (m-bip = 1-(3-(1H-imidazol-1-yl)phenyl)-1H-imidazole) displaying a bandgap (2.99 eV) have been reported displaying a semiconducting behavior. These MOFs with semiconducting properties will turn into an active research area as wide functionalities are available for modifications rather than the conventional methodologies. Few important applications of W-S-Cu clusters such as nonlinear optics, gas absorption, and solvatochromic effect have made them one of the most trusted areas of advanced material research.

3.4   EFFECT OF REACTION VARIABLES ON MOF SYNTHESIS

The strategic goal of crystal engineers is to synthesize MOFs with desired structures and useful properties. Apart from major ingredients for MOF synthesis (metal/ligand) ratios, their assembling strategies comprise of other reaction variables, such as nature of the solvent, temperature, and operating pH, also have an elusive role in the synthesis and crystal formation of MOFs and have an impact on their overall architecture and topology are briefed in the following section.

3.4.1 NATURE OF SOLVENT

The judicious choice of solvent is crucial in the synthesis of MOFs as it affects the coordination aspects of various metals and ligands. The precise mechanism of its impact is still not clear, though several studies propose their key role in introducing a variable coordination environment. Figure 3.9 outlines major chemical solvents used in the synthesis of MOFs. The solvent utilized for synthesis plays a significant function in MOF crystallization as it regulates reaction kinetics and thermo-dynamics throughout the coordination method. Since Yaghi et al. [66] have stated the effect of solvent on the dimensionality of MOF structure, several systematic studies have been reported.

3.4.1.1 Effect on MOF Topology

Solvents play a major role as they occupy the void spaces and serve as a guest in the crystal obtained [67,68]. Ordinarily, these solvents have a structure guiding role when they are not incorporated inside as synthesized MOFs, ultimately controlling

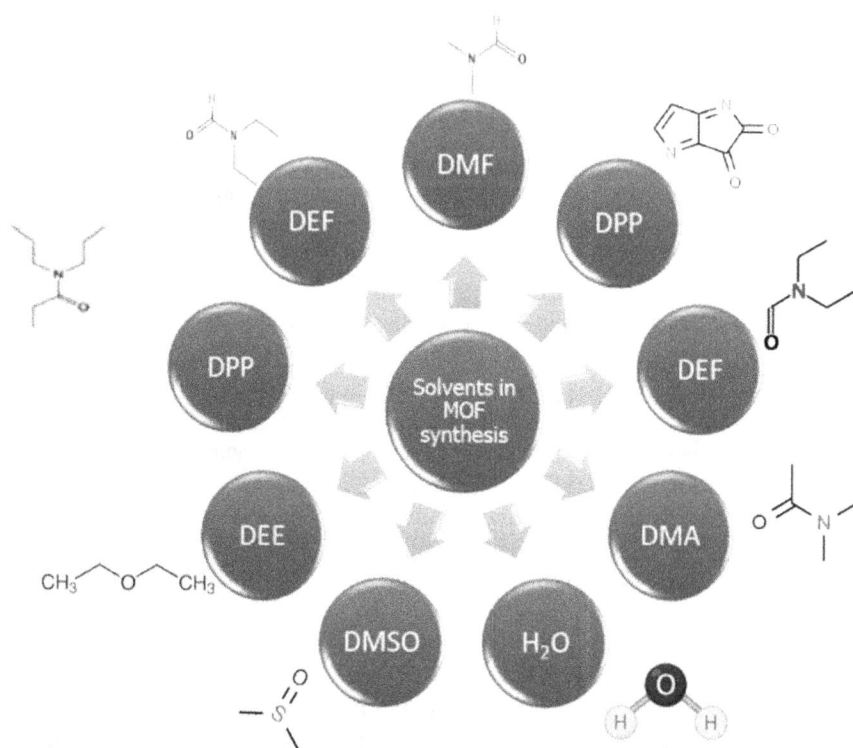

FIGURE 3.9 Various solvents used in MOF synthesis.

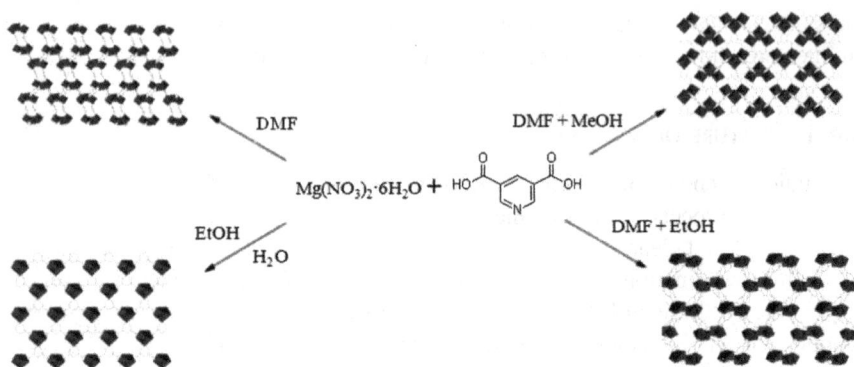

FIGURE 3.10 Variability in structure of Mg MOF obtained through solvent variability (Reproduced with permission Banerjee et al. 2011).

the deprotonation of organic ligands. Some commonly used solvents, di-methylformamide(DMF), dimethylformamide(DEF), and dimethylacetamide, etc., when utilized with carboxylate linkers, get transformed into their amines, leading to the deprotonation of carboxylates. The solvent-dependent behavior of Cd (II) MOF has been established by Li et al. [69] using three different solvents present, along with water-generating complexes.

Benarjee et al. have utilized the variable coordination behavior of solvents to the metal center for the synthesis of distinct MOF by utilizing varied mixtures of solvents (DMF, methanol, ethanol, etc.) and water, as shown in Figure 3.10.

The level of deprotonation of organic carboxylate ligands generally controls its coordination and can be regulated by altering the ionic concentration of the solvent. It has been noted that the capability of water molecules for coordinating with magnesium metal is quite superior in comparison with other polar solvents. Though the water has a high coordination with the Mg centers, followed by its interaction with DMF, other solvents like ethyl alcohol and methanol are likely, not co-ordinating to metal in presence of the water and DMF [70].

The structural variation ranges from a two-dimensional kgd sheet to a three-dimensional framework with Lonsdaleite (lon) topology [71]. The structurally distinct MOF has been obtained due to variation in coordination conditions of the metal ions and linking nodes of the ligand (tci herein) and is also influenced by solvent systems. The coordination geometries of metal (Zn, Mg) have been varied by utilization of different ratios of binary solvent mixtures such as water/DMA, and water/EtOH. The variation in the amount of ethanol from 0 to 100% has led to an increase in dimensionality of the framework from zero to three dimensions [72,73].

3.4.1.2 Effect on MOF Dimensionality

Molecular dimensions may lead to the growth of MOFs with distinct frameworks due to spatial arrangements of the linkers. A mixed solvent approach using DMF/H_2O and DEF/H_2O has been utilized for the synthesis of Y^{3+} center-based MOF with varied crystal structures and distinct arrangements of linkers [74]. Similarly,

Tzeng et al. have confirmed the formation of 1D/2D/3D forms ascribed to hindrances of the reaction solvents. Additional steric hindrance is noted for DMF, leading to the growth of a one-dimensional zigzag structure; however, the structures are less steric hindered using solvents CH_3CN and CH_3OH approve to build of two- to three-dimensional frameworks [75].

Recently, the mixed solvent method has become quite popular to achieve varied coordination modes. Cheng et al. [76] have confirmed the synthesis of two different zinc-based MOFs using two different solvent combinations (DMF-EtOH and MeOH). Anhydrous ethyl alcohol and dimethylformamide led to μ_2 along with μ_4 bridging with the organic ligand forming a 3D pillared structure while a 1D structure has been obtained through hydrogen bonding interactions between anhydrous methanol replacing DMF/EtOH mixture. The diversity in structures synthesized has been achieved through the difference in polarity of bridging organic ligands. Various combinations of solvent systems such as unary, binary, and ternary solvent have been investigated for preparation of four 2D coordination frameworks using zinc/cadmium as a metal source and a non-chiral diatopic ligand 2,6-bis (imidazole-1-yl) pyridine ($pyim_2$). In a methanol/DMF solvent mixture, a racemic mixture of zinc and cadmium MOFs have been obtained in the presence of water [77].

3.4.1.3　Effect on Porosity

Apart from the structural deviation, one common variability like pore dimension regulation can be attained by variation of different ratios of solvent mixtures. Different solvents [78] act as a template to generate new topologies with variable pore architecture along with the connectivity as they behave like solvent guest molecules having a structure-regulating effect during the process of self-assembly [79,80]. An isomorphic series of MOFs has been produced via solvothermal method using cadmium nitrate as metal salt and three different solvents [DMA (CH_3CN (CH_3)$_2$), C_5H_9NO (NMP), and $C_5H_{10}N_2O$ (DMI)]. The free void spaces of the channel framework have been obtained by coordinated and lattice solvent molecules. The solvent-accessible volume of the molecules can be altered by the difference in the size of the solvent. The extended organic anion has proven beneficial to design MOFs with larger porous sizes and advanced solvent-accessible volumes, as demonstrated in Figure 3.11.

A crystal to crystal transformation strategy has been approved to synthesize MOFs by post-synthetic alteration, not synthesized by conventional routes via altering the reaction solvents [81–84]. It has been discovered that the purity of solvent plays a crucial part in blending the coordination structures using carboxylic acids. Sporadically solvents endure hydrolysis in the air, finally generating the offset cations that play a role as a template to form the structural framework. Andrew D. Burrows and co-workers have reported the reaction of diethyl ammonium cation (NH_2Et^{2+}) and dimethylammonium cation (NH_2Me^{2+}), when present, along with DEF and DMF in the laboratory for several weeks.

Yao-Yu Wang and group carried out a reaction to produce two interpenetrated 3D microporous MOFs using water as a solvent and mixed solvents $DMF/CH_3OH/H_2O$. These frameworks have been described to be isomeric and nodal 3D-connected

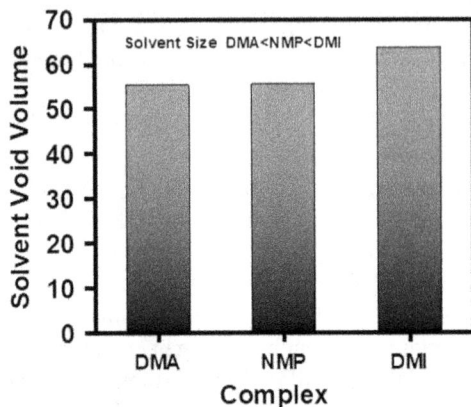

FIGURE 3.11 Occupancy of solvent void volume with respect to size of solvents.

structures. Further, the surface absorption will irreversibly translate the structure into a new phase. Therefore, two MOFs display similar network topology, but possess distinct properties [85]. The above-stated reports confirm that reactant sensitivity and solvent purity are highly imperative for synthesis.

3.4.2 TEMPERATURE

One of the key characteristics controlling the reaction of metal-organic frameworks (MOFs) synthesis is temperature. A notable impact of temperature over the growth and dimensional modifications in the structure of MOFs has been shown in Figure 3.12.

Empirically, the reaction temperature panes the MOF synthesis via various routes as it modifies the solubility of the organic linker; secondly, the presence of a

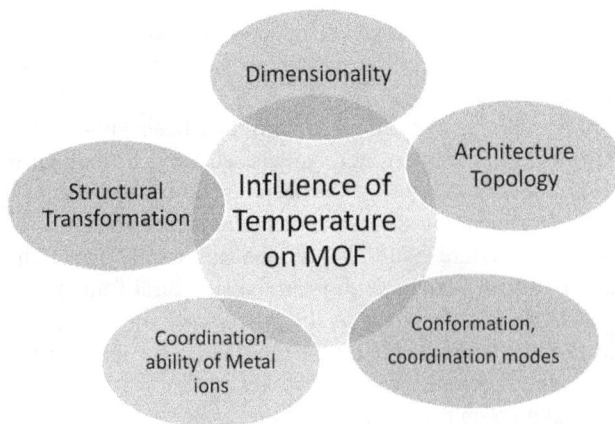

FIGURE 3.12 Effect of temperature on MOFs.

flexible organic linker inherently helps in obtaining diverse conformations under fluctuating temperatures, the reaction temperature may show a key part in varying the connecting manner of organic ligand, especially the carboxylate linker and convincingly, the temperature of reaction mixture alters the energy barrier existing within the framework. Subsequently, the structure directing role has been played by the reaction temperature and also in obtaining a multitudinal structure. Hydro/solvothermal methods have been well orchestrated for the production of MOFs. In a sealed reaction vessel, high temperature along with high pressure will modify the network assembly and ultimate framework of MOFs.

3.4.2.1   Effect on Topology

A comprehensive study has been conducted by Vivek et al. to evaluate that the topology of MOFs using microwave synthesis, proving the role of temperature and time [86]. A layer-by-layer assembly has been reported in SEM images. When the temperature is increased from 120 to 160°C, a transition of nanosheets to layered microrods can be observed. Continuous growth of microcrystals is obtained at a high temperature.

Urothermal synthesis, which utilizes urea as the derivative, has been recognized as an efficient technique for producing MOF structures with wide practical purposes. The shift of temperature during urothermal production in MOFs has been described [87]. Two differently structured MOFs, $[La_2(NDC)_{3(}urea)_3]$, at 120°C and $[La(NDC)1.5(e\text{-}urea)]$ at 140°C (H2NDC = 2,6-naphthalene dicarboxylic acid; e-urea = ethylene urea), have been reported. For instance, a high temperature tends to impart stability to the compounds. Different coordination forms of e-urea have been described by variable H-bonding occurring in between two distinct linkers, 2-Imidazolidinone hemihydrate ethylene urea and carboxylate groups present in NDC. Albeit thermal and morphological properties have been described to vary at different temperatures. Two cobalt MOFs have been prepared at varying temperatures (160°C and 220°C) from similar metal ion and organic linkers [88,89].

R. Q. Fan and the group have synthesized a group of lanthanide MOF using imidazole-4,5-dicarboxylic acid $(C_5H_4N_2O_4)$. They have reported that high-dimensional framework can be achieved at high temperatures analogous to low temperatures [90].

3.4.2.2   Effect on Dimensionality

Hydrothermal synthesis of MOFs [91,92] reports several advantages. On raising the reaction temperature, a rise in coordination to metal ion and framework dimensionality have been reported, along with a decrease in arrangement of solvent molecules.

Hydrothermal synthesis has been described as an effective route to produce MOFs with superior dimensionality. H.J. Mo et al. have analyzed the reaction of cobalt nitrate with a multifunctional linker 2,20 -benzimidazole (H2bbim) in 4-cyanopyridine using hydrothermal routes utilizing temperature as an independent entity, keeping the molar ratios and base reaction of the components constant [93]. Just after the temperature of reaction medium is raised to 130°C, the linker is monoprotonated to form a zero-dimensional complex. On raising the temperature from 140°C–200°C, a higher dimensionality in MOF structure has been reported. It is can be inferred that easier deprotonation at high temperature has led to the synthesis of high-dimensional MOFs.

G.E. Narda et al. have synthesized two Ho[III] succinate frameworks with varied dimensionality under mild hydrothermal conditions. The obtained compounds have reported varied thermal stability as well as magnetic properties. It has been observed that the thermal stability is enhanced for structures obtained from the hydrothermal synthesis in comparison to room temperature synthesis. Further, it has been established that the stable, open frameworks with empty channels can be synthesized by a hydrothermal route. Supramolecular isomerism has been reported in MOFs by Kanoo et al. [94].

On removal of water molecules at a high temperature, a high dimensionality in structural framework has been obtained by a thermodynamic entropy-driven approach. However, translating the framework from hydrated structure at low temperature towards the formation of metal hydroxide at an elevated temperature has proven the role of classical thermodynamic factors [95]. Though it has been described previously that elevated temperature promotes higher dimensionality, an exception to this has been reported [96]. The non-agreement on the effect of temperature on dimensionality is reported for three 3D homochiral cadmium camphorates formed from cadmium nitrate and Dextral-camphoric acid [97]. The extent of hydration has been regulated by temperature.

3.4.2.3 Effect on Porosity

Solid-to-solid structural (SCSC) transformation to synthesize different MOFs with alteration in temperature has been exploited for guest molecule desorption/ movement [98,99], and slipping of alternative layers, or breakage and formation of linkages [100]. A guest molecule persuades structural makeover with framework flexibility and dynamicity. Temperature variations cause reversible SCSC trans-formations that lead to the formation of pillared structures [101]. However, no excessive change in the 3D framework has been reported but a change in phase using slip and distortion is observed on heating/cooling. These frameworks react with the external stimuli and have a commercialization potential to be used as smart materials.

3.4.3 Effect of pH

Documented studies suggest that the crystal synthesis and growth of hybrid structures are extremely dependent on ionic condition of reaction medium (acidic/ basic). MOF synthesis has been labeled as an acid-base reaction requiring acidic metal and basic organic linkers (predominantly carboxylate, imidazolate, and pyridyl-based linkers). Too much reduction in pH averts the N-donor ligands from coordination, aided by protonation. Correspondingly, anionic ligands form MOFs at a lower rate at reduced pH due to a reduction in deprotonation of the carboxylic acid precursors. Besides, rapid deprotonation of carboxylic acid at higher pH produces a low-quality sample owing to reduced chances for self-healing [102,103]. Few major studies on effect of pH on MOF synthesis have been reported.

3.4.3.1 Effect on Topology

A structured study portraying the coordination capability of the free linker (H_2ADA) has shown various coordination predilections. Comparable results have been described for the synthesis of two 3D MOFs at varied pH (8.5 and at pH 4.5 respectively) using 2,4-Pyridinedicarboxylic acid [104].

Fractional deprotonation of multifunctional organic linker(tetrazole1-acetic acid) employed for the production of numerous coordination frameworks {[Cu (tza)$_2$(Htza)$_2$].2H$_2$O}, [Cu(tza)2]n (,{[Cu$_4$(tza)$_4$(CH$_3$COO)$_2$(l3-OH)$_2$(H2O)$_2$]2H$_2$O}n, {[Cu$_4$(tza)$_6$(l3-OH)$_2$]4H$_2$O}n. When pH of the medium is raised further (4–5), complete deprotonation of ligand occurs along with the coordination of hydroxyl group [105].

W.Y. Sun has verified the role of elevated pH to induce higher dimensionality in the crystal structure. A supramolecular 4-sulfocalixarene serving as a multifunctional organic linker on reacting with manganese in the presence of ammoniacal solution would lead to formation of a one-dimensional structural framework at pH 2 while a 2D framework has been reported at a pH 5. The conversion from one structure to another establish the role of pH in structural conformation [106,107].

Unlike the alteration in binding assemblies arising from the deprotonation of the organic linker over a pH range herein the structural variation of the frameworks has been observed due to differential generation of hydroxyl group at different pH. Also, at high pH range, coordination of water molecules with metal ions has been reported [108]. Seldom supramolecular isomeric assemblies possessing similar composition and building entities, yet disparate architectures is retrieved at a unique pH of the reaction medium [90].

3.4.3.2 Effect on Dimensionality

The pH of the medium has high prominence in controlling the dimension of the Cu/benzoate/bipyridine complexes obtained (ranging from zero to two dimensional). It also possesses an important role in the ratio of bipyridine/benzoate generated (Figure 3.13).

There is a substantial increase in the ratio of bipyridine to benzoate (1:1–1:2) with a rise in pH of the medium; further, the tendency for formation of a bridged structure is amplified with a rise in pH [109]. Z.M. Su and his group have determined the impact of base-induced hetero nucleation on the formation of different phases. A variation in the ratio of zinc nitrate hexahydrate and H$_3$BTC when dimethyl alkoxide is present, along with a discrete molar ratio of NaOH, ultimately form separate crystalline phases [110].

3.4.3.3 Effect on Crystal Color

Few reports have proven that on altering the pH, the change in color of MOF crystal is also observed. As the pH of the medium is tweaked over a range (5–9), a differentially colored Co-BTC-3,30,5,50- tetra(1H-imidazole-1-yl) -1,10-biphenyl MOF is obtained. BTC is partially deprotonated at pH = 5, while its full deprotonation takes both pH 7 and 9. Therefore, a coordination approach of the terephthalic acid ligand and color of the compounds can be adapted and tweaked by altering the pH [102].

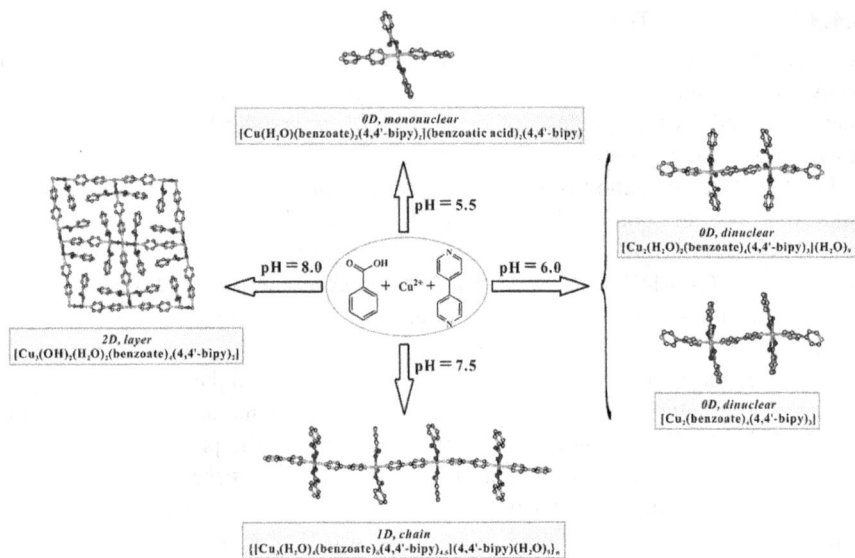

FIGURE 3.13 Systematic illustration of assembling tendencies at varied pH (Reproduced with permission [109].

3.5 CONCLUSION

Reticular synthesis of MOF structures to achieve desired features is the aim of crystal engineers. A tailored synthesis approach has been exploited to achieve specific topologies and properties. The framework assembly and dimensionalities can be modulated by the variation of reaction parameters and their linking methodologies. It can be concluded that the assembling strategies of ligand have a major role in coordination modes, architectural topologies, dimensionalities, and the transformation of the structure. The assembling strategy of the MOF is significantly dependent on the SBU, limiting the interpenetration and dimensionality of the framework. From the points discussed earlier, the effect of the solvent is visible in the variability of structures obtained. It is proposed that various solvent systems might control crystallization by divergent self-assembly processes, controlling dimensionality and porosity.

Temperature also plays a structure-guiding role by manipulating the connecting sites of the organic ligand and the coordination number possessed by a central position of a metal ion. At a high pH, a MOF with a high range of dimensionality and also varied coordination forms can be obtained. Though both methods and configurational factors are recognized to possess exigent effects over the ultimate framework and dimensionality of the structure obtained, however, it still remains a big challenge to construct the preferred structure via a true methodical approach as every factor has its unique role [111–115].

ACKNOWLEDGMENTS

The authors would like to extend their gratitude of thanks to Amity University, Noida Uttar Pradesh.

REFERENCES

[1] H.-C. Zhou, J. R. Long, and O. M. Yaghi, "Introduction to metal–organic frameworks," *Chem. Rev.*, vol. 112, no. 2, pp. 673–674, Feb. 2012.

[2] C. Wang, D. Liu, and W. Lin, "Metal–organic frameworks as a tunable platform for designing functional molecular materials," *J. Am. Chem. Soc.*, vol. 135, no. 36, pp. 13222–13234, Sep. 2013.

[3] F. Gándara, H. Furukawa, S. Lee, and O. M. Yaghi, "High methane storage capacity in aluminum metal–organic frameworks," *J. Am. Chem. Soc.*, vol. 136, no. 14, pp. 5271–5274, Apr. 2014.

[4] Y. He, W. Zhou, G. Qian, and B. Chen, "Methane storage in metal–organic frameworks," *Chem. Soc. Rev.*, vol. 43, no. 16, pp. 5657–5678, 2014.

[5] C. K. Brozek, L. Bellarosa, T. Soejima, T. V. Clark, N. López, and M. Dincă, "Solvent-dependent cation exchange in metal–organic frameworks," *Chem. – A Eur. J.*, vol. 20, no. 23, pp. 6871–6874, Jun. 2014.

[6] D. Cunha *et al.*, "Rationale of drug encapsulation and release from biocompatible porous metal–organic frameworks," *Chem. Mater.*, vol. 25, no. 14, pp. 2767–2776, Jul. 2013.

[7] Z. Dou *et al.*, "Luminescent metal–organic framework films as highly sensitive and fast-response oxygen sensors," *J. Am. Chem. Soc.*, vol. 136, no. 15, pp. 5527–5530, Apr. 2014.

[8] L. E. Kreno, K. Leong, O. K. Farha, M. Allendorf, R. P. Van Duyne, and J. T. Hupp, "Metal–organic framework materials as chemical sensors," *Chem. Rev.*, vol. 112, no. 2, pp. 1105–1125, Feb. 2012.

[9] X.-Q. Lü, J.-J. Jiang, C.-L. Chen, B.-S. Kang, and C.-Y. Su, "3D coordination polymers with nitrilotriacetic and 4,4'-bipyridyl mixed ligands: Structural variation based on dinuclear or tetranuclear subunits assisted by Na–O and/or O–H···O interactions," *Inorg. Chem.*, vol. 44, no. 13, pp. 4515–4521, Jun. 2005.

[10] S. Zhang, X. Liu, Q. Yang, Q. Wei, G. Xie, and S. Chen, "Mixed-metal–organic frameworks (M′MOFs) from 1D to 3D based on the 'organic' connectivity and the inorganic connectivity: Syntheses{,} structures and magnetic properties," *CrystEngComm*, vol. 17, no. 17, pp. 3312–3324, 2015.

[11] H. Robatjazi *et al.*, "Metal-organic frameworks tailor the properties of aluminum nanocrystals," *Sci. Adv.*, vol. 5, no. 2, p. eaav5340, Feb. 2019.

[12] M. Ahmadi *et al.*, "An investigation of affecting factors on MOF characteristics for biomedical applications: A systematic review," *Heliyon*, vol. 7, no. 4, p. e06914, 2021.

[13] N. Aljammal, C. Jabbour, S. Chaemchuen, T. Juzsakova, and F. Verpoort, "Flexibility in metal–organic frameworks: A basic understanding," *Catalysts*, vol. 9, no. 6, 2019, pp. 125–152.

[14] Z. Yin, Y.-L. Zhou, M.-H. Zeng, and M. Kurmoo, "The concept of mixed organic ligands in metal–organic frameworks: Design, tuning and functions," *Dalt. Trans.*, vol. 44, no. 12, pp. 5258–5275, 2015.

[15] A. Schneemann, V. Bon, I. Schwedler, I. Senkovska, S. Kaskel, and R. A. Fischer, "Flexible metal–organic frameworks," *Chem. Soc. Rev.*, vol. 43, no. 16, pp. 6062–6096, 2014.

[16] X. Lu et al., "Ligand effects on the structural dimensionality and antibacterial activities of silver-based coordination polymers," *Dalt. Trans.*, vol. 43, no. 26, pp. 10104–10113, 2014.

[17] M. J. Kalmutzki, N. Hanikel, and O. M. Yaghi, "Secondary building units as the turning point in the development of the reticular chemistry of MOFs," *Sci. Adv.*, vol. 4, no. 10, eaat9180, Oct. 2018.

[18] S. S. Park, C. H. Hendon, A. J. Fielding, A. Walsh, M. O'Keeffe, and M. Dincă, "The organic secondary building unit: Strong intermolecular π interactions define topology in MIT-25, a mesoporous MOF with proton-replete channels," *J. Am. Chem. Soc.*, vol. 139, no. 10, pp. 3619–3622, Mar. 2017.

[19] F. Yuan et al., "Effect of pH/metal ion on the structure of metal–organic frameworks based on novel bifunctionalized ligand 4'-carboxy-4,2':6',4''-terpyridine," *CrystEngComm*, vol. 15, no. 7, pp. 1460–1467, 2013.

[20] B. Zhang et al., "Solvent determines the formation and properties of metal–organic frameworks," *RSC Adv.*, vol. 5, no. 47, pp. 37691–37696, 2015.

[21] K. A. S. Usman et al., "Downsizing metal–organic frameworks by bottom-up and top-down methods," *NPG Asia Mater.*, vol. 12, no. 1, p. 58, 2020.

[22] M. O. Rodrigues, M. V. de Paula, K. A. Wanderley, I. B. Vasconcelos, S. Alves Jr., and T. A. Soares, "Metal organic frameworks for drug delivery and environmental remediation: A molecular docking approach," *Int. J. Quantum Chem.*, vol. 112, no. 20, pp. 3346–3355, Oct. 2012.

[23] M. Edgar, R. Mitchell, A. M. Z. Slawin, P. Lightfoot, and P. A. Wright, "Solid-state transformations of zinc 1,4-benzenedicarboxylates mediated by hydrogen-bond-forming molecules," *Chem. – A Eur. J.*, vol. 7, no. 23, pp. 5168–5175, Dec. 2001.

[24] H. Li, M. Eddaoudi, M. O'Keeffe, and O. M. Yaghi, "Design and synthesis of an exceptionally stable and highly porous metal-organic framework," *Nature*, vol. 402, no. 6759, pp. 276–279, 1999.

[25] P. Mahata, and S. Natarajan, "Pyridine- and Imidazoledicarboxylates of zinc: Hydrothermal synthesis, structure, and properties," *Eur. J. Inorg. Chem.*, vol. 2005, no. 11, pp. 2156–2163, Jun. 2005.

[26] L. Pan, T. Frydel, M. B. Sander, X. Huang, and J. Li, "The effect of pH on the dimensionality of coordination polymers," *Inorg. Chem.*, vol. 40, no. 6, pp. 1271–1283, Mar. 2001.

[27] J. B. Baruah, "Naphthalenedicarboxylate based metal organic frameworks: Multifaceted material," *Coord. Chem. Rev.*, vol. 437, p. 213862, 2021.

[28] F. Chen, Y. Wang, D. Bai, M. He, X. Gao, and Y. He, "Selective adsorption of C2H2 and CO2 from CH4 in an isoreticular series of MOFs constructed from unsymmetrical diisophthalate linkers and the effect of alkoxy group functionalization on gas adsorption," *J. Mater. Chem. A*, vol. 6, no. 8, pp. 3471–3478, 2018.

[29] Y. Sun, S. Zhao, H. Ma, Y. Han, K. Liu, and L. Wang, "Positional isomerism-driven two 3D pillar-layered metal-organic frameworks: Syntheses, topological structures and photoluminescence properties," *J. Solid State Chem.*, vol. 238, pp. 284–290, 2016.

[30] C.-Y. Zhao and C.-X. Jia, "Significant positional isomeric effect of zwitterionic ligands on the construction of two different 2D and 3D Pb(II) coordination polymers," *Inorg. Chem. Commun.*, vol. 48, pp. 52–56, 2014.

[31] B. Liu, H.-F. Zhou, L. Hou, J.-P. Wang, Y.-Y. Wang, and Z. Zhu, "Structural diversity of cadmium(II) coordination polymers induced by tuning the coordination sites of isomeric ligands," *Inorg. Chem.*, vol. 55, no. 17, pp. 8871–8880, Sep. 2016.

[32] L.-L. Liu, C.-X. Yu, Y.-R. Li, J.-J. Han, F.-J. Ma, and L.-F. Ma, "Positional isomeric effect on the structural variation of Cd(ii) coordination polymers based on flexible linear/V-shaped bipyridyl benzene ligands," *CrystEngComm*, vol. 17, no. 3, pp. 653–664, 2015.

[33] Q.-F. He, D.-S. Li, J. Zhao, X.-J. Ke, C. Li, and Y.-Q. Mou, "Positional isomeric effect of phenylenediacetate on the construction of mixed-ligand CdII coordination frameworks," *Inorg. Chem. Commun.*, vol. 14, no. 4, pp. 578–583, 2011.

[34] C.-J. Liu, T.-T. Zhang, W.-D. Li, Y.-Y. Wang, and S.-S. Chen, "Coordination assemblies of Zn(II) coordination polymers: Positional isomeric effect and optical properties," *Crystals*, vol. 9, no. 12, pp. 664–647, 2019.

[35] F.-P. Huang *et al.*, "A case study of the ZnII-BDC/bpt mixed-ligand system: Positional isomeric effect, structural diversification and luminescent properties," *CrystEngComm*, vol. 12, no. 4, pp. 1269–1279, 2010.

[36] J. Cisterna, C. Araneda, P. Narea, A. Cárdenas, J. Llanos, and I. Brito, "The positional isomeric effect on the structural diversity of Cd(II) coordination polymers, using flexible positional isomeric ligands containing pyridyl, triazole, and carboxylate fragments," *Molecules*, vol. 23, no. 10, pp. 2634–2648, 2018.

[37] N. N. Adarsh, D. K. Kumar, and P. Dastidar, "Metal–organic frameworks derived from bis-pyridyl-bis-amide ligands: Effect of positional isomerism of the ligands, hydrogen bonding backbone, counter anions on the supramolecular structures and selective crystallization of the sulfate anion," *CrystEngComm*, vol. 11, no. 5, pp. 796–802, 2009.

[38] L. Gou *et al.*, "An investigation of the positional isomeric effect of terpyridine derivatives: Self-assembly of novel cadmium coordination architectures driven by N-donor covalence and $\pi \cdots \pi$ non-covalent interactions," *Polyhedron*, vol. 27, no. 5, pp. 1517–1526, 2008.

[39] L. Fan *et al.*, "A series of metal–organic framework isomers based on pyridinedicarboxylate ligands: Diversified selective gas adsorption and the positional effect of methyl functionality," *Inorg. Chem.*, vol. 60, no. 4, pp. 2704–2715, Feb. 2021.

[40] J. Zhu, P. Zhu, J. Mei, J. Xie, J. Guan, and K.-L. Zhang, "Proton conduction and luminescent sensing property of two newly constructed positional isomer-dependent redox-active Mn(II)-organic frameworks," *Polyhedron*, vol. 200, p. 115139, 2021.

[41] P. Hao, Y. Xu, J. Shen, and Y. Fu, "Effect of positional isomerism on electron-transfer photochromism and photoluminescence of two pyromellitic diimide-based organic molecules," *Dye. Pigment.*, vol. 186, p. 108941, 2021.

[42] S. Lin *et al.*, "Modulation of topological structures and adsorption properties of copper-tricarboxylate frameworks enabled by the effect of the functional group and its position," *Inorg. Chem.*, vol. 60, no. 11, pp. 8111–8122, Jun. 2021.

[43] M. Du, X.-J. Jiang, and X.-J. Zhao, "Molecular tectonics of mixed-ligand metal–organic frameworks: Positional isomeric effect, metal-directed assembly, and structural diversification," *Inorg. Chem.*, vol. 46, no. 10, pp. 3984–3995, May 2007.

[44] H. Hahm, K. Yoo, H. Ha, and M. Kim, "Aromatic substituent effects on the flexibility of metal–organic frameworks," *Inorg. Chem.*, vol. 55, no. 15, pp. 7576–7581, Aug. 2016.

[45] T. Liu *et al.*, "Positional isomeric and substituent effect on the assemblies of a series of d10 coordination polymers based upon unsymmetric tricarboxylate acids and nitrogen-containing ligands," *CrystEngComm*, vol. 15, no. 27, pp. 5476–5489, 2013.

[46] Y. Ju *et al.*, "Interpenetration control in thorium metal–organic frameworks: Structural complexity toward iodine adsorption," *Inorg. Chem.*, vol. 60, no. 8, pp. 5617–5626, Apr. 2021.

[47] A. Torrisi, C. Mellot-Draznieks, and R. G. Bell, "Impact of ligands on CO_2 adsorption in metal-organic frameworks: First principles study of the interaction of CO_2 with functionalized benzenes. II. Effect of polar and acidic substituents," *J. Chem. Phys.*, vol. 132, no. 4, p. 44705, Jan. 2010.

[48] C. Wang et al., "MOFs constructed with the newly designed imidazole dicarboxylate bearing a 2-position aromatic substituent: Hydro(solvo)thermal syntheses, crystal structures and properties," *Dalt. Trans.*, vol. 42, no. 5, pp. 1715–1725, 2013.

[49] Z. Zhang et al., "R-substituent-induced structural diversity and single-crystal to single-crystal transformation of coordination polymers: Synthesis, luminescence, and magnetic behaviors," *Cryst. Growth Des.*, vol. 21, no. 9, pp. 5086–5099, Sep. 2021.

[50] P.-P. Cui, X.-D. Zhang, Y.-S. Kang, Y. Zhao, and W.-Y. Sun, "Cobalt-based metal–organic frameworks for adsorption of CO_2 and C_2 hydrocarbons: Effect of auxiliary ligands with different functional groups," *Inorg. Chem.*, vol. 60, no. 4, pp. 2563–2572, Feb. 2021.

[51] R. M. Cedeno et al., "Bandgap modulation in Zr-based metal–organic frameworks by mixed-linker approach," *Inorg. Chem.*, vol. 60, no. 12, pp. 8908–8916, Jun. 2021.

[52] P. Zhou, L. Yue, X. Wang, L. Fan, D.-L. Chen, and Y. He, "Improving ethane/ethylene separation performance of isoreticular metal–organic frameworks via substituent engineering," *ACS Appl. Mater. Interfaces*, vol. 13, no. 45, pp. 54059–54068, Nov. 2021.

[53] Y.-R. Xi et al., "The substituent effect on the luminescent properties of a set of 4-amino-4H-1,2,4-triazole: Syntheses, crystal structures and Hirshfeld analyses," *J. Mol. Struct.*, vol. 1243, p. 130893, 2021.

[54] Y.-J. Yin, C. Chen, Y.-H. Luo, and B.-W. Sun, "Three new co-crystals of 2,3,5,6-tetramethyl pyrazin with different substituted aromatic compounds _ crystal structure, spectroscopy and Hirshfeld analysis," *J. Mol. Struct.*, vol. 1241, p. 130580, 2021.

[55] A. Rossin et al., "Synthesis, characterization and CO_2 uptake of a chiral Co(ii) metal–organic framework containing a thiazolidine-based spacer," *J. Mater. Chem.*, vol. 22, no. 20, pp. 10335–10344, 2012.

[56] W. Wu, J. Wang, C. Shi, L. Lu, B. Xie, and Y. Wang, "Assembly of co coordination polymers tuned by the N-donor ligands with different spacer: Syntheses, structures and photocatalytic properties," *Inorganica Chim. Acta*, vol. 514, p. 119995, 2021.

[57] M. Dai et al., "Spacer length effect on the formation of different zinc coordination polymers of 1,4-benzenedicarboxylate and flexible bipyrazolyl ligands," *Inorg. Chem. Commun.*, vol. 29, pp. 70–75, 2013.

[58] L. Qin, J.-S. Hu, Y.-Z. Li, and H.-G. Zheng, "Three new coordination polymers based on one reduced symmetry tripodal linker," *Cryst. Growth Des.*, vol. 11, no. 7, pp. 3115–3121, Jul. 2011.

[59] J.-Z. Gu, Y. Cai, M. Wen, Z.-F. Shi, and A. M. Kirillov, "A new series of Cd(ii) metal–organic architectures driven by soft ether-bridged tricarboxylate spacers: Synthesis, structural and topological versatility, and photocatalytic properties," *Dalt. Trans.*, vol. 47, no. 40, pp. 14327–14339, 2018.

[60] D. Singh and C. M. Nagaraja, "Auxiliary ligand-assisted structural variation of Cd (II) metal–organic frameworks showing 2D → 3D polycatenation and interpenetration: Synthesis, structure, luminescence properties, and selective sensing of trinitrophenol," *Cryst. Growth Des.*, vol. 15, no. 7, pp. 3356–3365, Jul. 2015.

[61] J.-N. Rebilly, J. Bacsa, and M. J. Rosseinsky, "1 D tubular and 2 D metal–organic frameworks based on a flexible amino acid derived organic spacer," *Chem. – An Asian J.*, vol. 4, no. 6, pp. 892–903, Jun. 2009.

[62] C. Wang et al., "A series of anionic MOFs with cluster-based, pillared-layer and rod-spacer motifs: Near-sunlight white-light emission and selective dye capture," *CrystEngComm*, vol. 22, no. 5, pp. 878–887, 2020.

[63] C.-X. Chen *et al.*, "Dynamic spacer installation for multirole metal–organic frameworks: A new direction toward multifunctional MOFs achieving ultrahigh methane storage working capacity," *J. Am. Chem. Soc.*, vol. 139, no. 17, pp. 6034–6037, May 2017.

[64] E.-C. Yang, H.-K. Zhao, B. Ding, X.-G. Wang, and X.-J. Zhao, "Four novel three-dimensional triazole-based zinc(II) metal–organic frameworks controlled by the spacers of dicarboxylate ligands: Hydrothermal synthesis, crystal structure, and luminescence properties," *Cryst. Growth Des.*, vol. 7, no. 10, pp. 2009–2015, Oct. 2007.

[65] D.-D. Zheng, Y. Zhang, L. Wang, M. Kurmoo, and M.-H. Zeng, "A rod-spacer mixed ligands MOF [Mn₃(HCOO)₂(D-cam)₂(DMF)₂]ₙ as coating material for gas chromatography capillary column," *Inorg. Chem. Commun.*, vol. 82, pp. 34–38, 2017.

[66] C.-X. Chen *et al.*, "Precise modulation of the breathing behavior and pore surface in Zr-MOFs by reversible post-synthetic variable-spacer installation to fine-tune the expansion magnitude and sorption properties," *Angew. Chemie*, vol. 128, no. 34, pp. 10086–10090, 2016.

[67] C.-X. Chen *et al.*, "Stepwise engineering of pore environments and enhancement of CO₂/R22 adsorption capacity through dynamic spacer installation and functionality modification," *Chem. Commun.*, vol. 53, no. 83, pp. 11403–11406, 2017.

[68] E. Ilyes, M. Florea, A. M. Madalan, I. Haiduc, V. I. Parvulescu, and M. Andruh, "A robust metal–organic framework constructed from alkoxo-bridged binuclear nodes and hexamethylenetetramine spacers: Crystal structure and sorption studies," *Inorg. Chem.*, vol. 51, no. 15, pp. 7954–7956, Aug. 2012.

[69] M. Eddaoudi *et al.*, "Modular chemistry: Secondary building units as a basis for the design of highly porous and robust metal-organic carboxylate frameworks," *Acc. Chem. Res.*, vol. 34, no. 4, pp. 319–330, 2001.

[70] D. Yang, M. Babucci, W. H. Casey, and B. C. Gates, "The surface chemistry of metal oxide clusters: From metal–organic frameworks to minerals," *ACS Cent. Sci.*, vol. 6, no. 9, pp. 1523–1533, Sep. 2020.

[71] J. L. C. Rowsell and O. M. Yaghi, "Metal-organic frameworks: A new class of porous materials," *Microporous Mesoporous Mater.*, vol. 73, pp. 3–14, 2004.

[72] A. A. Yakovenko, Z. Wei, M. Wriedt, J.-R. Li, G. J. Halder, and H.-C. Zhou, "Study of guest molecules in metal–organic frameworks by powder X-ray diffraction: Analysis of difference envelope density," *Cryst. Growth Des.*, vol. 14, no. 11, pp. 5397–5407, Nov. 2014.

[73] K. Akhbari and A. Morsali, "Effect of the guest solvent molecules on preparation of different morphologies of ZnO nanomaterials from the [Zn₂(1,4-bdc)₂(dabco)] metal-organic framework," *J. Coord. Chem.*, vol. 64, no. 20, pp. 3521–3530, Oct. 2011.

[74] L. Li, S. Wang, T. Chen, Z. Sun, J. Luo, and M. Hong, "Solvent-dependent formation of Cd(II) coordination polymers based on a C2-symmetric tricarboxylate linker," *Cryst. Growth Des.*, vol. 12, no. 8, pp. 4109–4115, Aug. 2012.

[75] D. Banerjee *et al.*, "Synthesis and structural characterization of magnesium based coordination networks in different solvents," *Cryst. Growth Des.*, vol. 11, no. 6, pp. 2572–2579, Jun. 2011.

[76] P. Cui *et al.*, "Two solvent-dependent zinc(II) supramolecular isomers: Rare kgd and lonsdaleite network topologies based on a tripodal flexible ligand," *Cryst. Growth Des.*, vol. 11, no. 12, pp. 5182–5187, Dec. 2011.

[77] F.-K. Wang, S.-Y. Yang, R.-B. Huang, L.-S. Zheng, and S. R. Batten, "Control of the topologies and packing modes of three 2D coordination polymers through variation of the solvent ratio of a binary solvent mixture," *CrystEngComm*, vol. 10, no. 9, pp. 1211–1215, 2008.

[78] M. Mazaj, T. Birsa Čelič, G. Mali, M. Rangus, V. Kaučič, and N. Zabukovec Logar, "Control of the crystallization process and structure dimensionality of Mg–Benzene–1,3,5-Tricarboxylates by tuning solvent composition," *Cryst. Growth Des.*, vol. 13, no. 8, pp. 3825–3834, Aug. 2013.

[79] B.-X. Dong, X.-J. Gu, and Q. Xu, "Solvent effect on the construction of two microporous yttrium–organic frameworks with high thermostability viain situ ligand hydrolysis," *Dalt. Trans.*, vol. 39, no. 24, pp. 5683–5687, 2010.

[80] B.-C. Tzeng, H.-T. Yeh, T.-Y. Chang, and G.-H. Lee, "Novel coordinated-solvent induced assembly of Cd(II) coordination polymers containing 4,4′-Dipyridylsulfide," *Cryst. Growth Des.*, vol. 9, no. 6, pp. 2552–2555, Jun. 2009.

[81] M.-L. Cheng, E. Zhu, Q. Liu, S.-C. Chen, Q. Chen, and M.-Y. He, "Two coordinated-solvent directed zinc(II) coordination polymers with rare gra topological 3D framework and 1D zigzag chain," *Inorg. Chem. Commun.*, vol. 14, no. 1, pp. 300–303, 2011.

[82] S. Tripathi, R. Srirambalaji, N. Singh, and G. Anantharaman, "Chiral and achiral helical coordination polymers of zinc and cadmium from achiral 2,6-bis(imidazol-1-yl)pyridine: Solvent effect and spontaneous resolution," *J. Chem. Sci.*, vol. 126, no. 5, pp. 1423–1431, 2014.

[83] S. Biswas *et al.*, "A cubic coordination framework constructed from benzobis-triazolate ligands and zinc ions having selective gas sorption properties," *Dalt. Trans.*, no. 33, pp. 6487–6495, 2009.

[84] A. K. Chaudhari, S. Mukherjee, S. S. Nagarkar, B. Joarder, and S. K. Ghosh, "Bi-porous metal–organic framework with hydrophilic and hydrophobic channels: Selective gas sorption and reversible iodine uptake studies," *CrystEngComm*, vol. 15, no. 45, pp. 9465–9471, 2013.

[85] P.-Z. Li *et al.*, "Co(II)-tricarboxylate metal–organic frameworks constructed from solvent-directed assembly for CO_2 adsorption," *Microporous Mesoporous Mater.*, vol. 176, pp. 194–198, 2013.

[86] T. Zhang, Y. Lu, Z. Zhang, Q. Lan, D. Liu, and E. Wang, "Single-crystal to single-crystal transformation from a hydrophilic–hydrophobic metal–organic framework to a layered coordination polymer," *Inorganica Chim. Acta*, vol. 411, pp. 128–133, 2014.

[87] P. Wang *et al.*, "Two-/three-dimensional open lanthanide–organic frameworks containing rigid/flexible dicarboxylate ligands: Synthesis, crystal structure and photoluminescent properties," *CrystEngComm*, vol. 15, no. 10, pp. 1931–1949, 2013.

[88] H. J. Park, Y. E. Cheon, and M. P. Suh, "Post-synthetic reversible incorporation of organic linkers into porous metal–organic frameworks through single-crystal-to-single-crystal transformations and modification of gas-sorption properties," *Chem. – A Eur. J.*, vol. 16, no. 38, pp. 11662–11669, Oct. 2010.

[89] K. Davies, S. A. Bourne, and C. L. Oliver, "Solvent- and vapor-mediated solid-state transformations in 1,3,5-benzenetricarboxylate metal–organic frameworks," *Cryst. Growth Des.*, vol. 12, no. 4, pp. 1999–2003, Apr. 2012.

[90] B. Liu, L.-Y. Pang, L. Hou, Y.-Y. Wang, Y. Zhang, and Q.-Z. Shi, "Two solvent-dependent zinc(ii) supramolecular isomers: Structure analysis, reversible and non-reversible crystal-to-crystal transformation, highly selective CO_2 gas adsorption, and photoluminescence behaviors," *CrystEngComm*, vol. 14, no. 19, pp. 6246–6251, 2012.

[91] P. Sarawade, H. Tan, and V. Polshettiwar, "Shape- and morphology-controlled sustainable synthesis of Cu, Co, and in metal organic frameworks with high CO_2 capture capacity," *ACS Sustain. Chem. Eng.*, vol. 1, no. 1, pp. 66–74, Jan. 2013.

[92] D.-C. Hou, G.-Y. Jiang, Z. Zhao, and J. Zhang, "Temperature-dependent urothermal synthesis of two distinct La(III)-naphthalenedicarboxylate frameworks," *Inorg. Chem. Commun.*, vol. 29, pp. 148–150, 2013.

[93] Y. Wei, Y. Yu, R. Sa, Q. Li, and K. Wu, "Two cobalt(II) coordination polymers $[Co_2(H_2O)_4(Hbidc)_2]_n$ and $[Co(Hbidc)]_n$ (Hbidc = 1H-benzimidazole-5,6-dicarboxylate): Syntheses, crystal structures, and magnetic properties," *CrystEngComm*, vol. 11, no. 6, pp. 1054–1060, 2009.

[94] S.-M. Zhang, T.-L. Hu, J.-L. Du, and X.-H. Bu, "Tuning the formation of copper(I) coordination architectures with quinoxaline-based N,S-donor ligands by varying terminal groups of ligands and reaction temperature," *Inorganica Chim. Acta*, vol. 362, no. 11, pp. 3915–3924, 2009.

[95] M.-L. Han, X.-H. Chang, X. Feng, L.-F. Ma, and L.-Y. Wang, "Temperature and pH driven self-assembly of Zn(ii) coordination polymers: Crystal structures, supramolecular isomerism, and photoluminescence," *CrystEngComm*, vol. 16, no. 9, pp. 1687–1695, 2014.

[96] C. Wang, H. Jing, P. Wang, and S. Gao, "Series metal–organic frameworks constructed from 1,10-phenanthroline and 3,3′,4,4′-biphenyltetracarboxylic acid: Hydrothermal synthesis, luminescence and photocatalytic properties," *J. Mol. Struct.*, vol. 1080, pp. 44–51, 2015.

[97] E. Yang, H.-Y. Li, Z.-S. Liu, and Q.-D. Ling, "Urothermal synthesis of two photoluminescent cadmium coordination polymers," *Inorg. Chem. Commun.*, vol. 30, pp. 152–155, 2013.

[98] H.-J. Mo, Y.-R. Zhong, M.-L. Cao, Y.-C. Ou, and B.-H. Ye, "Hydrothermal syntheses and structural diversity of cobalt complexes with 2,2′-bibenzimidazole ligand by temperature tuning strategy," *Cryst. Growth Des.*, vol. 9, no. 1, pp. 488–496, Jan. 2009.

[99] P. Kanoo, K. L. Gurunatha and T. K. Maji, "Temperature-controlled synthesis of metal-organic coordination polymers: Crystal structure, supramolecular isomerism, and porous property," *Cryst. Growth Des.*, vol. 9, no. 9, pp. 4147–4156, Sep. 2009.

[100] P. Cui *et al.*, "Temperature-controlled chiral and achiral copper tetrazolate metal–organic frameworks: Syntheses, structures, and I_2 adsorption," *Inorg. Chem.*, vol. 51, no. 4, pp. 2303–2310, Feb. 2012.

[101] W.-X. Chen, S.-T. Wu, L.-S. Long, R.-B. Huang, and L.-S. Zheng, "Construction of a three-fold parallel interpenetration network and bilayer structure based on copper (II) and trimesic acid," *Cryst. Growth Des.*, vol. 7, no. 6, pp. 1171–1175, Jun. 2007.

[102] J. Zhang *et al.*, "Urothermal synthesis of crystalline porous materials," *Angew. Chemie Int. Ed.*, vol. 49, no. 47, pp. 8876–8879, Nov. 2010.

[103] C.-D. Wu and W. Lin, "Highly porous, homochiral metal–organic frameworks: Solvent-exchange-induced single-crystal to single-crystal transformations," *Angew. Chemie Int. Ed.*, vol. 44, no. 13, pp. 1958–1961, Mar. 2005.

[104] T. K. Maji, G. Mostafa, R. Matsuda, and S. Kitagawa, "Guest-induced asymmetry in a metal–organic porous solid with reversible single-crystal-to-single-crystal structural transformation," *J. Am. Chem. Soc.*, vol. 127, no. 49, pp. 17152–17153, Dec. 2005.

[105] J. Sun *et al.*, "Dimerization of a metal complex through thermally induced single-crystal-to-single-crystal transformation or mechanochemical reaction," *Angew. Chemie Int. Ed.*, vol. 50, no. 31, pp. 7061–7064, Jul. 2011.

[106] S. Kitagawa and K. Uemura, "Dynamic porous properties of coordination polymers inspired by hydrogen bonds," *Chem. Soc. Rev.*, vol. 34, no. 2, pp. 109–119, 2005.

[107] L. Luo, G.-C. Lv, P. Wang, Q. Liu, K. Chen, and W.-Y. Sun, "pH-dependent cobalt (ii) frameworks with mixed 3,3′,5,5′-tetra(1H-imidazol-1-yl)-1,1′-biphenyl and 1,3,5-benzenetricarboxylate ligands: Synthesis, structure and sorption property," *CrystEngComm*, vol. 15, no. 45, pp. 9537–9543, 2013.

[108] M. Chen *et al.*, "pH dependent structural diversity of metal complexes with 5-(4H-1,2,4-Triazol-4-yl)benzene-1,3-dicarboxylic acid," *Cryst. Growth Des.*, vol. 11, no. 5, pp. 1901–1912, May 2011.

[109] F.-J. Meng, H.-Q. Jia, N.-H. Hu, and J.-W. Xu, "pH-controlled synthesis of two new coordination polymers modeled by pyridine-2,4-dicarboxylic acid," *Inorg. Chem. Commun.*, vol. 21, pp. 186–190, 2012.

[110] Q. Yu *et al.*, "pH-dependent Cu(II) coordination polymers with tetrazole-1-acetic acid: Synthesis, crystal structures, EPR and magnetic properties," *Cryst. Growth Des.*, vol. 8, no. 4, pp. 1140–1146, Apr. 2008.

[111] B. Liu, G.-P. Yang, Y.-Y. Wang, R.-T. Liu, L. Hou, and Q.-Z. Shi, "Two new pH-controlled metal–organic frameworks based on polynuclear secondary building units with conformation-flexible cyclohexane-1,2,4,5-tetracarboxylate ligand," *Inorganica Chim. Acta*, vol. 367, no. 1, pp. 127–134, 2011.

[112] Q. Chu, G.-X. Liu, T. Okamura, Y.-Q. Huang, W.-Y. Sun, and N. Ueyama, "Structure modulation of metal–organic frameworks via reaction pH: Self-assembly of a new carboxylate containing ligand N-(3-carboxyphenyl)iminodiacetic acid with cadmium(II) and cobalt(II) salts," *Polyhedron*, vol. 27, no. 2, pp. 812–820, 2008.

[113] Z. Guo *et al.*, "A series of cadmium(II) coordination polymers synthesized at different pH values," *Eur. J. Inorg. Chem.*, vol. 2007, no. 5, pp. 742–748, 2007.

[114] S.-T. Wu, L.-S. Long, R.-B. Huang, and L.-S. Zheng, "pH-dependent assembly of supramolecular architectures from 0D to 2D networks," *Cryst. Growth Des.*, vol. 7, no. 9, pp. 1746–1752, Sep. 2007.

[115] H.-N. Wang, G.-S. Yang, X.-L. Wang, and Z.-M. Su, "pH-induced different crystalline behaviors in extended metal–organic frameworks based on the same reactants," *Dalt. Trans.*, vol. 42, no. 18, pp. 6294–6297, 2013.

4 Promising Functional Metal-Organic Frameworks for Gas Adsorption, Separation and Purification

Shabnam Khan, Fahmina Zafar, Farasha Sama, M. Shahid, and Mohammad Yasir Khan

CONTENTS

4.1 INTRODUCTION

The numerous aspects prevailing in the human society like environmental protection, use of energy, and industrial manufacture have a major impact on gas storage

DOI: 10.1201/9781003252061-4

and separation. Most importantly, hydrogen (H_2) and methane (CH_4) storage are vital for widespread clean energy usage; carbon dioxide (CO_2) separation is important to the increasing greenhouse effect; storage and separation of lethal gases, such as ammonia (NH_3) and nitrogen dioxide (NO_2), are crucial for pollution control and industrial chemicals synthesis.

Energy is required for today's world to function and for that purpose energy sources or carriers like natural gas and biogas significantly contribute to the production of energy [1]. With regard to energy production, the rapidly increasing global demand for fossil fuels has resulted in substantial environmental challenges, particularly the climate change produced by carbon emissions. Gas fuels are more ecologically benign than liquid petroleum and solid coal due to fewer carbon releases and high gravimetric energy densities. For instance, the gravimetric heat of combustion for H_2 and CH_4 are 123 and 55.7 MJ kg^{-1}, respectively, which is comparatively higher than gasoline (47.2 MJ kg^{-1}) [2]. Because of their exceptionally lower boiling point, low density, high diffusivity, and high critical pressure, the fundamental problem for gas fuels is their transit, storage, and transformation, which generally require severe conditions and use huge quantities of energy. Conventional gas fuel storage options include liquefaction at low temperature (526 K for liquefied hydrogen) or compression at high pressure under ambient temperature, both of which need a large facility. An alternate option for achieving affordable and safe storage and conveyance is to use porous materials for adsorption storage arrangements that function under mild circumstances.

Furthermore, separation and purification operations are vital for the contemporary chemical industry because they remove pure constituents from chemical mixtures that account for almost half of industrial energy demand. For example, olefins [primarily ethylene (C_2H_2), propylene (C_3H_6) and butadiene (C_4H_6)] constitute the foundation of numerous essential industrialized processes, with a global production reaching 2.0×10^8 metric tonnes, or nearly 30 kilos for every person on the globe [3]. Separation and purification steps are necessary prior to immediate use of gas commodities, that are traditionally performed by repetitive distillation–compression cycles of the combination. High-temperature separation techniques use up to ten times the energy of membrane-based and/or adsorptive separation methods [3]. Realizing that separation and purification operations typically consume > 40% of overall energy intake in the chemical industry, implementing sophisticated adsorptive separation methods centered on porous materials can result in significant savings of energy [3].

Numerous common substances, like zeolites, charcoal and ceramics have been profoundly utilized as adsorbents' porous medium. In order to handle the gases, the fabrication of porous materials showing a huge capacity for storage and great separation efficiency is crucial [4], particularly with high porosity and better modulation over varied dimensions. During the last two decades, a range of porous adsorbents has been introduced viz., metal-organic frameworks (MOFs) [5,6], covalent organic frameworks (COFs) [7–10] and hydrogen-bonded organic frameworks (HOFs) [11], in the context of this application. Among all the categories, MOFs are inorganic-organic hybrid materials that involve the organic spacers coordination to metal ions/clusters to form a self-assembled structure. The moderate linkage strength of about 90–350 kJ/mol (as compared to 1–170 kJ/mol and 300–600 kJ/mol in HOFs and COFs, respectively) and differing constituents impart the MOFs with remarkable porosity, high modularity, great crystallinity and varied functionality [11]. Overall, MOFs are displayed to be a promising platform in the area of gas storage and separation [12–14], optical [15–19], electric and magnetic materials [20], chemical sensing [21–25], catalysis [26,27] bio-medicine and much more.

Because of their extremely high porosity (surface area varying from 100 to 10,000 m^2/g) [28,29] configurable pore size of 3–100 Å, excellent thermal stability (up to 773 K), and also superior chemical stability, MOFs are widely known for gas storage and gas separation [6]. In the late 1990s, MOFs were discovered to have permanent porosity, which prompted their use as adsorbents [30–32]. This form of adsorbent appears to be particularly promising for gas processing, as evidenced by the fast renewing data of pore surface area. Significant advancements have been made by MOFs (Appendix 4.1) in the field of gas storage of CH_4, H_2, C_2H_2 and much more in comparison to zeolites and porous carbon.

Simultaneously, MOFs possessing uniformity in distribution of pore size, dec-orated with varied functional groups and modifiable pore size impart them the characteristics to act as promising candidates for gas separation. In principle, the use of gas chromatography in 2006 [33], fixed-bed breakthrough in 2006–2007 for the separation of a mixture of gases [34–37] and gas sorption site determination [38–40] by crystallographic technologies substantially facilitates MOFs in actual gas mixture separation. Since then, a lot of work (Appendix 4.2) has gone into the gas separation sector of MOFs [12,13,41–44], which has resulted in a lot of development, especially in the industrially important hydrocarbon separation.

Several MOFs have been utilized in the field of gas storage and separation up to this point. This dynamic field has emerged as the most significant in the subject of energy and materials chemistry. We hope to highlight the present situation of various MOFs in this chapter, providing comprehensive insight for distinguished readers and addressing contemporary issues in the area of storage and separation alongwith a brief overview on the mechanism involved in adsorptive separation and purification. In this chapter, we will focus on some great contributions made in the fied of gas storage (H_2, CH_4, C_2H_2), gas separation (C_2H_4/C_2H_6 separation, C_2H_2/C_2H_4 separation, C_2H_2/CO_2 separation) and harmful gas (NH_3, NO_2, SO_2) removal.

4.2 GAS STORAGE

Following the advent of their perpetual porosities, MOFs were predicted to be potential adsorbents for gas storage. MOFs have a number of structural benefits over conventional porous materials such as zeolites and activated carbons, including great porosity, large surface area, variable geometry and size of the pore, and a pore surface that is functionalizable. As a result, the advent of MOFs holds a lot of potential for storing essential gases, including H_2, CH_4, and C_2H_2 [45]. In the following section, the most recent developments in these three areas will be discussed.

4.2.1 H₂ STORAGE

Human society's demand for energy resources is expanding in tandem with its population growth and economic progress. The exploitation of fossil fuels has resulted in a variety of environmental challenges, including climate change, prompting people to look for new energy sources in order to reduce their carbon imprint. H_2 emerged to be an excellent alternative source of energy besides coal and gasoline owing to its extremely high gravimetric combustion heat and amiable combustion outcomes [46]. Most of the environmental issues associated with widespread use of fossil fuels could be surpassed by using H_2 as an automobile fuel [47]. This motivated many researchers towards the investigation on H_2 storage as it possesses highest energy per mass of any fuel while emitting negligible CO_2. The boiling point and critical temperature of H_2 are merely 20 K and 38 K, respectively, making it challenging to liquefy or compress. Hence, in order to store H_2, two probing techniques are the use of cryogenic liquid H_2 tanks and high-pressure tanks [48]. In the case of the former technique, an appreciably enormous amount of energy is a prerequisite for liquefying the H_2 and keeping the tanks cooler. While for the latter case, a high-pressure method demands the implication of tremendously heavy apparatus in order to prevent leakage of H_2, which ultimately decreases the 'actual gravimetric capacity' of the tanks. At present, the H_2 storage capacities of prevailing tanks are 3.4–4.7 wt% and 14–28 kg/m^2 [48]. In order to improve the storage capacities, efforts must need to be carried out for storing H_2 at a relatively low pressure and high temperature.

Yaghi et al. conveyed the very first example of MOF for H_2 storage (MOF5) in 2003 [49], following which several other MOFs have been discovered for their H_2 storage potential. According to research studies, it has been well established that gravimetric storage abilities of H_2 at high pressure and 77 K are found to be proportional to the MOF's pore volume and surface area. For example, NOTT112, NU111 and NU100/PCN610 having BET surface area of 3,800, 4,930 and 6,143 m^2/g, show total gravimetric uptake of H_2 about 10.0, 13.6 and 16.4 wt % , respectively, at 77 K and 70 bars [50–53]. H_2 storage must be done at near ambient temperatures for shipboard applications of hydrogen fuel cells. The MOFs and H_2 interactions are primarily weak van der Waals interactions, which makes this task extremely difficult. None of the MOFs have reached the DOE's gravimetric capacity targets thus far. Nonetheless, open metal site's immobilization can improve MOF-H_2 interactions and elevate the isosteric heat of H_2 adsorption to roughly 12 kJ/mol in some cases [2]. According to theoretical research, a minimum heat of H_2 adsorption of 15 kJ/mol is necessary to achieve substantial H_2 storage capacities at appropriate temperatures [54]. Volumetric capacity, in addition to gravimetric capacity, is a significant component to contemplate while assessing MOFs for storage of H_2, as the available volume in light-duty automobiles for a tank storing H_2 adsorbent is quite restricted, and is a crucial factor in shaping a vehicle's operating range. Regardless of the verity that MOFs with significant gravimetric capacities often have limited volumetric capacities owing to their wide diameter of pores and feeble interactions of MOF-H_2, our prognostications suggested that MOFs can reach both high gravimetric as well as volumetric H_2 densities. As a result, many potential MOFs were synthesized and their useful H_2 capacities were determined. IRMOF20 was expected to have excellent usable gravimetric (9.1 wt. %) and volumetric (51.0 g/L) capacities correspondingly, at a consolidated temperature and pressure shift to 160 K and 5 bars from 77 K and 100 bars), demonstrating the significance of exercising computational screening to assist experimental determinations toward MOFs with anticipated storage potential. Lately, Kapelewski investigated the accessible H_2 volumetric capacities of $Ni_2(m\text{-dobdc}^{4-} =$ 4,6-dioxido-1,3-benzenedicarboxylate), and observed volumetric capacity of 11.0 g/L at 298 K and 23.0 g/L through a temperature range of −75 to 298 K, rendering it the best physisorptive storage material to date [55]. The existence of open Ni(II) sites that are highly polarizing contributes to robust binding interactions as well as a compact H_2 packing within the MOF, resulting in a high capacity.

4.2.2 CH₄ STORAGE

Natural gas, which is primarily constituted of CH_4, is an extensive natural resource. Natural gas is appealing as a vehicle fuel for its high research octane number (RON) and little CO_2 emissions. This drives substantial research into effective CH_4 storage techniques [56]. Because the interactions amongst CH_4 and MOFs are mild, their storage could be done at ambient temperature and at a pressure that is relatively high, which is both practical and promising.

Substantial progress has been achieved in this sector since Kitagawa and Yaghi individualistically established the very first two instances of MOFs for the storage of CH_4 [30,57]. At high pressure, the gravimetric CH_4 storage competencies of MOFs were determined to be essentially correlated to their surface areas or volume of pores, based on extensive experimental data [58]. Bigger porosities suggest more capacity in MOFs, which can take in larger numbers of molecules of CH_4. Numerous empirical connections between rigid MOF pore volumes and CH_4 capacities have been found that insinuate valuable information in the search for potential MOF candidates for the storage of CH_4. Peng et al. in 2013 used a standardized measurement technique to assess the CH_4 storage capabilities of reported six MOFs with diverse kinds of structures (HKUST1, NiMOF74, PCN14, UTSA20, NU111, and NU125) [32]. Their findings demonstrated that the BET surface area was significantly proportional to pore volume, total gravimetric CH_4 adsorption (at 65 bar and 298 K), and inverse density of MOFs. Among them, NU111 had the maximum gravimetric uptake, which is about 0.36 g/g, with a calculated 4,930 m^2/g of BET surface area. As per the linear correlation, a theoretical MOF with surface area of 7,500 m^2/g and pore volume of 3.2 cm^3/g could fulfill the gravimetric target.

Given the restricted area available for CH_4 holding tanks on the vehicles, excessive volumetric CH_4 capabilities are much more critical than gravimetric storage capacities. Because MOFs with strong gravimetric CH_4 uptakes possess pores which are large and poor MOF-CH_4 interactions, they often have medium to lower volumetric capacities. According to research, optimum MOFs for large CH_4 volumetric capacities ought to have balance porosities and scaffold densities, along with greater densities of appropriate pore cages for CH_4 molecule identification [58]. To achieve high overall and operational (implementable amount between 5 and 65 bars) volumetric capacities, techniques such as maximizing pore spaces and incorporation of functional sites into MOFs were devised.

Lu et al. developed soc-MOF based on Fe, symbolized as Fe-pbpta (where H_4pbpta stands for 4,4',4'',4'''-(1,4-phenylenebis(pyridine-4,2–6-triyl))-tetrabenzoic acid) with an extremely high BET surface area of about 4,937 m^2/g and a high volume of pore of about 2.15 cm^3/g. In the DOE's operational storage settings (35 bars and 298 K), MOF had the maximum gravimetric uptake of 369 cm^3 (STP)/g and the peak volumetric delivery capacity of 192 cm^3/cm^3 at 298 K and 65 bars. Fe-pbpta is also a viable contender for on-board methane storage due to its strong thermal and aqueous stability [59].

Tu et al. recently synthesized iron-based MOFs, namely VNU21 and VNU22, as well as a fresh MOF called Fe-NDC, $Fe_3O(NDC)_2SO_4(HCO_2)(H_2O)_2$ (where NDC stands for 2,6-naphthalene dicarboxylate), all of which have one-dimensional (1D) channels of various shapes and sizes in their assemblies, and assessed their CH_4 storage (Figure 4.1a-b). Because of the structural pattern that uses mixed spacers to manage the width and length of rectangular 1D channels in the course of isoreticularly structural expanse. At 65 bars and 25°C, VNU21 (having long-narrow 1D channels of about 7.0 × 12.4 $Å^2$) outperformed VNU22, Fe-NDC, and typical MOFs with extensive 1D channels, exhibiting a total volumetric CH_4 absorption as high as 182 cm^3/cm^3 (DUT4 and DUT5). Despite its substantially reduced surface area, the CH_4 intake of VNU21 was equivalent to that of benchmark materials (e.g., MOF177, MOF205, MOF5, MOF210, and MOF905-NO2) for CH_4 storage at 65 bars. This approach will be used to innovate novel porous scaffolds, such as reticularly expanded structures of VNU21, that possess a large surface area as well as a significant CH_4-framework interaction for increased CH_4 storage capacity [60].

In order to increase CH_4 storage capacity, researchers used a combination technique of elongating organic linkers and inserting functional sites into MOFs.

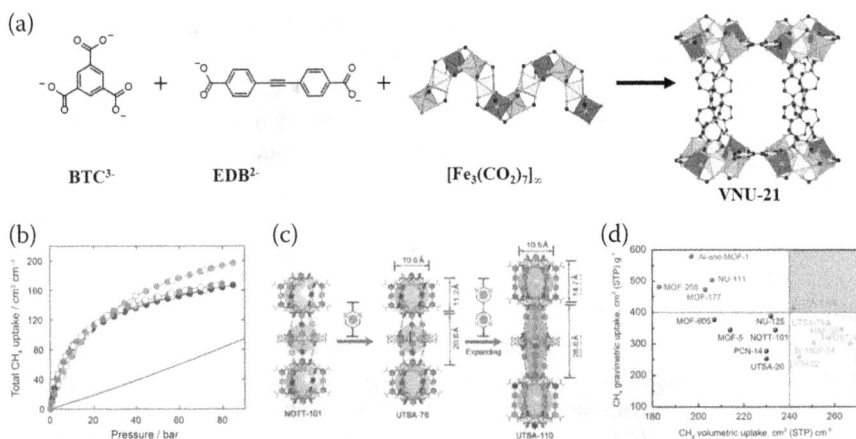

FIGURE 4.1 (a) Rod-shaped like SBUs are linked with organic spacers to form 3D scaffolds with the 1D channels. Color code of atoms: C, black; O, red; Fe, octahedra; and S, yellow tetrahedral (b) The total volumetric CH_4 uptake of VNU21 (red), Fe-NDC (green), and VNU22 (blue). Black curve represents methane's bulk density. (Reproduced with permission from Tu et al. 2019 [60]). (c) Comparison of the crystal structures of NOTT101, UTSA76, and UTSA110. (d) Comparison of UTSA110a's total gravimetric and volumetric CH_4 capacities at 65 bars and 298 K to other benchmark MOFs. (Reproduced with permission from Wen et al. 2018 [61]).

Wen et al. developed UTSA110a, a novel MOF with an increased density of functional nitrogen sites than UTSA76, in 2018 (Figure 4.1c-d) [61].

In comparison to UTSA76, UTSA110a has a greater surface area and functional N sites increased content (3,241 m^2/g and 3.94 mmol/cm^3, as compared to 2,820 m^2/g and 2.64 mmol/cm^3). The increased porosity could help with total gravimetric uptake; however, the functional nitrogen site's high density could aid in interactions between CH_4 and for better volumetric uptake capacity. At 298 K and 65 bars, UTSA110a shows great volumetric and gravimetric (241 cm^3 and 402 cm^3 (STP)/g, respectively) total CH_4 capacities, as envisaged.

4.2.3 C_2H_2 STORAGE

In modern industry, C_2H_2 is widely employed in the preparation of various electric and chemical materials. However, owing to its high explosiveness after compressing under pressures greater than 0.2 MPa, safe storage of C_2H_2 remains a challenge. C_2H_2 is currently stored in industry in specific cylinders packed with CH_3COCH_3 and functional inorganic materials, which results in acetone pollution and expensive costs. As a result, new ways for securely storing C_2H_2 are urgently needed. MOFs have been investigated as storage adsorbents for C_2H_2 [62]. To achieve high C_2H_2 storage capacities in MOFs, some techniques have been shown to be useful, including (1) integration of open metal sites, (2) tailoring the pore area, and (3) organic spacers functionalization to give particular desired sites for C_2H_2 molecules. After the discovery of very first MOF for C_2H_2 storage invented by Matsuda et al. in 2005 [38], Cu_2(pzdc)$_2$(pyz) (where pzdc stands for pyrazine-2,3-dicarboxylate and pyz for pyrazine), as C_2H_2 storage materials, a variety of different MOFs have been developed.

The use of open metal sites in MOFs for enhanced C_2H_2 absorption was initially described by Xiang et. al. in 2009 [63]. They looked at C_2H_2 storage qualities in six different MOF materials with distinct structures and different porosities (HKUST1, MOF5, MOF505, MOF508, MIL53 and ZIF8). HKUST1, for example, at 1 atm and 295 K has the highest C_2H_2 capacity of 201 cm^3/g, as per the findings. The strong binding sites, according to neutron powder diffraction analyzes, are open Cu(II) sites. Since these open Cu(II) sites have effective preferential binding with C_2H_2 molecules, they offer a large storage capacity for C_2H_2. The M-MOF74 series [M = Mn(II), Mg(II), Zn(II) and Co(II)] used the same method and also demonstrated large C_2H_2 uptakes, with Co-MOF74 attaining 197 cm^3/g at 1 bar and 298 K [64].

Organic spacers having functional groups/sites, such as amide groups and basic pyridyl sites, deliver more binding sites for C_2H_2 molecules, which is favorable for large C_2H_2 storage capacities [65,66]. Using a novel tetracarboxylic acid spacer with an accessible amino functional group H_4TTCA—NH_2(5-amino-3,3,5,5-tetracarboxylic acid), Qian et al. generated microporous MOF based on Cu [Cu_2L $(H_2O)_2$](DMF)$_{0.5}$(H_2O)$_7$, (ZJU195) as shown in Figure 4.2a-c. The activated ZJU195a was able to attain great C_2H_2 storages of 275.6 and 214.2 cm^3/g at 273 K and 298 K correspondingly under 1.0 bar [67]. Bai et al. developed a novel MOF505 counterpart with an amide functional moeity, i.e., [Cu_2(DBAI)(H_2O)$_2$]·4DMF·4H_2O

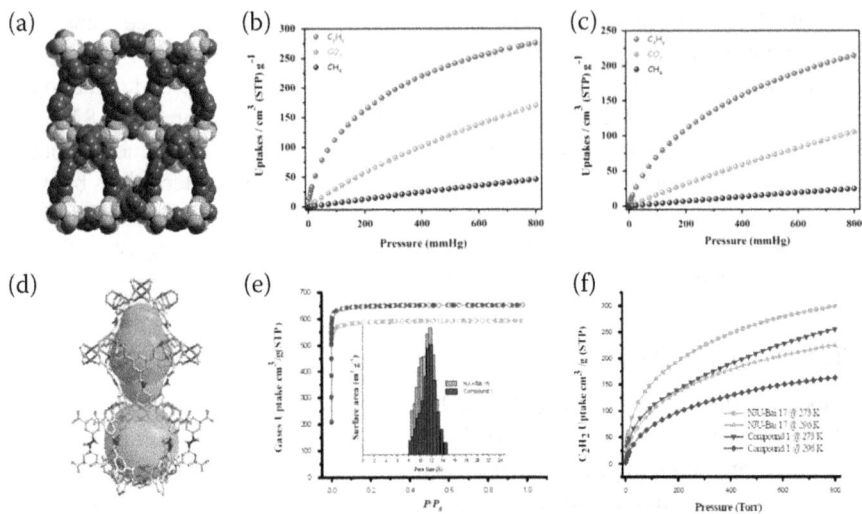

FIGURE 4.2 (a) X-ray single crystal structure representing the pore cages along *b* axis of ZJU195 (O, red; C, violet; Cu, azure; N, blue; for clarity, guest molecules and H atoms are excluded). (b) Single-component adsorption isotherms for CH_4 (blue), CO_2 (orange), C_2H_2 (violet) and of ZJU195a at 273 K and (c) 298 K, respectively. (Reproduced with permission from Qian et al. 2017 [67]). (d) NJU-Bai 17's X-ray single crystal structure. (e) Sorption isotherms of N_2 at 77 K (filled symbols symbolizes adsorption whereas vacant one symbolizes desorption); inset: distribution of pore size analyzed by NLDFT method. (f) Adsorption isotherms of C_2H_2 for NJU-Bai17 at 273 K and 296 K. (Reproduced with permission from Bai et al. 2016 [68]).

(DBAI, (5-(3,5-dicarboxybenzamido)isophthalic acid; NJU-Bai17, that has near best-ever C_2H_2 uptakes of 222.4 and 296 cm^3/g at 296 and 273 K, respectively, under 1 bar [68]. The addition of the novel and powerful adsorption site of the amide functional group towards C_2H_2 contributes significant C_2H_2 uptake in NJU-Bai 17 and the uptake of C_2H_2 at 25°C has been dramatically increased to 39% when compared to C≡C accordingly (Figure 4.2d-f).

The appropriate nanopore size/space is also a key component in the determination of MOF materials' C_2H_2 storage performances. Pang et al. designed the FJI-H8 MOF in 2015, which has adequate pore gaps (8, 12, and 15 Å in diameter, correspondingly) as well as metal sites that are open (3.59 mmol/g). FJI-H8 has the highest gravimetric uptake of C_2H_2 of about 224 cm^3/g (STP) and the volumetric capacity of 196 cm^3/cm^3 which is the second highest at 1 atm and 295 K (STP) [69]. Assuming that each open Cu(II) site will bind to one C_2H_2 molecule, the open Cu(II) sites in FJI-H8 could only provide 87 cm^3 of the overall 224 cm^3 storage capacity for C_2H_2. The remaining C_2H_2 uptake in FJI-H8 should indeed be attributable to the appropriate pore size, which contributes for more than 60% of total C_2H_2 absorption in FJI-H8. This research serves as an excellent example of another structural property which could be used to develop novel MOF materials with high C_2H_2 storage capacities.

4.3   SEPARATION OF GASES BY USING MOFS

Aside from storage of gases, a plethora of MOFs have showed considerable promise as adsorbents for gas separations owing to their unique pore chemistry. Over the last decade, the active field has advanced at an incredible rate [12,43,70–78].

Adsorptive separation of various gases may be done by distinguishing the adsorbate molecule sizes, morphologies, polarity, polarizabilities, coordinating capability, conformations, and so forth based on the size/shape of the adsorbent's pore and/or the binding ability to adsorption sites. Prior to separation process, one must gain a clear understanding of the basic mechanism behind adsorptive separation and purification processes to analyze the separation process through various solid porous materials (MOFs). It is widely known that gas adsorption is often achieved by one or more mechanisms [79]. The main four mechanisms involved in gas adsorptive separations are as follows.

- **Molecular sieving effect/Steric separation:** Some constituents of a gas mixture are forbidden to enter the adsorbent's pore due to size and/or shape exclusion, whilst other components are permitted to enter the pores and thus are eventually adsorbed. Size/shape exclusion, commonly referred to as steric separation, achieves selective adsorption depending on the adsorbate molecule's shape and cross-sectional size (kinetic or collision diameter). The shortest distance or diameter separating two molecules possessing zero kinetic energy that could approach each other and collide is referred to as the collision or kinetic diameter. It should also be noted that in certain circumstances, the temperature does have an effect on molecular sieving if the pore size is heat-sensitive [72]. Gas drying utilizing 3A zeolite and the separation of normal paraffin from iso-paraffins and some other hydrocarbons employing 5A zeolite are two of the most important uses of steric separation.
- **Equilibrium or dynamic separation:** In the case of equilibrium separation, when the pore of the adsorbent is sufficiently big to enable all of the constituent gases to penetrate, the interactions between the adsorbate and the surface of the adsorbent are critical in determining separation quality. It is noteworthy to mention that the intensity of this interaction is proportional to the adsorbate molecule's properties as well as the adsorbent's surface qualities such as magnetic susceptibility, polarizability, quadrupole moment, and permanent dipole moment [72].
- **Kinetic or partial molecular sieve separation:** This mechanism tends to play its role when equilibrium separation is not possible. The most critical aspect of kinetic separation is that the adsorbent's pore size must be precisely adjusted in between the kinetic diameters of the two molecules to be segregated. As per the literature, this mechanism is involved in the separation of methane from carbon dioxide via employing a carbon molecular sieve. Another common example involving this mechanism is air separation with the help of zeolite adsorbent by the pressure swing adsorption (PSA) method [72,80].

- **Quantum sieve effect:** The adsorption process occurring in the quantum sieve effect takes place due to the variations in the diffusion rates of the guest molecules and the comparability of the pore diameter well with de Broglie wavelength of such molecules. This mechanism is quite useful for the H_2/D_2 separation [72].

Under this section, significant advancements made in modern years in numerous key gas separation scenarios, including C_2H_4/C_2H_6, C_2H_2/C_2H_4 separation, C_2H_2/CO_2 separation will be discussed.

4.3.1 C_2H_4/C_2H_6 Separation

Since C_2H_4 is a key requisite in the petrochemical sector, C_2H_4/C_2H_6 separation is critical in the olefin-paraffin separation. Traditional methods rely heavily on energy-intensive cryogenic distillation.

Using adsorbents like MOFs in order to separate C_2H_4/C_2H_6 mixtures efficiently can be seen as a feasible alternative [81]. A variety of MOFs have been investigated and found to have excellent separations due to several of the below mentioned mechanistic pathways: (1) kinetic-based mechanism; (2) equilibrium-based mechanism; (3) gate-opening effect [82]; and (4) molecular sieving effect. MOF materials are classified as C_2H_4- or C_2H_6- selective adsorbents based on whether C_2H_4 or C_2H_6 is trapped by adsorbents.

4.3.1.1 C_2H_4 Selective Separation

Because C_2H_4 has a smaller molecular size and stronger framework-C_2H_4 interactions, C_2H_4-selective adsorbents remove C_2H_4 from C_2H_6/C_2H_4 mixtures [83]. A greater density of binding sites present in porous materials is a beneficial property for achieving strong framework-C_2H_4 interactions.

Yang et al. in 2014 synthesized a OH-functionalized Al-MOF (NOTT300) for highly selective C_2H_4/C_2H_6 separation [84]. At 293 K and 1.0 bar, the C_2H_4 and C_2H_6 adsorption quantities in NOTT300 were 4.28 and 0.85 mmol/g, correspondingly, resulting in a significant rise of 3.43 mmol/g and C_2H_4/C_2H_6 selectivity of about 48.7. According to neutron powder diffraction experiments and in-situ synchrotron X-ray, various groups in NOTT300 like aromatic-CH groups, M-OH groups, and C_6H_5-rings could result in feeble additive, supramolecular interactions with the molecules of C_2H_4 being adsorbed, where the C---HO distance between OH group and C_2H_4 molecule is reported as 4.62 Å. C_2H_6 is positioned far from the hydroxyl group (C---O = 5.07 Å), displayed poorer bindings with the NOTT-300 host. In 2020, Chen et al. developed a Th-MOF for C_2H_4 separation from binary C_2H_6/C_2H_4 and ternary $C_2H_4/C_2H_2/C_2H_6$ mixtures (Figure 4.3a-c). The developed MOF Azole-Th1 has UiO66-like moeties with fcu topology and a tetrazole-based linker. A O, N-donor ligand with excellent chemical stability connects such a significant structure. The as-synthesized showed exceptional C_2H_4 separation with 99.9% purity from not only a binary C_2H_6/C_2H_4 (1:9, v/v) mixture but also a ternary mixture of $C_2H_2/C_2H_4/C_2H_6$ (1:90:9, v/v/v) at 298 K and 100 kPa, having a working capacity of up to 1.13 and 1.34 mmol/g, correspondingly. This separation

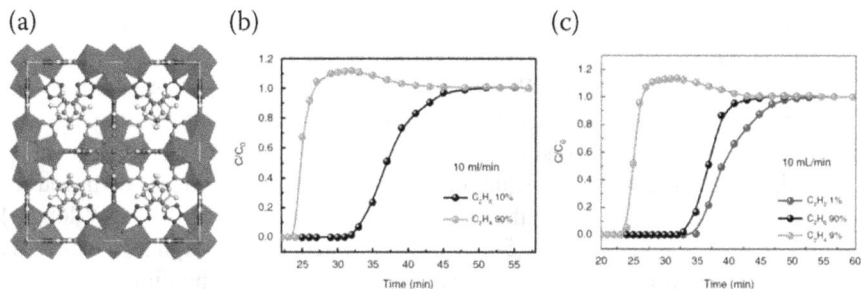

FIGURE 4.3 (a) Azole-Th1 crystal unit cell structure; experimental breakthrough on Azole-Th1 curves at 1 bar and 298 K for separation of (b) C_2H_6/C_2H_4 (10/90, v/v) binary mixture, (c) $C_2H_6/C_2H_4/C_2H_2$ (90/9/1, v/v/v) ternary mixture. (Reproduced with permission from Z. Xu et al. 2020 [85]).

is because of stronger van der Waals interaction amongst C_2H_6 and the MOF framework, according to the density functional theory calculations [85].

Another effective technique for selective separation is to precisely tune the binding sites of MOFs to isolate C_2H_4 from C_2H_6 via the molecular sieving effect. In 2018, Bao et al. created M-gallate (M= Ni, Mg, Co), a family of gallate-based MOFs, consisting of 3D interlinked zigzag channels with 3.47–3.69 Å aperture size. This results in efficient separation of C_2H_4 ($3.28 \times 4.18 \times 4.84$ Å^3) from C_2H_6 ($3.81 \times 4.08 \times 4.82$ Å^3) based on molecular sieving mechanism [86]. Co-gallate, in particular, has a strong C_2H_4/C_2H_6 IAST selectivity of 52 at 298 K and 1 bar, with a C_2H_4 absorption of 3.37 mmol/g, which is corroborated by direct breakthrough tests with equimolar C_2H_4/C_2H_6 mixes.

4.3.1.2 C_2H_6 Selective Separation

In MOF materials, C_2H_4 is usually adsorbed predominantly over C_2H_6 as unsaturated C_2H_4 exhibits greater interactions with the functional moieties owing to their high polarity and π-electron density. Furthermore, C_2H_4 products could be achieved by expelling the trapped C_2H_4 out from adsorbents, that requires a significant amount of energy. To come up with an energy-efficient solution, researchers have been looking out for adsorbents that preferentially extract C_2H_6 from the mixture, allowing pure C_2H_4 to be generated directly from the adsorbent fixed-bed outlet.

While designing and manufacturing of MOFs for selective adsorption of C_2H_6 over C_2H_4, many parameters for instance van der Waals interactions, hydrogen bonding and electrostatic interactions, can be utilized. Lin et al. reported in 2018 that manipulating the pore architectures in isoreticular ultramicroporous MOFs to widen the feebly polar pore surface of MOFs can improve their binding interactions to C_2H_6 and result in increased C_2H_6/C_2H_4 selectivities [87]. They discovered that in comparison to the typical Cu(ina)$_2$, its smaller-pore isoreticular equivalent Cu (Qc)$_2$ has a 237% (60.0/25.3 cm^3/cm^3) C_2H_6/C_2H_4 absorption ratio under ambient conditions; thus, significantly enhancing C_2H_6/C_2H_4 selectivity.

In 2018, Chen et al. stated that PCN250 may be used as a selective adsorbent for C_2H_6 for effective C_2H_4 purification from C_2H_4/C_2H_6 mixtures [88]. The effect of pore and the energy of van der Waals adsorption dictate the C_2H_6 separation over C_2H_4 for PCN250, notably at low pressures, according to a configurational biased grand canonical Monte Carlo simulation.

Another effective way for achieving C_2H_6/C_2H_4 separations is to introduce peroxo sites into MOFs. Li et al., in 2018, described a MOF, [Fe$_2$(O$_2$)(dobdc)] (where dobdc^{4-} stands for 2,5-dioxido-1,4-benzenedicarboxylate) which is microporous, featuring Fe-peroxo sites for preferred C_2H_6 binding over C_2H_4 (Figure 4.4a-d), and therefore accomplished extremely selective C_2H_6/C_2H_4 separation [89]. C_2H_2 binding affinity was initially seen in Fe$_2$(O$_2$)(dobdc) single-component sorption isotherms at 298 K, where C_2H_6 had a substantially greater adsorption than C_2H_4 (74.3 cm^3/g over 57 cm^3/g) at 1 bar. High-resolution neutron powder diffraction investigations on C_2D_6-loaded and C_2D_4-loaded Fe$_2$(O$_2$)(dobdc) samples were done to analyze how C_2H_4 and C_2H_6 are adsorbed in the scaffold resulting in "reversed C_2H_6/C_2H_4 adsorption". C_2H_6 preferentially binds to peroxo sites via C-D---O hydrogen bonding (D---O = 2.17–2.22 Å), with the D---O distance being substantially lesser in comparison to total of the oxygen (1.52 Å) and hydrogen (1.20 Å) van der Waals radii.

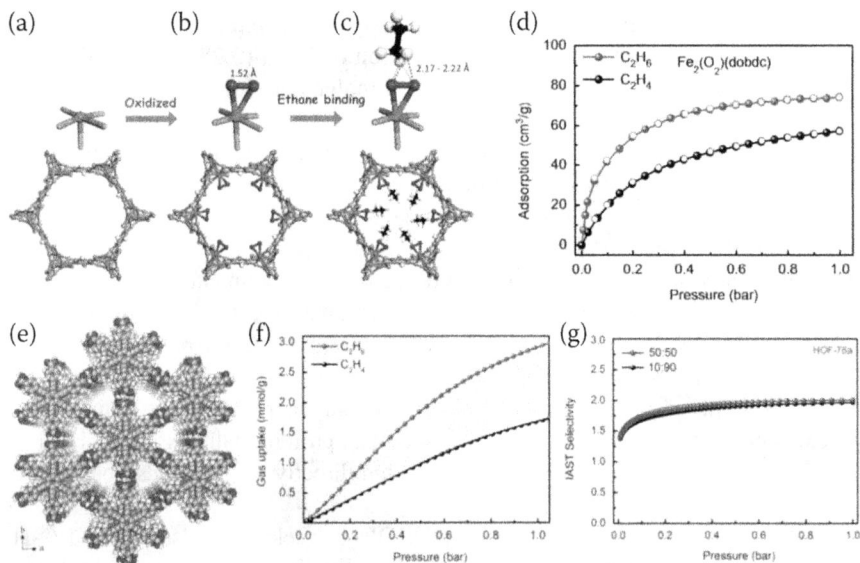

FIGURE 4.4 Determined structures from NPD studies. At 7 K, structures of (a) Fe$_2$(dobdc), (b) Fe$_2$(O$_2$)(dobdc), and (c) Fe$_2$(O$_2$)(dobdc)⊃C$_2$D$_6$. For the preferential binding of C_2H_6, note the change from the open Fe(II) site to the Fe(III)-peroxo site. Color code: C, dark gray; Fe, green; O, pink; H or D, O$_2^{2-}$, red white; C in C$_2$D$_6$, blue. (d) C_2H_4 and C_2H_6 adsorption isotherms of Fe$_2$(O$_2$)(dobdc). (Reproduced with permission from L. Li et al. 2018 [89]). (e) Representation of the porous scaffold of HOF76 (f) C_2H_6 (red) and C_2H_4 (black) adsorption isotherms for HOF76a at 296 K (d) HOF76a's IAST selectivity from C_2H_4/C_2H_6 (10/90 and 50/50) gas mixtures. (Reproduced with permission from X. Zhang et al. 2019 [90]).

Very recently, selective and effectual C_2H_4/C_2H_6 separation by the use of HOFs was demonstrated by Zhang et al. in 2019. Because HOFs are metal-free and lack highly polar groups, their pores are naturally endowed with nonpolar/inert surfaces, which could result in new C_2H_6-specific adsorbents. HOF-76a is synthesized, which is a new ultra-rugged HOF adsorbent with a BET surface area surpassing 1100 m^2/g and preferred C_2H_6 over C_2H_4 binding, allowing for extremely selective separation of C_2H_4/C_2H_6 (Figure 4.4e-g). The appropriate triangular channel-like pores and the nonpolar surface in HOF76a play a vital role in sterically "pairing" the nonplanar C_2H_6 molecule better as compared to planar C_2H_4, resulting in overall strong van der Waals interactions with C_2H_6. Breakthrough investigations on the HOF76a show that polymer-grade C_2H_4 gas can be generated easily from 50/50 (v/v) C_2H_4/C_2H_6 samples within the first adsorption cycle, with maximum production of 7.2 L/kg at 1.01 bar and 298 K and 18.8 L/kg at 5.0 bars and 298 K, correspondingly [90].

4.3.2 C_2H_2/C_2H_4 Separation

One of the vital yet difficult industrial-scale operations is separating C_2H_2 from C_2H_2/ C_2H_4 mixes. Partial hydrogenation and solvent extraction are two current commercial processes that are both expensive and energy-intensive. Adsorption-based separation using porous materials is a very effective way to achieve the desired separation.

Since Xiang et al. demonstrated the very first example of C_2H_2/C_2H_4 separation utilizing MOFs with the M'MOF series in 2011 [91], a great variety of functional materials can be used for their separation. Furthermore, functional materials have a trade-off among C_2H_2/C_2H_4 selectivity and C_2H_2 adsorption capacity as per literature. This could be relating with the example such as the open metal sites of Fe-MOF74 show significant increase in C_2H_2 absorption capacity, yet the huge nanopores result in a poor selectivity of 2.08 [92]. The identical case occurred with NOTT300 [93]. On the other hand, M'MOFs series has high selectivities up to 24, albeit very low C_2H_2 uptakes due to their restricted pore size [94].

In past few years, a lot of effort has gone into addressing the trade-off in C_2H_4/ C_2H_2 separations [95]. Yu et al. in 2020 designed Cr-MOF (HUST6) for the separation of C_2H_4/C_2H_2 after post-synthetic metalation of Fe-MOF (HUST5). As per IAST, the separation selectivity of HUST5 for samples formed of equimolar binary C_2H_4/C_2H_2 is 1.8 at 273K and starting pressure, whereas 3.8 for HUST6 at the same temperature and pressure. These values are greater than previously reported MOFs, such as TIFSIX2-Ni-i (6.2 for C_2H_2/CO_2), MOF74-Fe (1.8 for C_2H_4/C_2H_2) and NbU3-MnFe (2.7 for C_2H_4/C_2H_2), which are the most promising MOFs for C_2H_2/ C_2H_4 separations (Figure 4.5a-b). Because of the unique properties of chromium ions, the adsorption selectivity for C_2H_4/C_2H_2 has been significantly improved. At 290 K, dynamic breakthrough experiments for HUST5 and HUST6 upon actual separation processes of C_2H_4/C_2H_2 (1:99, v/v) mixtures indicated extremely effective separations for HUST6 in comparison to marginal separation performances for HUST5. Breakthrough time of C_2H_2 in C_2H_4/C_2H_2 mixture (1:99) is roughly 33.5 minutes, displayed 0.93 cm^3 of C_2H_2 is sustained/g HUST6 [96].

Du et al. recently developed a MOF, $Ni_2(BTEC)(bipy)_3$ (where H_4BTEC stands for 1,2,4,5-benzene tetracarboxylic acid and bipy for 4,4'-bipyridine), which is

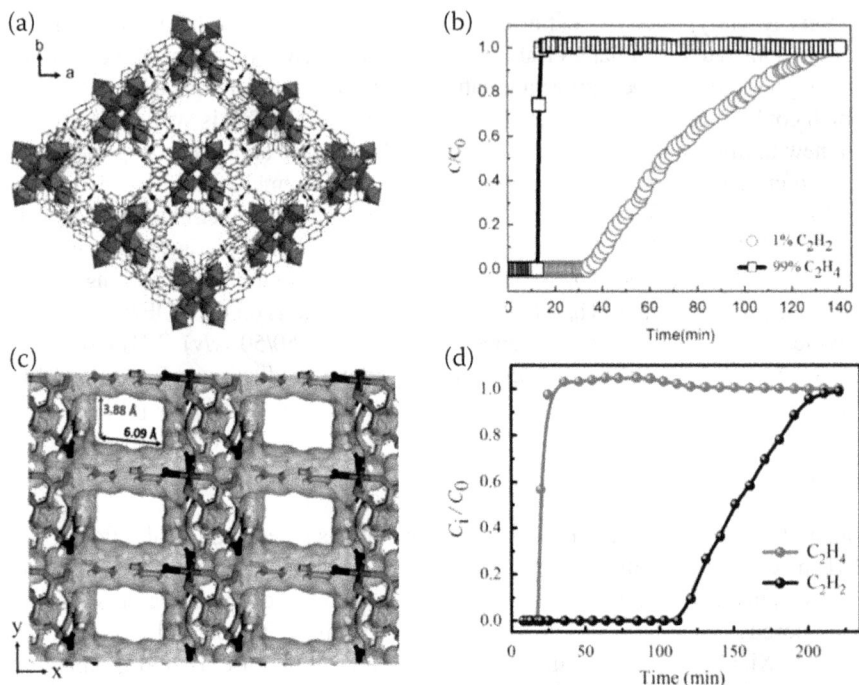

FIGURE 4.5 (a) A partial view of the Fe-MOF HUST-5's 3D porous structure along with c-axial direction. (b) Breakthrough curves for C_2H_2/C_2H_4 separations (1/99) for HUST6. (Reproduced with permission from Yu et al. 2020 [96]). (c) $Ni_2(BTEC)(bipy)_3$ pore structure (d) $Ni_2(BTEC)(bipy)_3$ experimental breakthrough curves of for the separation of C_2H_2/C_2H_4 (1:99, v/v) mixtures at 1 bar and 298 K. (Reproduced with permission from Du et al. 2021 [97]).

microporous, with abundant pyridine rings and carbonyl oxygens all around surface pore for excellent adsorptive C_2H_2 separation from the mixture of C_2H_2/C_2H_4 [97]. At 298 K and 1 bar, $Ni_2(BTEC)(bipy)_3$ displayed exceptional two-step C_2H_2 adsorption, with 76.8 cm^3/g adsorption capacity, considerably differentiating from C_2H_4 (7.9 cm^3/g) and CO_2 (13.0 cm^3/g). The significant adsorption difference of $Ni_2(BTEC)(bipy)_3$ between C_2H_2 and C_2H_4/CO_2 displayed high uptake ratios for C_2H_2/CO_2 (5.9) and C_2H_2/C_2H_4 (9.7) in robust MOFs at 1 atm and 298 K, as well as extremely high ideal adsorption solution selectivity (104, 33.5). Breakthrough studies confirmed the material's remarkable separation capability; it could efficiently separate C_2H_2/C_2H_4 mixtures with high C_2H_4 purity (> 99.999%) in one separation cycle (Figure 4.5c-d).

4.3.3 C_2H_2/CO_2 SEPARATION

Partially combusting CH_4 or cracking hydrocarbons yield C_2H_2, a crucial fuel gas and feedstock chemical. Because CO_2 is widely found in industrial operations, removing CO_2 from C_2H_2/CO_2 mixtures is critical for producing high-purity C_2H_2.

While most MOFs prefer to adsorb C_2H_2 over CO_2, certain MOFs prefer to adsorb CO_2. The discharge of the trapped C_2H_2 produces pure C_2H_2, which necessitates a significant amount of energy. Adsorbents must selectively adsorb CO_2 over C_2H_2 in an idealized C_2H_2/CO_2 separation model, allowing pure C_2H_2 to be created instantly in the course of the adsorption process.

Foo et al. demonstrated in 2016 that a novel flexible MOF [Mn(bdc)(dpe)] (where H_2bdc stands for 1,4-benzenedicarboxylic acid and dpe for 1,2-di(4-pyridyl) ethylene) exhibited preferred CO_2 adsorption over C_2H_2 [98]. With no gate-opening action, this MOF seems to have a sufficient CO_2 uptake of around 48 cm^3/g at 273 K and 1 bar. On the other hand, the C_2H_2 adsorption isotherm showed a gate-opening effect at 273 K, with a sharp step above 1 bar. The minimal C_2H_2 uptake at 1 bar (10 cm^3/g) indicates that this MOF may separate carbon dioxide from C_2H_2 at 273 K, as predicted through experimental and simulated breakthrough experiments.

In 2021, Gong et al. created MOF by reticulating prescreened NH_2- and OH-functionalities in isostructural ultramicroporous chiral MOFs—Ni_2(l-mal)$_2$(bpy) (MOF-OH) and Ni_2(l-asp)$_2$(bpy) (MOF-NH_2)—to target effective C_2H_2 uptake and C_2H_2/CO_2 separation, which surpasses most benchmark substances. Clearly, MOF-OH adsorbs a large quantity of C_2H_2 at ambient temperatures, with a storage density of about 0.81 g/mL, which is greater than the C2H2's solid density at 189 K. Furthermore, MOF-OH had an IAST selectivity of 25 toward an equimolar mixture of C_2H_2/CO_2, approximately double as high as MOF-NH_2. Interestingly, at zero converge the adsorption enthalpies for C_2H_2 in both MOFs are astonishingly low (as low as 17.5 kJ mol^{-1} and 16.7 kJ/mol for MOF-OH and MOF-NH_2 correspondingly), the lowest among effective rigid C_2H_2 sorbents within our knowledge. Multicycle breakthrough experiments confirm the efficiency of both MOFs for C_2H_2 /CO_2 separation [99].

Chen et al. recently developed an aluminum MOF (CAU10-H) that is economical, stable, and easily scalable for extremely competent CO_2/C_2H_2 separation (Figure 4.6). The appropriate pore confinement in CAU10-H allows for both multipoint binding interactions with C_2H_2 and dense packing of C_2H_2 within the pores. This substance has one of the maximum C_2H_2 storage densities (392 g/L) and

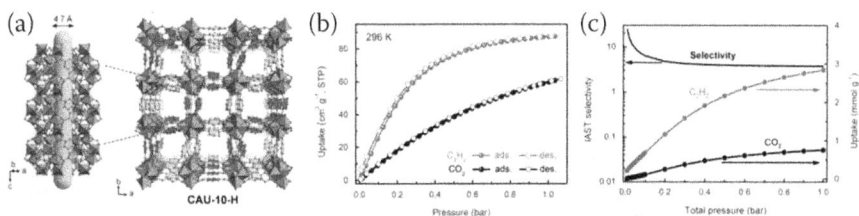

FIGURE 4.6 (a) CAU10-H 1D channel and helices configuration as seen through the b-axis. The AlO_6-polyhedra is epitomized by orange, C by gray, and O by red (b) CAU10-H gas adsorption isotherms for CO_2 and C_2H_2 at 296 K and up to 1 bar. (c) CAU10-H selectivity (black) and mixture adsorption isotherms (red and blue represents C_2H_2 and CO_2, respectively) predicted using the IAST method for a 50/50 C_2H_2/CO_2 mixture at 296 K. (Reproduced with permission from B. Chen et al. 2021 [100]).

extremely selective adsorption of C_2H_2 over CO_2 because of its low C_2H_2 adsorption enthalpy (27 kJ/mol), under normal conditions. Breakthrough studies show that it can separate C_2H_2/CO_2 mixtures with a 3.4 high separation factor and a 3.3 mmol/g substantial C_2H_2 uptake. For CO_2/C_2H_2 separation, CAU10-H meets the ideal balance of performance, robustness, and affordability [100].

4.4 MOFS FOR HARMFUL GAS REMOVAL

In recent years, MOFs have garnered considerable attention for their ability to adsorb and separate harmful gases. NH_3, SO_2, H_2S, Cl_2, NO and NO_2 are some of the most frequently detected poisonous gases These gases are discharged from industrial activities in the form of waste, causing serious environmental problems and harming human health. In comparison to H_2/CH_4 storage and CO_2 separation, the adsorption/removal of harmful toxic gases is comparatively with minimal investigation, which could be possibly related to the partial MOF stability toward these destructive gases. Hence, few investigations on MOFs with appreciable uptakes with recyclability for toxic gases have been reported. Even at ppm quantities, toxic gases like CO, NH_3, or H_2S pose an immediate threat to life. According to the U.S. Occupational Safety and Health Administration, the permissible limit of NH_3 in the industrial sector is 25 ppm [101]. Consequently, the development of adsorption material like MOFs with high porosity could offer a great support in the environment and public health protection. It is widely believed that relying just on the porosity of the materials is insufficient for successful gas/vapor extraction. When the pore surface integrates sites for coordination, acid–base interactions, electrostatic interactions, л–л stacking, or H-bonding, the adsorption degree and capability for gas molecules can be improved. A suitable porous material for reusable NH_3 adsorption should have interactions that are neither very strong nor too weak, enabling for both capture and release. These reversible contacts have been created in molecular chemistry via Lewis acid-base adducts by introducing steric demands, resulting in the "frustrated Lewis pairs" [102].

This section will provide information how several MOFs can be utilized for the effective removal of these lethal gases.

4.4.1 NH₃ Removal

Ammonia is the main prevalent gaseous alkaline pollutant. This gas has a very negative impact on animal and human health [103], and also reacts with SO_2, NOx, and other pollutants to generate $PM_{2.5}$ (fine particulate matter having less than 2.5 m aerodynamic diameters), which pose a potential menace to public safety [104]. The global need for NH_3 for chemical and fertilizer synthesis, refrigeration, and CO_2 capture, however, remains high [105–107]. NH_3 is also the carrier of sole carbon-free chemical energy (H_2), and it has a promising future as a hydrogen carrier in transportation [108,109]. In this way, there is a need to recycle, adsorb, and use NH_3 produced by ammonia emissions. MOFs employed in the field of NH_3 collection and utilization have also advanced in recent years.

Rieth et al. developed $M_2Cl_2(BTDD)$ (where M = Co, Mn, Ni) series materials with an uptake of 12 mmol of NH_3 at 1 bar [110]. Covalent organic frameworks (COF10) have a substantial adsorption of NH_3 (of about 15 mmol/g at 1 bar and 25°C), according to Yaghi [111]. Long has successfully removed NH_3 from the air using Bronsted acidic porous polymers [112]. Despite their great adsorption capacity, these materials demand an elevated regeneration temperature throughout the recycling process and exhibit performance degradation. Furthermore, for other common MOF materials utilized for adsorption and uptake applications of NH_3 (MOF5, HKUST1, and so on), ammonia and H_2O can easily affect their structures, causing bond deformation and permanent structural damage.

Chen et al. used a hydrothermal-solvothermal technique to synthesis MIL series MOFs [MIL53 (Al), NH_2-MIL53 (Al), MIL100 (Al), and MIL101 (Cr)] to examine their potential applicability in NH_3 adsorption (Figure 4.7a-b). According to the findings, MIL100 and MIL101 had significant NH_3 uptakes of 8 and 10 mmol/g, separately, and the altered NH_2-functional moieties boost the NH_3 adsorption

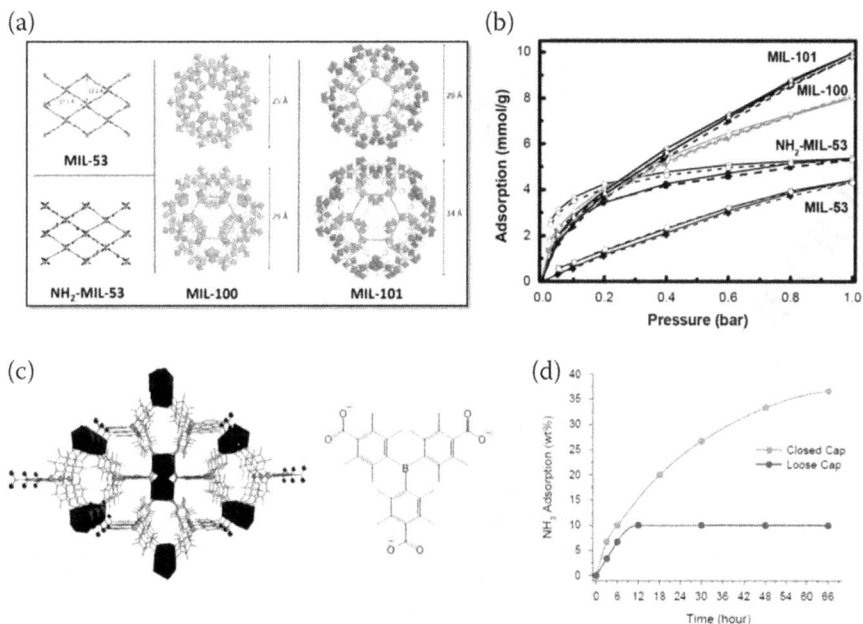

FIGURE 4.7 (a) Structures of MIL53, NH_2-MIL53, MIL100 and MIL101 (MIL100 and MIL101 both have large and small cages, symbolized as L-cage and as S-cage, correspondingly). (b) Obtained two cycle NH_3 adsorption-desorption isotherms at 25°C for MIL53, NH_2-MIL53, MIL100 and MIL101 samples (solid point represent point of adsorption, hollow point signifies point of desorption; solid line represents first adsorption-desorption curve and the dotted one signifies second adsorption-desorption curve). (Reproduced with permission from Y. Chen et al. 2018 [113]). (c) (left) SION105-Eu structure, with Eu ions represented as red polyhedral and for clarity H atoms are omitted; (right) ligand tctb^{3-}. (d) SION105-Eu NH_3 adsorption curve as a function of time. (Reproduced with permission from Nguyen et al. 2020 [114]).

ability in NH_2-MIL53. It is noteworthy to mention that the four MIL materials are potential ammonia adsorption materials since they exhibit reusable ammonia uptake and outstanding ammonia (and NH_3/H_2O) stability [113].

Nguyen et al. synthesized SION105-Eu, a very stable MOF with the chemical formula [Eu(tctb)$_3$(H$_2$O)] (Figure 4.7c-d). The connecting spacer is functionalized with a sterically constrained Lewis acidic boron (B) center, permitting effective NH_3 uptake and recycling facile MOF by heating at < 75°C. The MOF adsorbs 5.7 mmol/g NH_3 at 1 bar and 303 K. The utilization of Ln^{3+} ions offer hard Lewis's acids that firmly connect with hard donor atoms like O from carboxylate-based spacers, enabling stability in the presence of reactant molecules for ammonia. On the Lewis acidic B center, the presence of bulky duryl groups within the tctb^{3-} spacer also prevent significant acid-base B-N interactions in the presence of NH_3 that makes the MOF structure robust. The B presence on the organic spacers, on the other hand, makes the pores electrophilic, causing NH_3 to attract electrostatically and thus be captured [114].

4.4.2 NO$_2$ REMOVAL

NO_2 and nitrogen oxides in general are some of the most regularly encountered pollutants on a yearly basis. NO_2 is a typical commodity chemical that is produced mostly by combustion. Sorbents are necessary for the removal of NOx-based chemicals, whether in flue gas scrubbing operations or in protection-based applications, due to their widespread use and production.

In comparison to activated carbons, Peterson et al. developed the MOF UiO-66-NH_2, which eliminates the hazardous gas NO_2 at remarkable rates while limiting NOx by-product generation. At a relative humidity (RH) of 80%, UiO-66-NH_2 had a saturation NO_2 removal capacity of more than 1.4 g NO_2 g^{-1} as compared to 0.9 g NO_2 g^{-1} at 0% RH. The increase in water content, which assists in the interaction of NO_2 with MOF, was mostly responsible for the dramatically increased loading at 80% RH [115].

Wang et al. proved for the first time in 2020 that nanoporous dicarboxylic acid-functionalized MOF-808 analogs outperformed conventional MOF-808 counterparts in terms of NO_2 removal in both dry and wet environments (Figure 4.8). The MOF-808 framework was efficiently integrated with a series of dicarboxylic acid ligands, including pyridine-3,5-dicarboxylic acid, isophthalic acid, 5-aminoisophthalic acid, 5-hydroxyisophthalic acid and 5-nitroisophthalic acid, by substituting formate ligands. When compared to pristine MOF-808, the 5-aminoisophthalic acid-modified MOF-808 (808-NH_2IPA) demonstrated a considerable increase in NO_2 capacity (155 mg/g and 148 mg/g at 0% and 80% RH, respectively) and the most notable reduction in harmful NO by-product emission as per microbreakthrough tests [116].

4.4.3 SO$_2$ REMOVAL

The substantial harm to human well-being and the surroundings posed by SO_2 emissions produced by the ignition of inferior-quality fossil fuels in power plants has drawn global consideration [117]. Even a tiny amount of SO_2 in industry will have a significant impact on the subsequent elimination of other dangerous

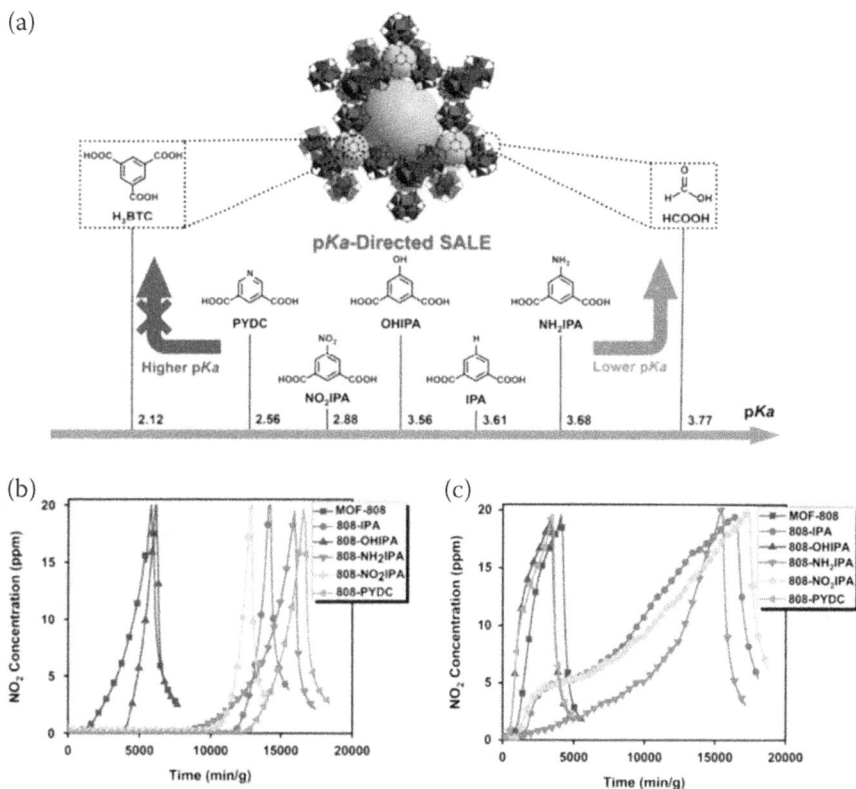

FIGURE 4.8 (a) Scheme of incorporation of dicarboxylic acid X into the MOF-808 framework via a pKa-directed SALE (solvent-assisted ligand exchange) strategy. (b) NO$_2$ microbreakthrough curves for MOF-808 and 808-X samples under dry conditions (c) under moist conditions. (Reproduced with permission from X. Wang et al. 2020 [116]).

constituents in flue gases, such as carbon dioxide and nitrogen oxides. Till this point, traditional flue gas desulfurization (FGD) techniques have relied on CaCO$_3$ scrubbing and wet sulfuric acid processes to remove 95% of sulfur dioxide from flue gases [118]; however, the exhaust gas still contains about 400 ppm SO$_2$ [119].

Numerous cooperative interactions of functionalized organic spacers with sulfur dioxide molecules have recently received a lot of attention. For instance, Schroder et al. described MFM601, a robust Zr-MOF with an all-time high SO$_2$ capacity of around 12.3 mmol/g at 298 K and 1.0 bar, as well as remarkable CO$_2$/SO$_2$ (32) and N$_2$/SO$_2$ (255) selectivities [118].

Zhang et al. recently established a microporous MOF, Cu$_2$(pzdc)$_2$(pyz) [CPL1, 22 pzdc = 2,3-pyrazinedicarboxylate; pyz = pyrazine] for effective sulfur dioxide removal through molecular sieving [120]. At ambient conditions, CPL1's appropriate size of around 4.1 × 6.2 Å and resilient sulfur dioxide trapping sites result in a high adsorption capacity of sulfur dioxide (44.8 cm^3/g) with such a record-high selectivity for N$_2$/SO$_2$ (368) and CH$_4$/SO$_2$ (74.3) separation (Figure 4.9).

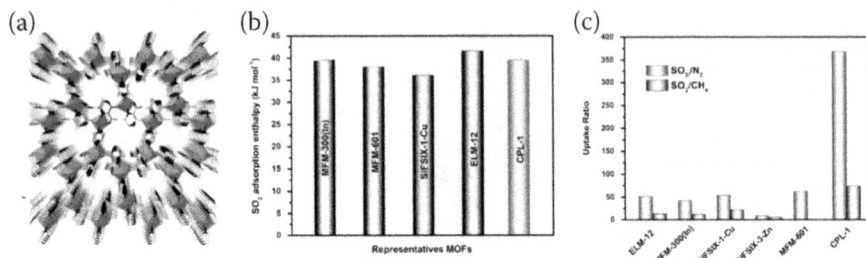

FIGURE 4.9 (a) Three-dimensional view of CPL1 framework (color code: Cu atoms are represented by light blue; O by red; N by blue; C by gray). (b) SO_2 adsorption enthalpy comparison on several MOFs. (c) Comparison of uptakes ratio amid representative MOFs. (Reproduced with permission from Y. Zhang et al. 2019 [120]).

In addition, the adsorption process and actual separation performance of SO_2 are vividly explained using a combination of first-principles DFT simulations and dynamic breakthrough tests. The dual binding sites are investigated as the strong interactions of (i) $S^{\delta+}$---$O^{\delta-}$ electrostatic interaction and (ii) $O^{\delta-}$---$H^{\delta+}$ hydrogen bonds.

4.4.4 REMOVAL OF OTHER POLLUTANTS

Particulate matter (PMs) pollution is a serious problem in some Asian countries as a result of rapid urbanization and industrialization. PMs are made up of several chemical compounds like C, NO_3^-, SO_4^{2-}, SiO_4^{4-}, and NH_4^+ ions, resulting from different sources such as soil dust, automotive emissions, coal/biomass burning, industrial pollution, and so forth [121–127]. PMs can also be produced by secondary processes involving nitrogen oxides, sulfur oxides, ammonia and other volatile organic compounds (VOCs), among other things. PM10 and PM2.5 are PMs with sizes less than 10 and 2.5 m, separately, and are often well defined by size (instead of component). PMs have an adverse effect on the human body, ranging from respiratory illness to death, specifically when they are small.

Wang et al. have demonstrated that putting various MOFs, such as ZIF8, onto air filters can significantly improve their efficiency through electrospinning [128], templated freeze drying [129], and hot pressing [130]. Following Wang's pioneering work, various MOFs (with or without functional moieties like NH_2) were placed onto filters or substrates to increase PM removal efficacy, including ZIF8 [131–136], ZIF67 [137,138], UiO-66 [139,140], MIL53 [141], MIL101 [142], HKUST1 [143], and PCP1/PCP1-L [144].

Very recently, Yoo et al. developed air filter for PM2.5 and PM10 by using cotton coated with MOFs (i.e., ZIF8s) and inferred for the first time that PM filtration does not only rely considerably on ZIF-8s pore size (Figure 4.10).

Among all the developed ZIF8s, DEA-ZIF-8/EC was found to exhibit a maximum removal efficiency of 92.3% (PM2.5) and 93.1% (PM10). The proposed mechanism for PM removal was the electrostatic interactions between small active sites on PMs and porous materials [145].

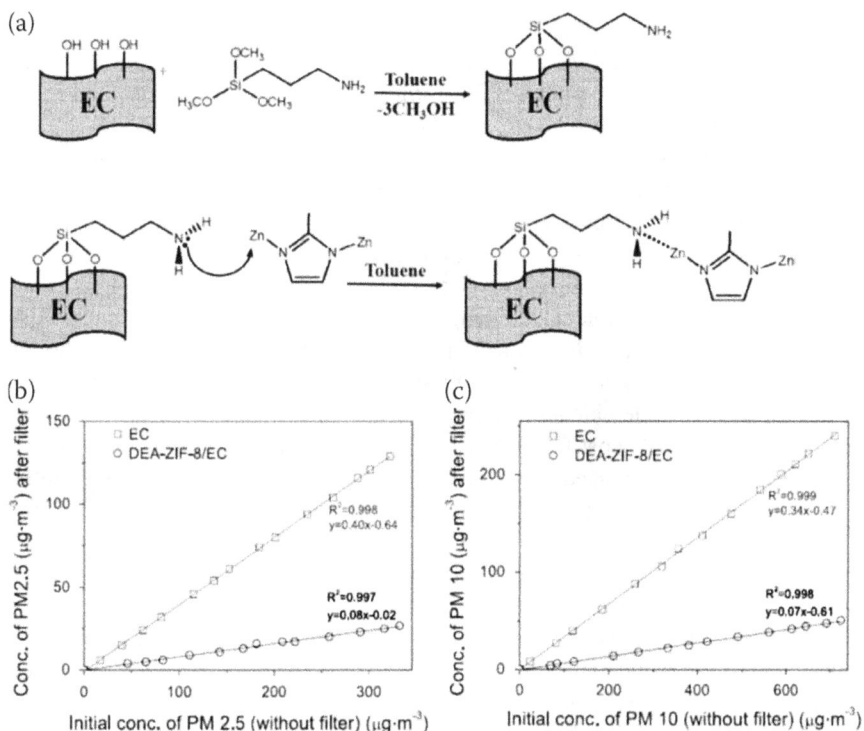

FIGURE 4.10 (a) A brief overview of the ZIF8 coating on the EC; concentration of PMs. (b) PM2.5 and (c) PM10 after filtering with EC and DEA-ZIF8/EC. The x-axes represent PMs concentrations prior to filtration. (Reproduced with permission from Yoo, Woo, and Jhung. 2021 [145]).

4.5 CONCLUSION AND OUTLOOK

Scientists and engineers have been fascinated by the evolution of functional materials to meet the requirements of storage and separation of gases, and great progress has been done in this arena. MOFs are more favorable for high-density fuel gas storage than standard porous adsorbents like zeolites and activated carbons because of their exclusive structural properties, including extraordinary porosity and high modularity. MOFs have much greater absorption capabilities than typical porous adsorbents for hydrogen, methane, and acetylene. Furthermore, much work need to be done before MOF-gas boarding platforms can be considered economically viable. During gas charging and discharging, thermal handling should be considered, bearing in mind the related exothermic and endothermic phenomena; bulk MOF adsorbent packing efficiency should be increased for compact concerns, to name a few. Ideal adsorbents with high volumetric storage and gravimetric capabilities under normal conditions would be excellent for gas storage. If allowed, the tremendous energy density of these clean fuel gases could result in enormous global benefits.

The advent of energy-efficient adsorptive systems embodied by functional materials has transformed the field of gas separation. MOFs' unique pore characteristics make them a suitable platform for testing new adsorbent materials. The recent decade, particularly the last 3 years, has seen a rapid evolution of MOFs for gas separation. MOFs with both a high adsorption capacity and a good selectivity represent the potentiality in terms of high separation efficiency. As a result, combining the functional moieties and tunable porosity in a single MOF, referred to as dual functionality, not only maximizes interactions with target gas molecules and moreover excludes competing ones, opening up new research opportunities. Waterproofing, purity tolerance, long lifetime, renewal, adsorption kinetics, mechanical features, huge implementation, processability, and flaw in membrane manufacturing are only a few of the hurdles that need to be surmounted before further industrial implementation. Another requirement is to scale up MOF production at a reasonable cost for large-scale implementation. Numerous MOFs have been conveniently produced on a large scale, employing simple precipitation reactions in aqueous solution under ambient circumstances or solvent-free reactions.

With a better knowledge of the association between MOF porosity/gas storage/functionalization/structure and separation capabilities, it is envisaged that in the coming years, MOF materials that are far superior for the separation and storage of gases would be strategically created and aimed. Substantial fundamental research in this area will undoubtedly emerge in the future, paving the way for the real-world application of new absorbent materials for purification of air and related applications.

ACKNOWLEDGMENTS

S. K. acknowledges the Council of Scientific and Industrial Research, New Delhi for granting Research Associateship [CSIR-RA, File No.#09/112(0665)/2020-EMR-I]. Authors also acknowledge the support from the Department of Chemistry, Aligarh Muslim University, Aligarh, India.

REFERENCES

[1] Chu, Steven, Yi Cui, and Nian Liu. 2017. "The Path towards Sustainable Energy." *Nature Materials* 16 (1): 16–22.
[2] Suh, Myunghyun Paik, Hye Jeong Park, Thazhe Kootteri Prasad, and Dae-Woon Lim. 2012. "Hydrogen Storage in Metal–Organic Frameworks." *Chemical Reviews* 112 (2): 782–835.
[3] Sholl, David S., and Ryan P. Lively. 2016. "Seven Chemical Separations to Change the World." *Nature* 532: 435–437.
[4] Kitagawa, Susumu. 2015. "Porous Materials and the Age of Gas." *Angew. Chem. Int. Ed* 54: 10686–10687.
[5] Li, Bin, Hui-Min Wen, Yuanjing Cui, Wei Zhou, Guodong Qian, and Banglin Chen. 2016. "Emerging Multifunctional Metal–Organic Framework Materials." *Advanced Materials* 28 (40): 8819–8860.

[6] Furukawa, Hiroyasu, Kyle E. Cordova, Michael O'Keeffe, and Omar M. Yaghi. 2013. "The Chemistry and Applications of Metal-Organic Frameworks." *Science* 341 (6149): 1230444.

[7] Ding, San-Yuan, and Wei Wang. 2013. "Covalent Organic Frameworks (COFs): From Design to Applications." *Chemical Society Reviews* 42 (2): 548–568.

[8] Feng, Xiao, Xuesong Ding, and Donglin Jiang. 2012. "Covalent Organic Frameworks." *Chemical Society Reviews* 41 (18): 6010–6022.

[9] Jiang, Juncong, Yingbo Zhao, and Omar M. Yaghi. 2016. "Covalent Chemistry beyond Molecules." *Journal of the American Chemical Society* 138 (10): 3255–3265.

[10] Diercks, Christian S., and Omar M. Yaghi. 2017. "The Atom, the Molecule, and the Covalent Organic Framework." *Science* 355 (6328): eaal1585.

[11] Lin, Rui-Biao, Yabing He, Peng Li, Hailong Wang, Wei Zhou, and Banglin Chen. 2019. "Multifunctional Porous Hydrogen-Bonded Organic Framework Materials." *Chemical Society Reviews* 48 (5): 1362–1389.

[12] Bao, Zongbi, Ganggang Chang, Huabin Xing, Rajamani Krishna, Qilong Ren, and Banglin Chen. 2016. "Potential of Microporous Metal–Organic Frameworks for Separation of Hydrocarbon Mixtures." *Energy & Environmental Science* 9 (12): 3612–3641.

[13] Adil, Karim, Youssef Belmabkhout, Renjith S. Pillai, Amandine Cadiau, Prashant M. Bhatt, Ayalew H. Assen, Guillaume Maurin, and Mohamed Eddaoudi. 2017. "Gas/Vapour Separation Using Ultra-Microporous Metal–Organic Frameworks: Insights into the Structure/Separation Relationship." *Chemical Society Reviews* 46 (11): 3402–3430.

[14] Bachman, Jonathan E., Matthew T. Kapelewski, Douglas A. Reed, Miguel I. Gonzalez, and Jeffrey R. Long. 2017. "M_2 (m-Dobdc)(M= Mn, Fe, Co, Ni) Metal–Organic Frameworks as Highly Selective, High-Capacity Adsorbents for Olefin/Paraffin Separations." *Journal of the American Chemical Society* 139 (43): 15363–15370.

[15] Cui, Yuanjing, Yanfeng Yue, Guodong Qian, and Banglin Chen. 2012. "Luminescent Functional Metal–Organic Frameworks." *Chemical Reviews* 112 (2): 1126–1162.

[16] Sun, Chun-Yi, Xin-Long Wang, Xiao Zhang, Chao Qin, Peng Li, Zhong-Min Su, Dong-Xia Zhu, Guo-Gang Shan, Kui-Zhan Shao, and Han Wu. 2013. "Efficient and Tunable White-Light Emission of Metal–Organic Frameworks by Iridium-Complex Encapsulation." *Nature Communications* 4 (1): 1–8.

[17] Alezi, Dalal, Youssef Belmabkhout, Mikhail Suyetin, Prashant M. Bhatt, Łukasz J. Weseliński, Vera Solovyeva, Karim Adil, Ioannis Spanopoulos, Pantelis N. Trikalitis, and Abdul-Hamid Emwas. 2015. "MOF Crystal Chemistry Paving the Way to Gas Storage Needs: Aluminum-Based Soc-MOF for CH_4, O_2, and CO_2 Storage." *Journal of the American Chemical Society* 137 (41): 13308–13318.

[18] Verma, Gaurav, Sanjay Kumar, Harsh Vardhan, Junyu Ren, Zheng Niu, Tony Pham, Lukasz Wojtas, Sydney Butikofer, Jose C. Echeverria Garcia, and Yu-Sheng Chen. 2021. "A Robust Soc-MOF Platform Exhibiting High Gravimetric Uptake and Volumetric Deliverable Capacity for on-Board Methane Storage." *Nano Research* 14 (2): 512–517.

[19] Fang, Han, Bin Zheng, Zong-Hui Zhang, Hong-Xin Li, Dong-Xu Xue, and Junfeng Bai. 2021. "Ligand-Conformer-Induced Formation of Zirconium-Organic Framework for Methane Storage and MTO Product Separation." *Angewandte Chemie* 133(30): 16657–16664.

[20] Talin, A Alec, Andrea Centrone, Alexandra C. Ford, Michael E. Foster, Vitalie Stavila, Paul Haney, R Adam Kinney, Veronika Szalai, Farid El Gabaly, and Heayoung P. Yoon. 2014. "Tunable Electrical Conductivity in Metal-Organic Framework Thin-Film Devices." *Science* 343 (6166): 66–69.

[21] Kreno, Lauren E., Kirsty Leong, Omar K. Farha, Mark Allendorf, Richard P. Van Duyne, and Joseph T. Hupp. 2012. "Metal–Organic Framework Materials as Chemical Sensors." *Chemical Reviews* 112 (2): 1105–1125.

[22] Raizada, Mukul, Farasha Sama, Mo Ashafaq, M. Shahid, Mohd Khalid, Musheer Ahmad, and Zafar A. Siddiqi. 2018. "Synthesis, Structure and Magnetic Studies of Lanthanide Metal–Organic Frameworks (Ln–MOFs): Aqueous Phase Highly Selective Sensors for Picric Acid as Well as the Arsenic Ion." *Polyhedron* 139: 131–141.

[23] Ahamad, M Naqi, M. Shahid, Musheer Ahmad, and Farasha Sama. 2019. "Cu (II) MOFs Based on Bipyridyls: Topology, Magnetism, and Exploring Sensing Ability toward Multiple Nitroaromatic Explosives." *ACS Omega* 4 (4): 7738–7749.

[24] Xu, Ruoyu, Youfu Wang, Xiaopin Duan, Kuangda Lu, Daniel Micheroni, Aiguo Hu, and Wenbin Lin. 2016. "Nanoscale Metal–Organic Frameworks for Ratiometric Oxygen Sensing in Live Cells." *Journal of the American Chemical Society* 138 (7): 2158–2161.

[25] Campbell, Michael G., Dennis Sheberla, Sophie F. Liu, Timothy M. Swager, and Mircea Dincă. 2015. "Cu3 (Hexaiminotriphenylene) 2: An Electrically Conductive 2D Metal–Organic Framework for Chemiresistive Sensing." *Angewandte Chemie International Edition* 54 (14): 4349–4352.

[26] Yang, Qihao, Qiang Xu, and Hai-Long Jiang. 2017. "Metal–Organic Frameworks Meet Metal Nanoparticles: Synergistic Effect for Enhanced Catalysis." *Chemical Society Reviews* 46 (15): 4774–4808.

[27] Xu, Hai-Qun, Jiahua Hu, Dengke Wang, Zhaohui Li, Qun Zhang, Yi Luo, Shu-Hong Yu, and Hai-Long Jiang. 2015. "Visible-Light Photoreduction of CO2 in a Metal–Organic Framework: Boosting Electron–Hole Separation via Electron Trap States." *Journal of the American Chemical Society* 137 (42): 13440–13443.

[28] Farha, Omar K., Ibrahim Eryazici, Nak Cheon Jeong, Brad G. Hauser, Christopher E. Wilmer, Amy A. Sarjeant, Randall Q. Snurr, SonBinh T. Nguyen, A. Özgür Yazaydın, and Joseph T. Hupp. 2012. "Metal–Organic Framework Materials with Ultrahigh Surface Areas: Is the Sky the Limit?" *Journal of the American Chemical Society* 134 (36): 15016–15021.

[29] Furukawa, Hiroyasu, Nakeun Ko, Yong Bok Go, Naoki Aratani, Sang Beom Choi, Eunwoo Choi, A. Özgür Yazaydin, Randall Q. Snurr, Michael O'Keeffe, and Jaheon Kim. 2010. "Ultrahigh Porosity in Metal-Organic Frameworks." *Science* 329 (5990): 424–428.

[30] Kondo, Mitsuru, Tomomichi Yoshitomi, Hiroyuki Matsuzaka, Susumu Kitagawa, and Kenji Seki. 1997. "Three-Dimensional Framework with Channeling Cavities for Small Molecules:{[M$_2$(4, 4'-bpy)$_3$(NO$_3$)$_4$]·xH$_2$O}$_n$ (M= Co, Ni, Zn)." *Angewandte Chemie International Edition* 36 (16): 1725–1727.

[31] Li, Hailian, Mohamed Eddaoudi, Thomas L. Groy, and O.M. Yaghi. 1998. "Establishing Microporosity in Open Metal–Organic Frameworks: Gas Sorption Isotherms for Zn (BDC) (BDC= 1, 4-Benzenedicarboxylate)." *Journal of the American Chemical Society* 120 (33): 8571–8572.

[32] Peng, Yang, Vaiva Krungleviciute, Ibrahim Eryazici, Joseph T. Hupp, Omar K. Farha, and Taner Yildirim. 2013. "Methane Storage in Metal–Organic Frameworks: Current Records, Surprise Findings, and Challenges." *Journal of the American Chemical Society* 135 (32): 11887–11894.

[33] Chen, Banglin, Chengdu Liang, Jun Yang, Damacio S. Contreras, Yvette L. Clancy, Emil B. Lobkovsky, Omar M. Yaghi, and Sheng Dai. 2006. "A Microporous Metal–Organic Framework for Gas-chromatographic Separation of Alkanes." *Angewandte Chemie International Edition* 45 (9): 1390–1393.

[34] Mueller, U., M. Schubert, F. Teich, H. Puetter, K. Schierle-Arndt, and J. Pastre. 2006. "Metal–Organic Frameworks—Prospective Industrial Applications." *Journal of Materials Chemistry* 16 (7): 626–636.

[35] Bárcia, Patrick S., Fatima Zapata, José A C Silva, Alírio E Rodrigues, and Banglin Chen. 2007. "Kinetic Separation of Hexane Isomers by Fixed-Bed Adsorption with a Microporous Metal– Organic Framework." *The Journal of Physical Chemistry B* 111 (22): 6101–6103.

[36] Hayashi, Hideki, Adrien P. Cote, Hiroyasu Furukawa, Michael O'Keeffe, and Omar M. Yaghi. 2007. "Zeolite A Imidazolate Frameworks." *Nature Materials* 6 (7): 501–506.

[37] Banerjee, Rahul, Anh Phan, Bo Wang, Carolyn Knobler, Hiroyasu Furukawa, Michael O'Keeffe, and Omar M. Yaghi. 2008. "High-Throughput Synthesis of Zeolitic Imidazolate Frameworks and Application to CO_2 Capture." *Science* 319 (5865): 939–943.

[38] Matsuda, Ryotaro, Ryo Kitaura, Susumu Kitagawa, Yoshiki Kubota, Rodion V. Belosludov, Tatsuo C. Kobayashi, Hirotoshi Sakamoto, Takashi Chiba, Masaki Takata, and Yoshiyuki Kawazoe. 2005. "Highly Controlled Acetylene Accommodation in a Metal–Organic Microporous Material." *Nature* 436 (7048): 238–241.

[39] Rowsell, Jesse L. C., Elinor C. Spencer, Juergen Eckert, Judith A. K. Howard, and Omar M. Yaghi. 2005. "Gas Adsorption Sites in a Large-Pore Metal-Organic Framework." *Science* 309 (5739): 1350–1354.

[40] Vaidhyanathan, Ramanathan, Simon S. Iremonger, George K. H. Shimizu, Peter G. Boyd, Saman Alavi, and Tom K. Woo. 2010. "Direct Observation and Quantification of CO_2 Binding within an Amine-Functionalized Nanoporous Solid." *Science* 330 (6004): 650–653.

[41] Wu, Haohan, Qihan Gong, David H. Olson, and Jing Li. 2012. "Commensurate Adsorption of Hydrocarbons and Alcohols in Microporous Metal Organic Frameworks." *Chemical Reviews* 112 (2): 836–868.

[42] Li, B., H-M Wen, Y. Yu, Yuanjing Cui, Wei Zhou, B. Chen, and Guodong Qian. 2018. "Nanospace within Metal–Organic Frameworks for Gas Storage and Separation." *Materials Today Nano* 2: 21–49.

[43] Zhao, Xiang, Yanxiang Wang, Dong-Sheng Li, Xianhui Bu, and Pingyun Feng. 2018. "Metal–Organic Frameworks for Separation." *Advanced Materials* 30 (37): 1705189.

[44] Lin, Rui-Biao, Shengchang Xiang, Huabin Xing, Wei Zhou, and Banglin Chen. 2019. "Exploration of Porous Metal–Organic Frameworks for Gas Separation and Purification." *Coordination Chemistry Reviews* 378: 87–103.

[45] Getman, Rachel B., Youn-Sang Bae, Christopher E. Wilmer, and Randall Q. Snurr. 2012. "Review and Analysis of Molecular Simulations of Methane, Hydrogen, and Acetylene Storage in Metal–Organic Frameworks." *Chemical Reviews* 112 (2): 703–723.

[46] Berg, Annemieke W C van den, and Carlos Otero Areán. 2008. "Materials for Hydrogen Storage: Current Research Trends and Perspectives." *Chemical Communications* 6: 668–681.

[47] Schlapbach, Louis. 2002. "Hydrogen as a Fuel and Its Storage for Mobility and Transport." *MRS Bulletin* 27 (9): 675–679.

[48] Rowsell, Jesse L. C., and Omar M. Yaghi. 2005. "Strategies for Hydrogen Storage in Metal–Organic Frameworks." *Angewandte Chemie International Edition* 44 (30): 4670–4679.

[49] Kaye, Steven S., Anne Dailly, Omar M. Yaghi, and Jeffrey R. Long. 2007. "Impact of Preparation and Handling on the Hydrogen Storage Properties of $Zn_4O(1,4$-Benzenedicarboxylate$)_3$ (MOF-5)." *Journal of the American Chemical Society* 129 (46): 14176–14177.

[50] Yan, Yong, Xiang Lin, Sihai Yang, Alexander J. Blake, Anne Dailly, Neil R. Champness, Peter Hubberstey, and Martin Schröder. 2009. "Exceptionally High H_2 Storage by a Metal–Organic Polyhedral Framework." *Chemical Communications* 9: 1025–1027.

[51] Farha, Omar K., Christopher E. Wilmer, Ibrahim Eryazici, Brad G. Hauser, Philip A. Parilla, Kevin O'Neill, Amy A. Sarjeant, SonBinh T. Nguyen, Randall Q. Snurr, and Joseph T. Hupp. 2012. "Designing Higher Surface Area Metal–Organic Frameworks: Are Triple Bonds Better than Phenyls?" *Journal of the American Chemical Society* 134 (24): 9860–9863.

[52] Farha, Omar K., A Özgür Yazaydın, Ibrahim Eryazici, Christos D. Malliakas, Brad G. Hauser, Mercouri G. Kanatzidis, SonBinh T. Nguyen, Randall Q. Snurr, and Joseph T. Hupp. 2010. "De Novo Synthesis of a Metal–Organic Framework Material Featuring Ultrahigh Surface Area and Gas Storage Capacities." *Nature Chemistry* 2 (11): 944–948.

[53] Yuan, Daqiang, Dan Zhao, Daofeng Sun, and Hong-Cai Zhou. 2010. "An Isoreticular Series of Metal–Organic Frameworks with Dendritic Hexacarboxylate Ligands and Exceptionally High Gas-uptake Capacity." *Angewandte Chemie* 122 (31): 5485–5489.

[54] Bhatia, Suresh K., and Alan L. Myers. 2006. "Optimum Conditions for Adsorptive Storage." *Langmuir* 22 (4): 1688–1700.

[55] Kapelewski, Matthew T., Tomče Runčevski, Jacob D. Tarver, Henry Z. H. Jiang, Katherine E. Hurst, Philip A. Parilla, Anthony Ayala, Thomas Gennett, Stephen A. FitzGerald, and Craig M. Brown. 2018. "Record High Hydrogen Storage Capacity in the Metal–Organic Framework Ni_2(m-Dobdc) at near-Ambient Temperatures." *Chemistry of Materials* 30 (22): 8179–8189.

[56] He, Yabing, Wei Zhou, Guodong Qian, and Banglin Chen. 2014. "Methane Storage in Metal–Organic Frameworks." *Chemical Society Reviews* 43 (16): 5657–5678.

[57] Eddaoudi, Mohamed, Jaheon Kim, Nathaniel Rosi, David Vodak, Joseph Wachter, Michael O'Keeffe, and Omar M. Yaghi. 2002. "Systematic Design of Pore Size and Functionality in Isoreticular MOFs and Their Application in Methane Storage." *Science* 295 (5554): 469–472.

[58] Li, Bin, Hui-Min Wen, Wei Zhou, and Banglin Chen. 2014. "Porous Metal–Organic Frameworks for Gas Storage and Separation: What, How, and Why?" *The Journal of Physical Chemistry Letters* 5 (20): 3468–3479.

[59] Lu, Kuangda, Theint Aung, Nining Guo, Ralph Weichselbaum, and Wenbin Lin. 2018. "Nanoscale Metal–Organic Frameworks for Therapeutic, Imaging, and Sensing Applications." *Advanced Materials* 30 (37): 1707634.

[60] Tu, Thach N., Huong T. D. Nguyen, and Nhung Thi Tran. 2019. "Tailoring the Pore Size and Shape of the One-Dimensional Channels in Iron-Based MOFs for Enhancing the Methane Storage Capacity." *Inorganic Chemistry Frontiers* 6 (9): 2441–2447.

[61] Wen, Hui-Min, Bin Li, Libo Li, Rui-Biao Lin, Wei Zhou, Guodong Qian, and Banglin Chen. 2018. "A Metal–Organic Framework with Optimized Porosity and Functional Sites for High Gravimetric and Volumetric Methane Storage Working Capacities." *Advanced Materials* 30 (16): 1704792.

[62] Zhang, Zhangjing, Shengchang Xiang, and Banglin Chen. 2011. "Microporous Metal–Organic Frameworks for Acetylene Storage and Separation." *CrystEngComm* 13 (20): 5983–5992.

[63] Xiang, Shengchang, Wei Zhou, Jose M. Gallegos, Yun Liu, and Banglin Chen. 2009. "Exceptionally High Acetylene Uptake in a Microporous Metal– Organic Framework with Open Metal Sites." *Journal of the American Chemical Society* 131 (34): 12415–12419.

[64] Xiang, Shengchang, Wei Zhou, Zhangjing Zhang, Mark A. Green, Yun Liu, and Banglin Chen. 2010. "Open Metal Sites within Isostructural Metal–Organic Frameworks for Differential Recognition of Acetylene and Extraordinarily High Acetylene Storage Capacity at Room Temperature." *Angewandte Chemie* 122 (27): 4719–4722.

[65] Rao, Xingtang, Jianfeng Cai, Jiancan Yu, Yabing He, Chuande Wu, Wei Zhou, Taner Yildirim, Banglin Chen, and Guodong Qian. 2013. "A Microporous Metal–Organic Framework with Both Open Metal and Lewis Basic Pyridyl Sites for High C_2H_2 and CH_4 Storage at Room Temperature." *Chemical Communications* 49 (60): 6719–6721.

[66] Moreau, Florian, Ivan Da Silva, Nada H. Al Smail, Timothy L. Easun, Mathew Savage, Harry G. W. Godfrey, Stewart F. Parker, Pascal Manuel, Sihai Yang, and Martin Schröder. 2017. "Unravelling Exceptional Acetylene and Carbon Dioxide Adsorption within a Tetra-Amide Functionalized Metal-Organic Framework." *Nature Communications* 8 (1): 1–9.

[67] Zhang, Ling, Ke Jiang, Yanping Li, Dian Zhao, Yu Yang, Yuanjing Cui, Banglin Chen, and Guodong Qian. 2017. "Microporous Metal–Organic Framework with Exposed Amino Functional Group for High Acetylene Storage and Excellent C_2H_2/ CO_2 and C_2H_2/CH_4 Separations." *Crystal Growth & Design* 17 (5): 2319–2322.

[68] Zhang, Mingxing, Bin Li, Yunzhi Li, Qian Wang, Wenwei Zhang, Banglin Chen, Shuhua Li, Yi Pan, Xiaozeng You, and Junfeng Bai. 2016. "Finely Tuning MOFs towards High Performance in C_2H_2 Storage: Synthesis and Properties of a New MOF-505 Analogue with an Inserted Amide Functional Group." *Chemical Communications* 52 (45): 7241–7244.

[69] Pang, Jiandong, Feilong Jiang, Mingyan Wu, Caiping Liu, Kongzhao Su, Weigang Lu, Daqiang Yuan, and Maochun Hong. 2015. "A Porous Metal-Organic Framework with Ultrahigh Acetylene Uptake Capacity under Ambient Conditions." *Nature Communications* 6 (1): 7575.

[70] Li, Jian-Rong, Julian Sculley, and Hong-Cai Zhou. 2012. "Metal–Organic Frameworks for Separations." *Chemical Reviews* 112 (2): 869–932.

[71] Herm, Zoey R., Eric D. Bloch, and Jeffrey R. Long. 2014. "Hydrocarbon Separations in Metal–Organic Frameworks." *Chemistry of Materials* 26 (1): 323–338.

[72] Li, Jian-Rong, Ryan J. Kuppler, and Hong-Cai Zhou. 2009. "Selective Gas Adsorption and Separation in Metal–Organic Frameworks." *Chemical Society Reviews* 38 (5): 1477–1504.

[73] Yoon, Ji Woong, Hyunju Chang, Seung-Joon Lee, Young Kyu Hwang, Do-Young Hong, Su-Kyung Lee, Ji Sun Lee, Seunghun Jang, Tae-Ung Yoon, and Kijeong Kwac. 2017. "Selective Nitrogen Capture by Porous Hybrid Materials Containing Accessible Transition Metal Ion Sites." *Nature Materials* 16 (5): 526–531.

[74] Li, Bin, Xili Cui, Daniel O'Nolan, Hui-Min Wen, Mengdie Jiang, Rajamani Krishna, Hui Wu, Rui-Biao Lin, Yu-Sheng Chen, and Daqiang Yuan. 2017. "An Ideal Molecular Sieve for Acetylene Removal from Ethylene with Record Selectivity and Productivity." *Advanced Materials* 29 (47): 1704210.

[75] Gao, Junkuo, Xuefeng Qian, Rui-Biao Lin, Rajamani Krishna, Hui Wu, Wei Zhou, and Banglin Chen. 2020. "Mixed Metal–Organic Framework with Multiple Binding Sites for Efficient C_2H_2/CO_2 Separation." *Angewandte Chemie International Edition* 59 (11): 4396–4400.

[76] Cadiau, Amandine, Karim Adil, P.M. Bhatt, Youssef Belmabkhout, and Mohamed Eddaoudi. 2016. "A Metal-Organic Framework–Based Splitter for Separating Propylene from Propane." *Science* 353 (6295): 137–140.

[77] Li, Libo, Rui-Biao Lin, Rajamani Krishna, Xiaoqing Wang, Bin Li, Hui Wu, Jinping Li, Wei Zhou, and Banglin Chen. 2017. "Flexible–Robust Metal–Organic Framework for Efficient Removal of Propyne from Propylene." *Journal of the American Chemical Society* 139 (23): 7733–7736.

[78] Zhang, Ling, Ke Jiang, Lifeng Yang, Libo Li, Enlai Hu, Ling Yang, Kai Shao, Huabin Xing, Yuanjing Cui, and Yu Yang. 2021. "Benchmark C_2H_2/CO_2 Separation in an Ultra-Microporous Metal–Organic Framework via Copper (I)-Alkynyl Chemistry." *Angewandte Chemie* 133 (29): 16131–16138.

[79] Keller, Jürgen U, and Reiner Staudt. 2005. *Gas Adsorption Equilibria: Experimental Methods and Adsorptive Isotherms*. Springer Science & Business Media.

[80] Yang, Ralph T. 2003. *Adsorbents: Fundamentals and Applications*. John Wiley & Sons.

[81] Liao, Pei-Qin, Wei-Xiong Zhang, Jie-Peng Zhang, and Xiao-Ming Chen. 2015. "Efficient Purification of Ethene by an Ethane-Trapping Metal-Organic Framework." *Nature Communications* 6 (1): 1–9.

[82] Gucuyener, Canan, Johan Van Den Bergh, Jorge Gascon, and Freek Kapteijn. 2010. "Ethane/Ethene Separation Turned on Its Head: Selective Ethane Adsorption on the Metal– Organic Framework ZIF-7 through a Gate-Opening Mechanism." *Journal of the American Chemical Society* 132 (50): 17704–17706.

[83] Wang, Qing Min, Dongmin Shen, Martin Bülow, Miu Ling Lau, Shuguang Deng, Frank R. Fitch, Norberto O. Lemcoff, and Jessica Semanscin. 2002. "Metallo-Organic Molecular Sieve for Gas Separation and Purification." *Microporous and Mesoporous Materials* 55 (2): 217–230.

[84] Yang, Sihai, Anibal J. Ramirez-Cuesta, Ruth Newby, Victoria Garcia-Sakai, Pascal Manuel, Samantha K. Callear, Stuart I. Campbell, Chiu C. Tang, and Martin Schröder. 2015. "Supramolecular Binding and Separation of Hydrocarbons within a Functionalized Porous Metal–Organic Framework." *Nature Chemistry* 7 (2): 121–129.

[85] Xu, Zhenzhen, Xiaohong Xiong, Jianbo Xiong, Rajamani Krishna, Libo Li, Yaling Fan, Feng Luo, and Banglin Chen. 2020. "A Robust Th-Azole Framework for Highly Efficient Purification of C_2H_4 from a $C_2H_4/C_2H_2/C_2H_6$ Mixture." *Nature Communications* 11 (1): 1–9.

[86] Bao, Zongbi, Jiawei Wang, Zhiguo Zhang, Huabin Xing, Qiwei Yang, Yiwen Yang, Hui Wu, Rajamani Krishna, Wei Zhou, and Banglin Chen. 2018. "Molecular Sieving of Ethane from Ethylene through the Molecular Cross-Section Size Differentiation in Gallate-based Metal–Organic Frameworks." *Angewandte Chemie* 130 (49): 16252–16257.

[87] Lin, Rui-Biao, Hui Wu, Libo Li, Xiao-Liang Tang, Zhiqiang Li, Junkuo Gao, Hui Cui, Wei Zhou, and Banglin Chen. 2018. "Boosting Ethane/Ethylene Separation within Isoreticular Ultramicroporous Metal–Organic Frameworks." *Journal of the American Chemical Society* 140 (40): 12940–12946.

[88] Chen, Yongwei, Zhiwei Qiao, Houxiao Wu, Daofei Lv, Renfeng Shi, Qibin Xia, Jian Zhou, and Zhong Li. 2018. "An Ethane-Trapping MOF PCN-250 for Highly Selective Adsorption of Ethane over Ethylene." *Chemical Engineering Science* 175: 110–117.

[89] Li, Libo, Rui-Biao Lin, Rajamani Krishna, Hao Li, Shengchang Xiang, Hui Wu, Jinping Li, Wei Zhou, and Banglin Chen. 2018. "Ethane/Ethylene Separation in a Metal-Organic Framework with Iron-Peroxo Sites." *Science* 362 (6413): 443–446.

[90] Zhang, Xu, Libo Li, Jia-Xin Wang, Hui-Min Wen, Rajamani Krishna, Hui Wu, Wei Zhou, Zhong-Ning Chen, Bin Li, and Guodong Qian. 2019. "Selective Ethane/ Ethylene Separation in a Robust Microporous Hydrogen-Bonded Organic Framework." *Journal of the American Chemical Society* 142 (1): 633–640.

[91] Xiang, Sheng-Chang, Zhangjing Zhang, Cong-Gui Zhao, Kunlun Hong, Xuebo Zhao, De-Rong Ding, Ming-Hua Xie, Chuan-De Wu, Madhab C. Das, and Rachel Gill. 2011. "Rationally Tuned Micropores within Enantiopure Metal-Organic Frameworks for Highly Selective Separation of Acetylene and Ethylene." *Nature Communications* 2 (1): 1–7.

[92] Bloch, Eric D., Wendy L. Queen, Rajamani Krishna, Joseph M. Zadrozny, Craig M. Brown, and Jeffrey R. Long. 2012. "Hydrocarbon Separations in a Metal-Organic Framework with Open Iron (II) Coordination Sites." *Science* 335 (6076): 1606–1610.

[93] Li, Peng, Qishui Chen, Timothy C. Wang, Nicolaas A. Vermeulen, B Layla Mehdi, Alice Dohnalkova, Nigel D. Browning, Dengke Shen, Ryther Anderson, and Diego A. Gómez-Gualdrón. 2018. "Hierarchically Engineered Mesoporous Metal-Organic Frameworks toward Cell-Free Immobilized Enzyme Systems." *Chem* 4 (5): 1022–1034.

[94] Das, Madhab C., Qunsheng Guo, Yabing He, Jaheon Kim, Cong-Gui Zhao, Kunlun Hong, Shengchang Xiang, Zhangjing Zhang, K Mark Thomas, and Rajamani Krishna. 2012. "Interplay of Metalloligand and Organic Ligand to Tune Micropores within Isostructural Mixed-Metal Organic Frameworks (M'MOFs) for Their Highly Selective Separation of Chiral and Achiral Small Molecules." *Journal of the American Chemical Society* 134 (20): 8703–8710.

[95] Cui, Xili, Kaijie Chen, Huabin Xing, Qiwei Yang, Rajamani Krishna, Zongbi Bao, Hui Wu, Wei Zhou, Xinglong Dong, and Yu Han. 2016. "Pore Chemistry and Size Control in Hybrid Porous Materials for Acetylene Capture from Ethylene." *Science* 353 (6295): 141–144.

[96] Yu, Fan, Bing-Qian Hu, Xiao-Ning Wang, Yu-Meng Zhao, Jia-Luo Li, Bao Li, and Hong-Cai Zhou. 2020. "Enhancing the Separation Efficiency of a C_2H_2/C_2H_4 Mixture by a Chromium Metal–Organic Framework Fabricated via Post-Synthetic Metalation." *Journal of Materials Chemistry A* 8 (4): 2083–2089.

[97] Du, Yadan, Yang Chen, Yong Wang, Chaohui He, Jiangfeng Yang, Libo Li, and Jinping Li. 2021. "Optimized Pore Environment for Efficient High Selective C_2H_2/C_2H_4 and C_2H_2/CO_2 Separation in a Metal-Organic Framework." *Separation and Purification Technology* 256: 117749.

[98] Foo, Maw Lin, Ryotaro Matsuda, Yuh Hijikata, Rajamani Krishna, Hiroshi Sato, Satoshi Horike, Akihiro Hori, Jingui Duan, Yohei Sato, and Yoshiki Kubota. 2016. "An Adsorbate Discriminatory Gate Effect in a Flexible Porous Coordination Polymer for Selective Adsorption of CO_2 over C_2H_2." *Journal of the American Chemical Society* 138 (9): 3022–3030.

[99] Gong, Wei, Hui Cui, Yi Xie, Yingguo Li, Xianhui Tang, Yan Liu, Yong Cui, and Banglin Chen. 2021. "Efficient C_2H_2/CO_2 Separation in Ultramicroporous Metal–Organic Frameworks with Record C_2H_2 Storage Density." *Journal of the American Chemical Society* 143 (36): 14869–14876.

[100] Chen, Banglin, Jiyan Pei, Hui-Min Wen, Xiao-Wen Gu, Quan-Li Qian, Yu Yang, Yuanjing Cui, Bin Li, and Guodong Qian. 2021. "Dense Packing of Acetylene in a Stable and Low-Cost Metal–Organic Framework for Efficient C_2H_2/CO_2 Separation." *Angewandte Chemie.* 133 (47): 25272–25278.

[101] Organization, World Health. 2016. "Ambient Air Pollution: A Global Assessment of Exposure and Burden of Disease." World Health Organization.

[102] Stephan, Douglas W. 2016. "The Broadening Reach of Frustrated Lewis Pair Chemistry." *Science* 354 (6317): aaf7229.

[103] Xie, Lili, Qiuming Gao, Chundong Wu, and Juan Hu. 2005. "Rapid Hydrothermal Synthesis of Bimetal Cobalt Nickel Phosphate Molecular Sieve CoVSB-1 and Its Ammonia Gas Adsorption Property." *Microporous and Mesoporous Materials* 86 (1–3): 323–328.

[104] Chen, Yang, Libo Li, Jinping Li, Kun Ouyang, and Jiangfeng Yang. 2016. "Ammonia Capture and Flexible Transformation of M-2 (INA)(M= Cu, Co, Ni, Cd) Series Materials." *Journal of Hazardous Materials* 306: 340–347.

[105] Zhang, Jun, Dan Yue, Tifeng Xia, Yuanjing Cui, Yu Yang, and Guodong Qian. 2017. "A Luminescent Metal-Organic Framework Film Fabricated on Porous Al_2O_3 Substrate for Sensitive Detecting Ammonia." *Microporous and Mesoporous Materials* 253: 146–150.

[106] Erisman, Jan Willem, Mark A. Sutton, James Galloway, Zbigniew Klimont, and Wilfried Winiwarter. 2008. "How a Century of Ammonia Synthesis Changed the World." *Nature Geoscience* 1 (10): 636–639.

[107] Makhlouf, Ali, Tayeb Serradj, and Hamza Cheniti. 2015. "Life Cycle Impact Assessment of Ammonia Production in Algeria: A Comparison with Previous Studies." *Environmental Impact Assessment Review* 50: 35–41.

[108] Wang, Li, Yanhui Yi, Yue Zhao, Rui Zhang, Jialiang Zhang, and Hongchen Guo. 2015. "NH_3 Decomposition for H_2 Generation: Effects of Cheap Metals and Supports on Plasma–Catalyst Synergy." *ACS Catalysis* 5 (7): 4167–4174.

[109] Klerke, Asbjørn, Claus Hviid Christensen, Jens K. Nørskov, and Tejs Vegge. 2008. "Ammonia for Hydrogen Storage: Challenges and Opportunities." *Journal of Materials Chemistry* 18 (20): 2304–2310.

[110] Rieth, Adam J., Yuri Tulchinsky, and Mircea Dincă. 2016. "High and Reversible Ammonia Uptake in Mesoporous Azolate Metal–Organic Frameworks with Open Mn, Co, and Ni Sites." *Journal of the American Chemical Society* 138 (30): 9401–9404.

[111] Doonan, Christian J., David J. Tranchemontagne, T Grant Glover, Joseph R. Hunt, and Omar M. Yaghi. 2010. "Exceptional Ammonia Uptake by a Covalent Organic Framework." *Nature Chemistry* 2 (3): 235–238.

[112] Barin, Gokhan, Gregory W. Peterson, Valentina Crocellà, Jun Xu, Kristen A. Colwell, Aditya Nandy, Jeffrey A. Reimer, Silvia Bordiga, and Jeffrey R. Long. 2017. "Highly Effective Ammonia Removal in a Series of Brønsted Acidic Porous Polymers: Investigation of Chemical and Structural Variations." *Chemical Science* 8 (6): 4399–4409.

[113] Chen, Yang, Feifei Zhang, Yong Wang, Chengyin Yang, Jiangfeng Yang, and Jinping Li. 2018. "Recyclable Ammonia Uptake of a MIL Series of Metal-Organic Frameworks with High Structural Stability." *Microporous and Mesoporous Materials* 258: 170–177.

[114] Nguyen, Tu N., Ian M. Harreschou, Jung-Hoon Lee, Kyriakos C. Stylianou, and Douglas W. Stephan. 2020. "A Recyclable Metal–Organic Framework for Ammonia Vapour Adsorption." *Chemical Communications* 56 (67): 9600–9603.

[115] Peterson, Gregory W., John J. Mahle, Jared B. DeCoste, Wesley O. Gordon, and Joseph A. Rossin. 2016. "Extraordinary NO_2 Removal by the Metal–Organic Framework UiO-66-NH_2." *Angewandte Chemie* 128 (21): 6343–6346.

[116] Wang, Xinbo, Zhenzhu Xu, Li Li, Yue Zhao, Ruyue Su, Guojie Liang, Bo Yang, Yefan Miao, Wei Meng, and Zhiqiang Luan. 2020. "NO_2 Removal under Ambient Conditions by Nanoporous Multivariate Zirconium-Based Metal–Organic Framework." *ACS Applied Nano Materials* 3 (11): 11442–11454.

[117] Mon, Marta, Estefanía Tiburcio, Jesús Ferrando-Soria, Rodrigo Gil San Millán, Jorge A. R. Navarro, Donatella Armentano, and Emilio Pardo. 2018. "A Post-Synthetic Approach Triggers Selective and Reversible Sulphur Dioxide Adsorption on a Metal–Organic Framework." *Chemical Communications* 54 (65): 9063–9066.

[118] Carter, Joseph H., Xue Han, Florian Y. Moreau, Ivan Da Silva, Adam Nevin, Harry G. W. Godfrey, Chiu C. Tang, Sihai Yang, and Martin Schröder. 2018. "Exceptional Adsorption and Binding of Sulfur Dioxide in a Robust Zirconium-Based Metal–Organic Framework." *Journal of the American Chemical Society* 140 (46): 15564–15567.

[119] Bruce, Nigel, Dan Pope, Eva Rehfuess, Kalpana Balakrishnan, Heather Adair-Rohani, and Carlos Dora. 2015. "WHO Indoor Air Quality Guidelines on Household Fuel Combustion: Strategy Implications of New Evidence on Interventions and Exposure–Risk Functions." *Atmospheric Environment* 106: 451–457.

[120] Zhang, Yan, Zhonghang Chen, Xing Liu, Ze Dong, Peixin Zhang, Jun Wang, Qiang Deng, Zheling Zeng, Shuhua Zhang, and Shuguang Deng. 2019. "Efficient SO_2 Removal Using a Microporous Metal–Organic Framework with Molecular Sieving Effect." *Industrial & Engineering Chemistry Research* 59 (2): 874–882.

[121] Liu, Hui, Chunyan Cao, Jianying Huang, Zhong Chen, Guoqiang Chen, and Yuekun Lai. 2020. "Progress on Particulate Matter Filtration Technology: Basic Concepts, Advanced Materials, and Performances." *Nanoscale* 12 (2): 437–453.

[122] Xiao, Jing, Jiachen Liang, Chen Zhang, Ying Tao, Guo-Wei Ling, and Quan-Hong Yang. 2018. "Advanced Materials for Capturing Particulate Matter: Progress and Perspectives." *Small Methods* 2 (7): 1800012.

[123] Li, Zhisheng, Qingmei Wen, and Ruilin Zhang. 2017. "Sources, Health Effects and Control Strategies of Indoor Fine Particulate Matter (PM2.5): A Review." *Science of the Total Environment* 586: 610–622.

[124] Zhang, Zhi-Hui, Andrey Khlystov, Leslie K. Norford, Zhen-Kang Tan, and Rajasekhar Balasubramanian. 2017. "Characterization of Traffic-Related Ambient Fine Particulate Matter (PM2. 5) in an Asian City: Environmental and Health Implications." *Atmospheric Environment* 161: 132–143.

[125] Zhang, Rufan, Chong Liu, Guangmin Zhou, Jie Sun, Nian Liu, Po-Chun Hsu, Haotian Wang, Yongcai Qiu, Jie Zhao, and Tong Wu. 2018. "Morphology and Property Investigation of Primary Particulate Matter Particles from Different Sources." *Nano Research* 11 (6): 3182–3192.

[126] Zhang, Xiaowei, Wei Zhang, Mingqiang Yi, Yingjie Wang, Pengjun Wang, Jun Xu, Fenglei Niu, and Feng Lin. 2018. "High-Performance Inertial Impaction Filters for Particulate Matter Removal." *Scientific Reports* 8 (1): 1–8.

[127] Khalid, Bilal, Xiaopeng Bai, Hehe Wei, Ya Huang, Hui Wu, and Yi Cui. 2017. "Direct Blow-Spinning of Nanofibers on a Window Screen for Highly Efficient PM2. 5 Removal." *Nano Letters* 17 (2): 1140–1148.

[128] Zhang, Yuanyuan, Shuai Yuan, Xiao Feng, Haiwei Li, Junwen Zhou, and Bo Wang. 2016. "Preparation of Nanofibrous Metal–Organic Framework Filters for Efficient Air Pollution Control." *Journal of the American Chemical Society* 138 (18): 5785–5788.

[129] Chen, Yifa, Fan Chen, Shenghan Zhang, Ya Cai, Sijia Cao, Siqing Li, Wenqi Zhao, Shuai Yuan, Xiao Feng, and Anyuan Cao. 2017. "Facile Fabrication of Multifunctional Metal–Organic Framework Hollow Tubes to Trap Pollutants." *Journal of the American Chemical Society* 139 (46): 16482–16485.

[130] Chen, Yifa, Shenghan Zhang, Sijia Cao, Siqing Li, Fan Chen, Shuai Yuan, Cheng Xu, Junwen Zhou, Xiao Feng, and Xiaojie Ma. 2017. "Roll-to-roll Production of Metal-organic Framework Coatings for Particulate Matter Removal." *Advanced Materials* 29(15): 1606221.

[131] Ma, Shanshan, Meiyun Zhang, Jingyi Nie, Jiaojun Tan, Bin Yang, and Shunxi Song. 2019. "Design of Double-Component Metal–Organic Framework Air Filters with PM2. 5 Capture, Gas Adsorption and Antibacterial Capacities." *Carbohydrate Polymers* 203: 415–422.

[132] Su, Zhiping, Meiyun Zhang, Zhaoqing Lu, Shunxi Song, Yongsheng Zhao, and Yang Hao. 2018. "Functionalization of Cellulose Fiber by in Situ Growth of Zeolitic Imidazolate Framework-8 (ZIF-8) Nanocrystals for Preparing a Cellulose-Based Air Filter with Gas Adsorption Ability." *Cellulose* 25 (3): 1997–2008.

[133] Zhu, Qiuyun, Xi Tang, Shasha Feng, Zhaoxiang Zhong, Jianfeng Yao, and Zhong Yao. 2019. "ZIF-8@ SiO_2 Composite Nanofiber Membrane with Bioinspired Spider Web-like Structure for Efficient Air Pollution Control." *Journal of Membrane Science* 581: 252–261.

[134] Li, Ting-Ting, Xixi Cen, Hai-Tao Ren, Liwei Wu, Hao-Kai Peng, Wei Wang, Bo Gao, Ching-Wen Lou, and Jia-Horng Lin. 2020. "Zeolitic Imidazolate Framework-8/Polypropylene–Polycarbonate Barklike Meltblown Fibrous Membranes by a Facile in Situ Growth Method for Efficient PM2. 5 Capture." *ACS Applied Materials & Interfaces* 12 (7): 8730–8739.

[135] Wang, Zhe, Youfang Zhang, Xiu Yun Daphne Ma, Jiaming Ang, Zhihui Zeng, Bing Feng Ng, Man Pun Wan, Shing-Chung Wong, and Xuehong Lu. 2020. "Polymer/ MOF-Derived Multilayer Fibrous Membranes for Moisture-Wicking and Efficient Capturing Both Fine and Ultrafine Airborne Particles." *Separation and Purification Technology* 235: 116183.

[136] Zhang, Kun, Qian Huo, Ying-Ying Zhou, Hong-Hong Wang, Gao-Peng Li, Yao-Wu Wang, and Yao-Yu Wang. 2019. "Textiles/Metal–Organic Frameworks Composites as Flexible Air Filters for Efficient Particulate Matter Removal." *ACS Applied Materials & Interfaces* 11 (19): 17368–17374.

[137] Hu, Min, Linghui Yin, Nicholas Low, Donghuan Ji, Yishui Liu, Jianfeng Yao, Zhaoxiang Zhong, and Weihong Xing. 2020. "Zeolitic-Imidazolate-Framework Filled Hierarchical Porous Nanofiber Membrane for Air Cleaning." *Journal of Membrane Science* 594: 117467.

[138] Bian, Ye, Rutao Wang, Shijie Wang, Chenyu Yao, Wei Ren, Chun Chen, and Li Zhang. 2018. "Metal–Organic Framework-Based Nanofiber Filters for Effective Indoor Air Quality Control." *Journal of Materials Chemistry A* 6 (32): 15807–15814.

[139] Feng, Shasha, Xingya Li, Shuaifei Zhao, Yaoxin Hu, Zhaoxiang Zhong, Weihong Xing, and Huanting Wang. 2018. "Multifunctional Metal Organic Framework and Carbon Nanotube-Modified Filter for Combined Ultrafine Dust Capture and SO_2 Dynamic Adsorption." *Environmental Science: Nano* 5 (12): 3023–3031.

[140] Yoo, Dong Kyu, and Sung Hwa Jhung. 2019. "Effect of Functional Groups of Metal–Organic Frameworks, Coated on Cotton, on Removal of Particulate Matters via Selective Interactions." *ACS Applied Materials & Interfaces* 11 (50): 47649–47657.

[141] Wang, Xiaoyu, Wenshi Xu, Xiaoying Yan, Yi Chen, Mengyu Guo, Guoqiang Zhou, Shengrui Tong, Maofa Ge, Ying Liu, and Chunying Chen. 2019. "MOF-Based Fibrous Membranes Adsorb PM Efficiently and Capture Toxic Gases Selectively." *Nanoscale* 11 (38): 17782–17790.

[142] Xie, Fan, Nan Zhang, Longhai Zhuo, Panliang Qin, Shanshan Chen, Yafang Wang, and Zhaoqing Lu. 2019. "'MOF-Cloth' Formed via Supramolecular Assembly of NH_2-MIL-101(Cr) Crystals on Dopamine Modified Polyimide Fiber for High Temperature Fume Paper-Based Filter." *Composites Part B: Engineering* 168: 406–412.

[143] Zhao, Xing, Liping Chen, Yi Guo, Xu Ma, Zhuoyi Li, Wen Ying, and Xinsheng Peng. 2019. "Porous Cellulose Nanofiber Stringed HKUST-1 Polyhedron Membrane for Air Purification." *Applied Materials Today* 14: 96–101.

[144] Gai, Shuang, Ruiqing Fan, Kai Xing, Ani Wang, Xubin Zheng, Xuesong Zhou, Ping Wang, and Yulin Yang. 2019. "Preparation of Composite Filters Based on Porous Coordination Polymers by Using a Vacuum Filtration Method for Highly Efficient Removal of Particulate Matters." *Chemistry–An Asian Journal* 14 (13): 2291–2301.

[145] Yoo, Dong Kyu, Ho Chul Woo, and Sung Hwa Jhung. 2021. "Removal of Particulate Matters by Using Zeolitic Imidazolate Framework-8s (ZIF-8s) Coated onto Cotton: Effect of the Pore Size of ZIF-8s on Removal." *ACS Applied Materials & Interfaces* 13 (29): 35214–35222.

APPENDIX

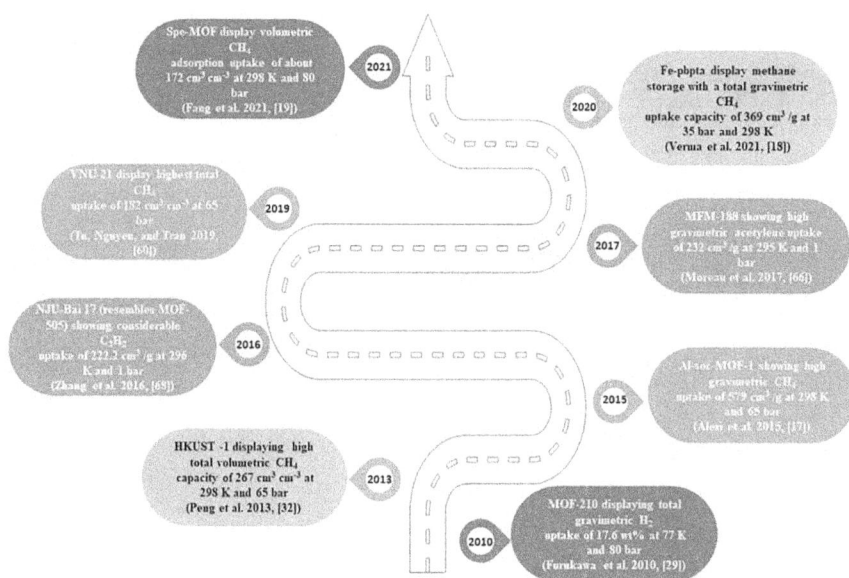

APPENDIX 4.1 Timeline of major developments in MOF-based gas storage.

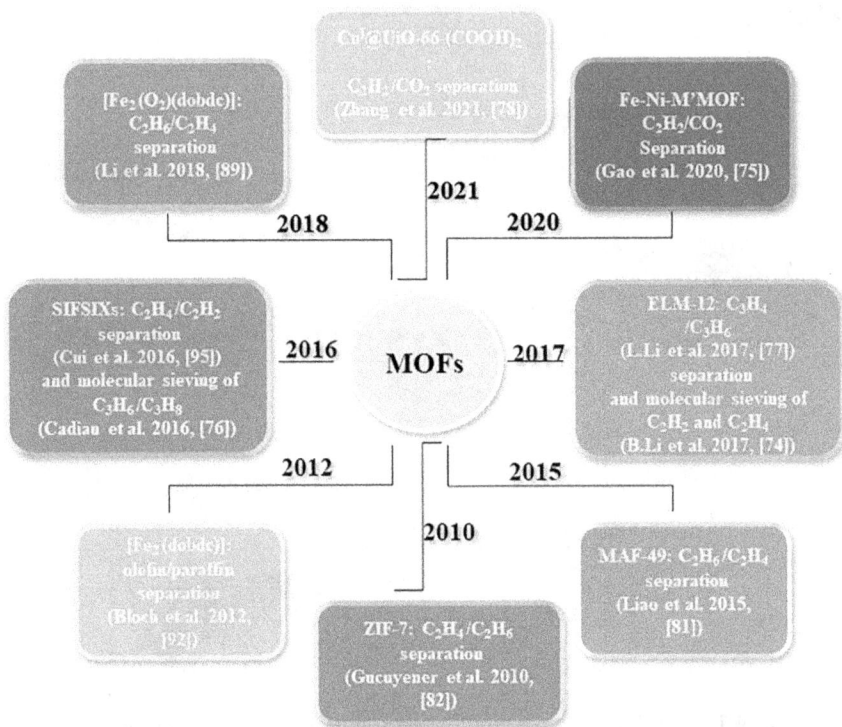

APPENDIX 4.2   Timeline of major developments in MOF-based gas separation.

5 Metal-Organic Frameworks in Heterogeneous Catalysis

A. Manjceevan and K. Velauthamurty

CONTENTS

5.1 INTRODUCTION

Metal-organic frameworks (MOFs) are developed by combining organic and inorganic units with coordination bonds and show high porosity [1,2]. MOFs consist of metal centers and organic links. The inorganic metal centers are considered secondary building units that act as 'joints', and the organic secondary building units act as 'struts'. The metal center, organic links, and topology of the framework are essential parts of MOFs [3]. The MOFs could be synthesized with a large porosity size. Careful modification of MOFs tropology and use of extended length of organic molecules provides higher porosity and larger pore size of MOFs. The great selection of metal centre, organic stunt and topology of MOF leads to flexible modification of physicochemical properties of MOFs (shown in Table 5.1).

The existence of metal ions or clusters and organic molecules by coordination bonds undergo self-assembly polymerization, leading to the three-dimensional

DOI: 10.1201/9781003252061-5

TABLE 5.1

Physical and Chemical Properties of MOFs

Metal atoms	Organic stunt	Formula	Physical and chemical properties	Ref.
$Cu(OAc)_2.H_2O$	H_3BTC	HKUST-1@AC	High conductivity of the MOF composite, the high-performance electrode for pseudo-capacitors	[4]
PtNi	UiO-67	PtNi@UiO-67, Ni@UiO-67	Intrinsic structural and catalytic properties, the PtNi synergistic effect enhanced the catalytic properties	[5]
$Cu(NO_3)_2.3H_2O$	H_3BTC	Pd/rGO@Cu-BTC	Uniform pore size, high surface area, catalyzed CO oxidation effectively, further catalytic performance affected by the presence of Pd	[6]
$Cu(NO_3)_2 \cdot 3H_2O$, $Co(NO_3)_2 \cdot 6H_2O$, $Fe(NO_3)_3 \cdot 9H_2O$	Benzenedicarboxylic acid	Cu-based UiO-66-, CuCo–UiO-66-, and CuFe–UiO-6	Best catalytic activity, CO oxidation in air and the COProx reaction, synergic effects of metals	[7]
Silver nitrate, zinc nitrate Hexahydrate, copper (II) chloride	4-mercaptophenol, 2-methylimidazole	Ag-MP@ZIF	Unconventional regioselectivity at the molecular level	[8]
palladium (II) nitrate, cerium (III) nitrate hexahydrate	trimesic acid (H_3BTC),	$([Ce(BTC)(H_2O)]$ $DMF)_n$	Better catalytic activity in CO oxidation	[9]
$Co(NO_3)_2$ $4H_2O$	1,4-benzene dicarboxylic acid	$[Co(pca)(bdc)$ $0.5(H_2O)2]$	Selective water uptake	[10]
$Co(NO_3)_2 \cdot 6H_2O$	(2,5-dihydroxyterephthalic acid	Co -MOF-74	High density of CUSs and remarkable porosity, efficient catalytic activity	[11]
$Cu(NO_3)_2 \cdot 3H_2O$, $Pd(NO_3)_2$	1,3,5-benzenetricarb-oxylic acid	1%Pd/ $Cu_3(BTC)_2$-P	Excellent catalysts for CO oxidation	[11]

crystal-like structure of MOFs [3]. The organic molecules with more than one nitrogen or oxygen atom may link metal centers and be useful for MOFs [3].

The great selection of metal ions, organic struts, and topology leads to infinite inorganic-organic MOFs, and the rational design of MOFs leads to enhanced physicochemical properties over the zeolites and porous carbon materials [12]. The regular structural motifs of MOFs exist and may be used to predict MOFs' structure [13]. The MOFs with higher porosity were successfully applied to many catalytic processes and attracted unprecedented scientific interest. The catalyst enhances the rate of chemical reactions without consumption. There are a vast number of catalytic reactions used in the laboratory and industrial-scale production of substances. Egyptian civilization, Greek, Roman, and Islamic empires used the catalytic process to produce alcohol, control synthesis of ammonia, etc. MOF shows excellent catalytic activities due to several characteristic features of MOF such as higher porosity, a vast number of structural diversities, possibilities for rational design, and the presence of open metal sites and accessible organic struts [2]. Further, the catalytic process by MOF differs to a great extent due to its structural features, coordination vacancies, charge transfer between ligand and metals, post-synthetic modifications, etc [2]. This chapter summarizes the characteristic features of MOFs in heterogeneous catalysis, characterization of MOFs and its applications.

5.2 CATALYTIC ACTIVITY OF MOFS

The catalytic activities of MOFs commonly arise due to open metal sites (OMS) existing without coordination with organic struts, defects in MOFs, and the catalytic organic species formed by pre- or post-synthesis modifications [2]. The vacant site of metal ions exists without coordination by organic strut, called an open metal site (OMS) of MOFs. The formation of OMS in MOF should be a rationalized design without collapsing the three-dimensional structure and the porosity of MOF (Figure 5.1).

There are many key applications associated with OMS of MOFs such as catalysis, sensing, gas separation and gas adsorption, etc. [12]. The OMS formation in MOFs generally plays with ligand removal, mainly achieved via solvent exchange in which the removal is termed *activation*. Removal of a solvent by heating under a vacuum may collapse the MOFs and lead to degradation or completely decompose [12].

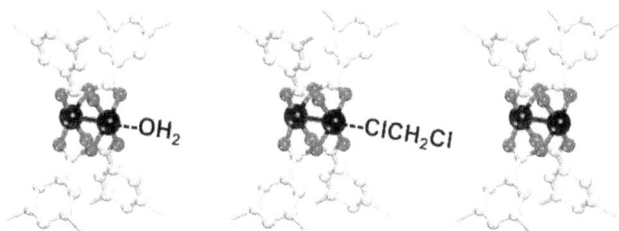

FIGURE 5.1 Schematic design of chemical activation of HKUST-1 (HKUST stands for Hong Kong University of Science and Technology) by Dichloromethane. Reprinted (adapted) with permission from [14] Copyright 2015, American Chemical Society.

There are several strategies performed to remove the coordinated solvents in the MOFs, such as solvent exchange and removal by thermal activation [10,15], chemical activation [14], and photothermal activations [12].

In a solvent exchange and removal by thermal activation process, the less volatile solvent molecules in MOFs exchange by a readily volatile solvent molecule and then are followed by a thermal process under proper conditions yield OMS. In the chemical activation process, after the exchange of less volatile solvent molecules in MOF, similar to the earlier process; the less volatile solvent and weakly coordinated molecules in MOFs are removed carefully by air drying at room temperature without utilizing the extra thermal energy.

If the solvent molecules exist as kinetically stable in MOF, they could be removed using the ultraviolet-visible (UV-Vis) irradiation process. This process is called photothermal activation [12,16]. The summary of OMS formed in MOFs by different techniques and its significance are tabulated in Table 5.2.

Defects of MOFs create nonstoichiometric nature between metal and ligand, occurring due to loss of metal clusters, strut, and coordinated solvents. Quantitative measures of defect could be performed experimentally via elemental analysis. The missing linker positions are occupied by the groups with oxygen, water, and chloride. The defects created on MOFs enhanced mechanical properties, porosity, and Bronsted and Lewis acidities [2]. The impact of the defect in MOFs is shown in Table 5.3.

The MOFs consist of porous structures constructs by inorganic and organic molecules in MOFs. Those structures are highly tuneable and properties could be optimized by rationalized design of MOFs. The functional group in MOFs tailors the MOFs properties to a great extent. The inclusion of organic molecules into the MOFs during the synthesis of MOFs was rather limited. The scope of expanded and versatile deployed of functional group could be performed by post synthesis

---

TABLE 5.2

Catalytic Sites Via Open Metal Site (OMS) in MOFs

MOFs Formula	Activation process	Ref.
HKUST-1 or Cu-MOF-2	Activate by methylene chloride (chemical route)	[14]
$[Zn_2(iso)_2(bpy)_2]_n$-300	Thermal treatment method	[17]
Co-MOF-74	Heat and vacuum process	[18]
Co-MOF-74	Induce structure defects	[19]
MIL-88A	Thermal activation	[20]
HKUST-1, UiO-66-NH_2, ZIF-67, CPO-27-M (M = Zn, Ni, and Mg), Fe-MIL-101-NH_2, and IRMOF-3,	Photothermal activation	[16]
MOF MIL-100(Cr)	Activation temperature	[21]
UMCM-1-OBnNO$_2$ UMCM-1-(OBnNO$_2$)$_2$	Photochemical activation	[22]

TABLE 5.3

Recent Publications of Defects Associated MOFs Used in Photocatalytic Properties

MOFs Formula	Defect on	Significance of having the defect	Ref.
UiO-66	Linker defects and missing cluster defects, open metal sites	Heterogeneous catalytic performances	[23]
UiO-67	The bipyridine replaced phenyl groups	The high degree of ammonia adsorption	[24]
MOF-199	During synthesis, defective linkers doped systematically	Modified bandgap, magnetism, porosity	[25]
UiO-66	Missing-linker defects	Greatly improved porosity strongly affects its gas adsorption behavior	[26]
MIL-53(Al)	Formation of defects by using an acetic acid modulator	The dramatic increase in both pore volume and surface area	[27]
Pt@UiO66-NH$_2$-100	Structural defects in MOFs form by molar equivalents of the acetic acid modulator	Structural defects possess the fastest relaxation kinetics and the highest charge separation efficiency	[28]

modifications (PSM). The success of the PSM process depends on a judicious and rationalized selection of reaction conditions and reagents to complete the PSM process without altering the structure, crystallinity, and porosity of MOFs [29]. In the PSM process in MOFs, the inclusion of functional groups occurs covalently and/or coordinatively.

PSM via coordination bond formation is achieved with the organic linker molecules having aldehyde, amine, azide-derivatives due to those functional groups are quite reactive and better selectivity towards targets [29]. PSM via coordination bond formation is used to hybridize the MOF with polymers towards the desired properties of catalytic and other applications of MOFs. Coordinative PSM was firstly observed in the coordination of solvent molecule tethering with open metal void of MOFs. Tethering of amine of the pyridine-based linker of MOFs and chelating ligands with carboxylates and phosphates functional groups selectively bind with the surface of MOFs. The surface modification performed in MOFs is tabulated in Table 5.4.

5.3 METAL NANOPARTICLES IN MOFS

Nanomaterials are defined as at least one of its dimensions existing on the nanometer-scale, typically as 1–100 nm. The small size of nanomaterials shows unique properties than the bulk materials such as high surface area, tunable bandgap, high excitation generation, and so on leads to the applicability to current and future desired applications. Further, MOFs have several desired features, such as structural diversity, high surface area, porosity, etc., leading to vast applications including catalyst, gas storage, sensing, energy applications, etc. Currently, the nano-MOFs had strong enthusiasm due to their unique properties, such as

TABLE 5.4

Some Recent Publications of Post-Treatment-Associated MOFs

PSM	Formula of MOFs	Significance of PSM	Ref.
Modified with pyrene-based chemodosimeter probe	UiO-66	MOF emission behavior changed	[30]
Post CO_2 atmosphere treatment	IRMOF-1 membranes	Higher CO_2 adsorption. Better selectivity over affinity of H_2 and reduced CO_2	[31]
Condensation with 1-pyrenecarboxaldehyde	UiO-66–NH_2	Biologically active anions sensing, fast response time, excellent selectivity and sensitivity	[32]
Modification by using $MoO_2(acac)_2$ and $VO(acac)_2$	$Cu_3(BTC)_2$	Mo-catalyst shows maximum activity in epoxidation	[33]
Dilute sulfonic acid functionalization	MIL-100(Fe)	Catalyze the methanol, oleic acid esterification reaction	[34]

accelerated adsorption/desorption kinetics, accessible to the inner side of MOF for good catalytic activity, suitable size for biomedical application, etc. [35]. Further, the MOFs exploit the size-selective catalytic process in which selectivity mainly depends on the size of metal nanoparticles. This is called selective structure reactions [2]. The catalytic process of MOFs depends on metal atoms and organic molecules. In MOFs, metal atoms can act as a Lewis acid. Metal atom and organic strut can be modified for optimized performance of the catalytic process. The optimization of metal is achieved via tuning the size of metal particles and organic molecules via using various functional groups such as -OH, -NH_2, -SO_3H, etc. The MOFs involve a different role in the nano-MOFs composite in the catalytic process. MOFs stabilize the metal nanoparticles within the pores and assist the selectivity of reaction based on size. Further, MOFs control the nanomaterial properties by controlling the electronic and charge transfer processes between MOFs and metal nanoparticles. The significance of nano-MOFs are shown in Table 5.5.

5.4 APPLICATIONS OF MOFS IN HETEROGENEOUS CATALYSIS

Attractive unique features of MOFs are ideal structures for real-world applications. The MOFs in heterogeneous catalyst uses in many catalytic processes such as CO oxidation, NO reduction, CO_2 hydrogenation, alcohol esterification, biofuel upgrade, CO_2 fixation, etc. (Figure 5.2).

The MOFs show superior performances and provide a better yield than the usage of conventional catalysts. The applications of MOFs are summarized as follows.

5.4.1 CO OXIDATION

Y. Zhao et al. reported that the copper-containing MOFs, such as $Cu_5(OH)_2(nip)_4$ shows higher catalytic activity and oxidized 100% of CO. The higher catalytic activity

TABLE 5.5
Recent Publications of Metal Nanoparticles in MOFs

Nanomaterial	Formula of MOFs	Significance of Nano-MOFs	Ref.
CuNPs	CuNPs@NU-901 and CuNPs@NU-907	Selective hydrogenation of acetylene to ethylene	[36]
CuO NPs	CuO@CuBDC	High catalytic activity in C-S cross-coupling reactions and photo-degradation of dye pollutant	[37]
PdCu	PtCu@MIL-101	High catalytic activity, better selectivity, recyclability	[38]
Pd	Pd@UiO-66-OH	Highest activity	[39]
Au	Au@UIO-66	Better catalytic activity and stability for CO oxidation	[40]
Au	Au-NP/Ni-Cu MOF	Chemical degradation, exceptionally high catalytic activity	[41]
Au	Au/MIL-101 (CD/PVP)	Enhanced catalytic activities in liquid-phase aerobic oxidation of a wide range of alcohols	[42]
Au	Ce-MOF	Exhibited ultrahigh turnover frequency for the reduction of 4-nitrophenol	[43]
Ag	Ag@NOTT-300(Al)	Highly efficient catalytic reduction of nitrophenols	[44]
Ag	Fe_3O_4@MIL-100(Fe)/Ag	Better catalytic activity and recyclable for 4-nitrophenol catalytic reduction	[45]

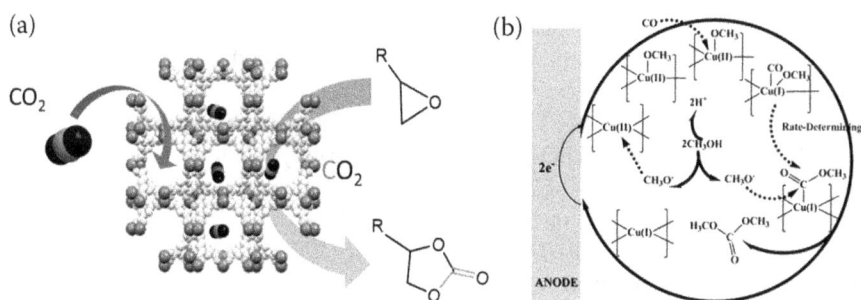

FIGURE 5.2 Heterogeneous catalyst: (a) Representation of cycloaddition reaction of CO_2, captured by MOFs, to epoxides. Reproduce under Creative Commons Attribution (CC BY) license [46]. (b) Reaction scheme of the electrosynthesis of dimethyl carbonate (DMC) over the Cu (II)-based MOF. Reprinted (adapted) with permission from, [47], Copyright 2013, Elsevier.

arises due to the dense and nanoporous MOF with a higher concentration of reactive metal site [48]. $Cu_3(BTC)_2$ (BTC: benzene-1,3,5-tricarboxylate) is act as good catalyst for oxidation of CO [49]. Jing-yun Ye and Chang-jun Liu in 2011, confirmed that the use of $Cu_3(BTC)_2$ catalyst enhances the CO oxidation capabilities. Further, they confirmed that the use of amorphized $Cu_3(BTC)_2$ and palladium (Pd) support shows enhanced in catalytic activities against the oxidation activity of CO [49]. Kim, J.Y. Kim, et al. synthesized redox-active Pd supported MOFs such as Pd@ra-MOF at room temperature and exploited to $CO_{(g)}$ oxidation reaction. During the process, the Pd@ra-MOFs transformed to catalytic active PdO_x-NiO_y/C nanocomposite in-situ [50].

In 2015, T. Kim, et. al., reported that Co-MOF-74 showed catalytic activities in CO reduction reactions due to the unrivaled Lewis acid character of the coordinatively unsaturated site [11]. In 2016, C. Zhang, et. al., reported that Co_3O_4-MOFs such as Co_3O_4-MA (MA = methylamine) and Co_3O_4-DMA (DMA = dimethylamine) shows better catalytic performance for CO oxidation. Here the Co_3O_4 exists in nanometer-scale particles. Both of the catalysts showed 100% conversion of CO [51]. In 2017, X. Zhang and co-workers showed the thermal activation of $Cu_3(BTC)_2$ at 240°C in the presence of different gases like Ar, O_2, H_2, or CO. The catalyst thermally activated in the presence of $O_{2(g)}$, little collapse in the structure and formed active CuO, and the thermal activation performed in the presence of $CO_{(g)}$ formed many pores and more effective Cu_2O form in the structure. However, the thermal activation in the presence of reductive H_2 destroys the structure, and the presence of inert Ar gas weakly influences the structure. As a result, the $Cu_3(BTC)_2$ thermally activated in the presence of H_2 and Ar, shows poor catalytic activity [52].

A. Lin, et al., reported that Pd-supported Ce-MOFs act as a catalyst for CO oxidation at modest temperatures and uptake CO_2 at a lower temperature. The high catalytic activity arises due to Ce(III) and Ce(IV) presence in the MOFs. Further, the existence of Pd on Ce-MOF stored the oxygen as Pd oxide. The stored oxygen is released during the oxidation at 373 K, and the catalyst significantly adsorbs $CO_{2(g)}$ [9]. In 2019, L.A. Lozano, et al., synthesized non-noble metal in cooperated MOF and showed that the MOF with non-noble metal shows catalytic activity. Cu-based UiO-66-(Universitetet i Oslo), bimetallic CuCo–UiO-66, and CuFe–UiO-6-derived materials showed catalytic activity to oxidize CO(g), which is a good result as the metals are a good alternative for Pd [7].

In 2020, RezaVakili et al., synthesized PtNi bimetallic nanoparticles supported on UiO-67 MOFs (PtNi@UiO-67). Synthesized catalyst active in CO oxidation reaction. Ni in the PtNi@UiO-67 showed different behavior than the Ni that exist as monometallic nanoparticles in the MOFs, and PtNi@UiO-67 showed better catalytic activity than Pt@UiO-67 and Ni@UiO-67. The improvement of catalytic behavior arises via stabilized shell NiO_x phase in PtNi@UiO-67 [5].

In 2021, H.M. Altass, et al., reported 3.0 wt.% of Pd/rGO@Cu-BTC effectively oxidize CO [6]. T_{50} and T_{100} values are 71 and 82°C, respectively. Further, the synthesized catalyst showed catalytic activity after six cycles also. Interestingly, the inclusion of Pd into the rGO@Cu-BTC, increases the catalytic activity drastically. However, the excess load of Pd on rGO@Cu-BTC accumulates and hinders the performance of the catalytic activity.

5.4.2 ALCOHOL OXIDATION

F. Carson, et. al. reported that the ruthenium trichloride was included in the aluminum doped MOF by post-synthetic modifications. MOF-253 has open bipyridine sites, and ruthenium occupied these open positions. The $PhI(OAc)_2$ oxidized primary, and secondary alcohols, including allylic alcohols, under very mild conditions in the presence of ruthenium complexation in the aluminum doped MOF. The efficient catalytic process was achieved via short reaction time, low catalyst loading, high selectivity, performing the reaction at room temperature, and reusability upto six cycles [53].

In 2016, A.K. Babahydarai, et. al. reported that PW-MOF showed superb catalytic activity and perform selective oxidation of the benzylic, linear and secondary alcohols to the corresponding aldehydes and ketones. Further, allylic alcohols were selectively converted as the aldehyde. Also, the catalyst can recover by filtration from the reaction mixture and reusable over several time [54]. In the same year B.R. Kim, et. al. reported that Cu@C/TEMPO/NMI catalyst (TEMPO: 2,2,6,6-tetramethylpiperidine-N-oxyl and NMI: Nmethylimidazole) shows excellent catalytic activity for aerobic oxidation of alcohol. Further, it shows excellent reusability could apply to broad scope for the substrate [55].

In 2017, L. Peng, et al. reported that Co-BTC MOFs showed (BTC: 1,3,5-benzenetricarboxylic acid) selectively oxidize the benzyl alcohol under air with high conversion (92.9%) and good selectivity (97.1%) [56]. A. Taher, et al. reported that copper containing MOFs catalysis a wide range of oxidization reactions of alcohols, including inactive hetero-aryl and long-chain alkyl units selectively oxidized to their corresponding aldehydes. Cu-containing MOFs with TEMPO showed high catalytic, selectively under base free conditions, broad substrate scope, usable under open-air conditions, and reusable more than six times without significant loss of catalytic activity [57].

The Zr-MOFs are applicable to benzyl alcohol oxidation to benzaldehyde selectively. It is an important reaction usable in a wide range of areas, including chemical synthesis, pharmaceutical and industries. Air-stable Fe-anchored MOF-808 (MOF-808: $Zr_6(\mu_3\text{-}O)_4(\mu_3\text{-}OH)_4(HCOO)_6(1,3,5\text{-benzenetricarboxylate})_2$) was synthesized by W. Jumpathong, et al. and used to selective oxidation of benzyl alcohol to benzaldehyde. The catalyst showed better catalytic activity, selectivity, and reusability [58].

In 2019, X. Wang, et al. synthesized isostructural mesoporous MOFs based on transition metals, lanthanides, and actinides (Zr, Hf, Ce, Th) and used as a support for vanadium catalyst. It catalyzes the aerobic oxidation reaction of 4-methoxybenzyl alcohol in which the V-Zr-NU-1200 showed ~20 times higher turnover frequency than V-Ce-NU-1200. Further, catalytic activity was correlated with electronegativity and the oxidation state of the support metals [59].

In 2021, Y. Chen, et al. analyzed the structure–activity relationship of MOFs catalyst in aerobic alcohol oxidation in which they used molybdenum(VI) deposited on mesoporous zirconium-based metal-organic framework (MOF) NU-1200 as an active layer for the oxidation reaction. They observed two distinct anchoring modes of molybdenum such as coordinated through one node of terminal oxygen atom in octahedral geometry and coordinated to two adjacent Zr_6 node oxygen atoms in a tetrahedral geometry, as shown in Figure 5.3.

The catalytic activity of Mo-NU-1200 was measured based on the oxidation of 4-methoxybenzyl alcohol. The two types of Mo-NU-1200 MOFs play differently in

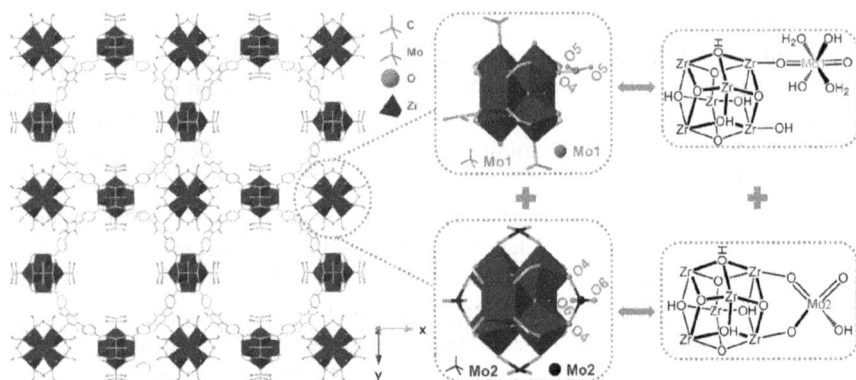

FIGURE 5.3 Mo-NU-1200 (NU stands for Northwestern University) Crystal structure showing two different anchoring modes of molybdenum. Reprinted (adapted) with permission from [60] Copyright 2021, American Chemical Society.

base and base-free conditions. The almost three times faster activity of catalyst under base free conditions than in base condition and reveal the presence of base catalytic alcohol oxidation process inhibited. Further, the two distinct anchoring modes are played differently based on reaction conditions [60].

5.4.3 STYRENE EPOXIDATION

In 2011, R. Sen, et al. synthesized lanthanides containing MOF heterogeneous catalysts by hydrothermal methods and tested them over epoxidation of olefins [61]. The compounds synthesized in their methods showed better catalytic performance in olefin epoxidation reactions [61]. Later in 2012, the cobalt-based MOFs (STA-12(Co)) were synthesized and tested against aerobic epoxidation of olefins in DMF. They reported that the STA-12(Co) showed remarkably high catalytic activities over the Co-doped zeolite catalysts. Further, the selectivity of epoxide reactions depends on the substrate in which styrene epoxidation shows low selectivity and (E)-stilbene epoxidation perform with high selectivity [62]. In 2014, J. Sun, et al. performed epoxidation of styrene in the presence of MOF catalysts. Their study synthesized MIL-101 MOF (MIL, Matérial Institut Lavoisier) with different metal ions against the styrene epoxidation in the presence of air, H_2O_2, and TBHP (tert-butyl hydroperoxide) oxidants. Fe and Cr in MOFs show better catalytic activities and epoxide selectivity in styrene conversion. The styrene epoxidation activity in various oxidants in Fe-MIL-101 are air > TBHP > H_2O_2 and in Cr-MIL-101 are H_2O_2 > air > TBHP. However, selectivity to styrene oxide are TBHP > air > H_2O_2 [63].

J. Tang, et al. synthesized Zr-MOF catalysts, which were modified Mo(VI) and used for epoxidations of olefins. In this study, stable porous Zr-MOF (UiO-66(NH$_2$)) modified the post-synthesis process by using salicylaldehyde, pyridine-2-aldehyde, or 2-pyridine chloride and Mo-based catalyst. Here, the MOFs act as carriers of the Mo(VI). The MOFs consist of the big pore size improve the contact

TABLE 5.6
MOFs in Catalytic Epoxidation

Epoxidation	Catalyst	Ref.
Epoxidation of alkenes with tert-BuOOH	UiO-66-NH$_2$-SA-Mo, UiO-66-NH$_2$-TC-Mo	[66]
Alkene epoxidation	NU-1000-supported Nb(V) catalysts	[67]
Styrene epoxidation	ZIF-67, Zn0.5Co1-ZIF, Zn1Co1-ZIF, Zn1.5Co1-ZIF, Zn2Co1-ZIF)	[68]
Olefin epoxidation	MnII-MOF-5	[69]
Enantioselective epoxidation	Chiral-NU-1000-Mo	[70]
Carboxylation of alkynes and epoxides	Metal-organic frameworks, zeolites, and alumina-supported nanoparticles	[71]
Selective oxidation of styrene	Cu$_x$-Co$_y$-MOF	[72]
Epoxidation of olefin with H$_2$O$_2$	Mn$_x$Cu$_{1-x}$-MOF	[73]

between substrate and catalytic active center, thus yielding excellent catalytic activity on epoxide formation reaction. Similarly, in their study, MoO$_2$(acac)$_2$ loaded Zr-MOF was also synthesized, and the catalytic activity of this molecule in epoxide formation reaction also showed good performances [64]. Later in 2015, J.W. Brown synthesized manganese porphyrin containing MOF, MOF-525-Mn and tested in epoxide formation reaction of alkene at mild conditions using molecular oxygen and showed that the catalyst shows better catalytic activity, reusability, maintain its structure, and minimal deactivation after several cycles of the catalytic process [65]. Likewise, several catalysts were used for epoxide formation reactions (shows in Table 5.6).

5.4.4 BIODIESEL PRODUCTION

Biodiesel, labeled as free fatty acid methyl ester, has been reported as an environmentally friendly alternative to fossil fuel. Biodiesel produces from edible or non-edible oils, commonly via using esterification or transesterification process. In the transesterification process, extracted oils reacted with methanol in the presence of the catalyst and produce biodiesel, and in the esterification process, the free fatty acid reacted with methanol and produced biodiesel [74]. In the esterification process, the nucleophilic attack of alcohol, after acidic site interaction of carbonyl oxygen of free fatty acid [75]. If the extracted oil consists higher amount of free fatty acid, the base cannot be used in this process. The presence of a base in higher free fatty acid content may lead to saponification [76]. Therefore, biodiesel produces via the transesterification process. Biodiesel production could be performed in the presence of a catalyst or without the use of a catalyst. The use of a catalyst is classified as alkali, acid, enzyme, or heterogeneous [76]. The use of the homogeneous catalysts may consume water for biodiesel purification [76]. The heterogeneous catalysts could be removed easily after the reaction. The MOFs are the most

TABLE 5.7

Summary of MOFs Use as Catalysis in Bio-Fuel Production

Catalytic reactions	MOFs Catalyst	Significance	Ref.
Biofuel upgrade	Pd/NPC-ZIF-8	Well-dispersed in water, electron-rich Pd sites, high surface area, hierarchical pores, favorable hydrophilicity	[77]
Model for biofuel upgrade reactions	Pd/PRGO/Ce MOF	Stable, uniform Pd nanoparticles dispersion, hydrogenolysis reaction promote by presence of acidic active sites.	[78]
Catalytic biofuel upgrade	Pd1Ni$_4$/C$_{16}$-CNC-3	High performance in the vanillin hydrodeoxygenation under mild conditions.	[79]
Upgrade of biofuels	Co$_8$Ni$_2$/NC nanoalloy catalyst	Excellent catalytic activity, high selectivity, better stability towards vanillin hydrodeoxygenation, synergetic effect between cobalt and nickel improved catalytic performance.	[80]
Biofuel production	UiO-66, UiO66(COOH)$_2$, and UiO-66(NH$_2$)	Active catalytic centers: uncoordinated COOH groups, defective sites on the cluster. higher surface area due to its open structure, better internal diffusion of the butyric acid	[81]
Biofuel upgrade with formic acid	Multilayered N-doped graphene (Co@NG)	Largest specific surface area, efficient catalysts, high surface area carbon skeleton, full exposure of catalytic active sites	[82]

promising characteristic features such as high surface area, modifiable pore size, tailorable functional groups, structural stability, etc. Two strategies could be used to catalyze biodiesel production, such as embedding acid, base and/or active constituents into the MOFs or in cooperating with the functional groups as secondary building units and linking active site via covalent bonds (Table 5.7).

It could be called the introduction of active constituent via impregnation or encapsulation process [76].

5.4.5 ESTERIFICATION

Ester is one of the abundant chemicals widely used in many applications, including pharmaceuticals, agriculture and food additives, etc. In conventional methods, the esters are prepared by using carboxylic acid or carboxylic acid derivatives. The multistep preparation methods yield undesired products. Further researchers are devoted to synthesis the products via benign and eco-friendly procedures for synthesis esters [83]. Oxidative direct oxidation of alcohol by molecular oxygen to yield esters is one of the most promising, sustainable, cost-effective, and green procedures for the synthesis of ester. Using heterogeneous catalytic oxidation of alcohol yield ester would be cost-effective, easy to separate catalyst, reusability of catalyst, etc. There are numerous researches done so far, synthesis of the ester by using heterogeneous catalysts. Some of the recently reported catalysts and the significance of such research works are tabulated in Table 5.8 as follows.

5.4.6 NO REDUCTION

NO_x is one of the air pollutants that evolve from industrial processes and the combustion of petroleum products, etc. It causes many issues such as greenhouse effect, acid rain, photochemical smog, etc. Selective catalytic reduction is one of the cost-effective methods for NO_x reduction. Nowadays, MOFs are synthesizing towards catalyst, storage, separation, etc. There is a lot of research that uses MOFs to reduce NO_x to nitrogen gases such as 5% MnO_x–ZnO, C_{ex}/Zr-CAU-24 (CAU stands for Christian-Albrechts-University, Kiel, Germany), Mn on a UiO-66, etc. [93–95].

In 2017, M. Zhang, et al. synthesized manganese-based catalysts on a UiO-66 carrier by in-situ deposition methods; 8.5 wt-% MnO_x loading shows the highest catalytic activity and reported that 90% NO conversion occurred [93]. Later in 2020, Y. Shi, et al. hydrothermally synthesized bimetallic Cu-Al-BTC MOFs with different ratios of Cu:Al content and observed complete NO conversion in the specific temperature range from 325°C to 400°C [96]. Porous cerium oxide clusters were synthesized by S. Smolders, et al. synthesized C_{ex}/Zr-CAU-24, and it shows catalytic reduction of $NO_{(g)}$ selectively in the presence of $CO_{(g)}$ [95]. In 2020, 5% MnOx–ZnO was used as a catalyst to reduce $NO_{(g)}$ [95]. The $NO_{(g)}$ reduces as $N_2O_{(g)}$ ferrous form of the MOF [97]. $Ni_{0.65}Mn_{0.35}$-MOF-74 catalyst reduced 100% of No at 175°C [98].

5.4.7 CO₂ METHANATION

The synthesis of CH_4 from CO_2 is called methanation. This reaction is thermo-dynamically favorable. However, in the kinetic perspective, the reaction is slow at low temperature, and the production cycle lengthen. In the MOFs, the immobilized metal nanoparticles and organic stunt confined the pores, prevents metal nano-particle agglomeration and shows the best photocatalytic process [99]. The CO_2 methanation process is tabulated in Table 5.9.

5.4.8 CO₂ FIXATION

Cyclic organic carbonates (COCs) are used as solvents and are used for several applications due to their attractive properties such as being stable, non-corrosive, and non-toxic. The conversion of CO_2 into carbonate called artificial CO_2 fixation. It uses in many industrial synthetic processes due to cost-effectiveness. Several types of catalyst are used in the cycloaddition reactions, including metal oxides and metal salts [99]. Although most of the catalyst required high temperature and

TABLE 5.8

Recent Publications Related to MOFs in Esterification Reactions

Catalytic reactions	MOFs Catalyst	Significance	Ref.
Esterification of amide	Zr-MOF-808-P	An efficient and versatile catalyst, applicable to primary, secondary and tertiary amides, reused for more than five consecutive cycles	[84]
Esterification	MIL-125(Ti) MOFs-derived nanoporous titanium dioxide–heteropoly acid (PW–TiO$_2$)	Excellent activity, high conversion, simply separated via centrifugation, reused for six cycles	[85]
Levulinic acid, ethanol esterification	Zr-containing UiO-66, UiO-66–NH$_2$ MOFs	Active catalyst. Catalytic activity depends on synthesis condition, particle size, and the concentration of defects. stable and reusable catalyst	[86]
Alcohols to esters direct oxidation with molecular oxygen	Co nanoparticles embedded N-doped graphite	Catalyze the oxidation of the aerobic alcohol in base-free conditions to esters at room temperature, catalyze, broad substrate scope: aromatic alcohols, aliphatic alcohols, diols, higher yields of ester	[83]
Esterification of ethanol	MIL-101(Cr)-PI, Mixed matrix membranes based on the polyimide Matrimid® (PI)	The use of synthesized porous nanostructured filler used to apply for separation element in a membrane reactor, increased in permeability than MOF, higher stability	[87]
Esterification reaction of palmitic acid	Zirconium(IV) based MOFs and H$_3$BTC linker (H3BTC= benzene-1,3,5-tricarboxylic acid)	High thermal stability, Zr-O bond: significant covalent character	[88]
Esterification	Cr$_3$(μ_3-O)(H$_2$O)$_3$(NDC(SO$_3$H$_{5/6}$)$_2$)$_3$ (BUT-8(Cr)-SO$_3$H, NDC(SO$_3$H)$_2$ $^{2-}$ = 4,8-disulfonaphthalene-2,6-dicarboxylatlate)	High chemical and thermal stability, excellent performance in various esterification reactions, homogeneously spread, high density Brønsted acidic –SO$_3$H catalytic active sites, good reusability	[89]
About oleic acid, methanol esterification	(MIL-101(Cr) @MBIAILs) where MBIAILs: 2-Mercaptobenzimidazole ILs	Superior catalytic behavior and excellent reusability	[90]

TABLE 5.8 (Continued)
Recent Publications Related to MOFs in Esterification Reactions

Catalytic reactions	MOFs Catalyst	Significance	Ref.
Acetic acid, isooctyl alcohol esterification	Acidic ionic liquid based catalysts: UiO-67–HSO$_4$, UiO-67–CF$_3$SO$_3$ and UiO-67–hifpOSO$_3$	Good catalytic activities, under optimized conditions shows maximum isooctyl alcohol conversion by UiO-67–CF$_3$SO$_3$, reusable five times without a significant decrease, excellent stability	[91]
Lauric acid, methanol esterification	HSiW-UiO-66	Good thermal stability, and high catalytic activity, excellent reusability, better textural properties	[92]

pressure to get better yield, it leads to increased production costs. MOFs are promising catalytic materials, widely used in many reactions, including CO_2 to COC conversion reactions. The defect sites of MOFs, nodes of MOFs, and linker molecules in MOFs are catalytic sites effectively involved in the catalytic process. The defects that exist in the MOFs produce some Lewis acid or Lewis base sites. The unsaturated catalytic sites in nodes of MOFs or the secondary building unit act as Lewis acid. Linkers in MOFs effectively bind the Lewis acid, Lewis base, and metal nodes and enhance the catalytic activities [99]. Some of the recent publications related to CO_2 fixation are tabulated in Table 5.10.

TABLE 5.9
MOFs Involved in CO$_2$ Methanation

Catalytic reactions	MOFs Catalyst	Significance	Ref.
CO$_2$ to CH$_4$ conversions	Ru impregnated UiO-66	Catalyst superiority, remarkable stability, and high selectivity	[100]
CO$_2$ to CH$_4$	Zr based UiO-66/Ni	Activated catalyst, better turnover frequency, high selectivity, high stability	
CO$_2$ to CH$_4$	Co/Al$_2$O$_3$/ ZIF-67	Highly-active catalyst, elevate photothemocatalytic performance good catalytic durability	[101]
CO$_2$ to CH$_4$	1 wt%Ru/UiO-66	Highly active catalyst	[102]
CO$_2$ to CH$_4$	Ni@MOF-5	Activity at low temperature, high stability	[103]
CO$_2$ to CH$_4$	Ru/Zr-MOF	Facilitated CO2 conversion, high selectivity,	[104]
	Ni-Co@CMOF-74	Best activity at moderate pressures, with stable performance	[105]

TABLE 5.10
MOFs Supported CO_2 Fixation

Catalytic reactions	MOFs catalyst	Significance	Ref.
CO_2 to cyclic carbonates	$(Cu_6(TATAB)_4(DABCO)_3 (H_2O)_3]\cdot24DMF)_n$	Recyclable catalytic activity	[106]
CO_2 into value-added chemicals	$(Cu_2[(C_{20}H_{12}N_2O_2)(COO)_4])_n$	High efficiency on CO_2 catalytic conversion and better size selectivity	[107]
CO_2 and epichlorohydrin	Zn-DPA$\cdot$2H$_2$O and Zn-DPA	Yield have shown a progressive increase	[108]
Carbon dioxide to epoxides	$Zn(atz)(bdc)_{0.5}$	High propylene epoxide conversion, selectivity, stability, defined pore system	[109]

5.5 MOFS AS ELECTROCATALYSTS

Redox processes of H_2O, H_2, and O_2 get great interest due to their applicability in many fields such as fuel cells, batteries and electrolysis of water [110]. The development of efficient catalytic reactions in a cost-effective way may provide the solution for the world facing global energy and environmental crises. The electrocatalytic process is used to prepare the artificial fuel via water splitting and it produces H_2 and O_2 gas. Further, the electrocatalytic process applicabe to the $CO_2(g)$ and the O_2 reduction reaction. It is reverse to the utilization of product fuel [2]. Electrocatalyst should be responsible for quick electron transport, mass diffusion, and surface reaction process. The heterogeneous gas-phase reactions show poor kinetics due to multi-step electron transfer and occur only in the triple-phase boundary region [111]. Many researches have been done so far to synthesize high-efficient and cost-effective electrocatalyst as an alternative for noble metal catalysts. MOFs are exceptional catalytic abilities due to their open molecular structure, high porosity and flexibility. However, the use of MOFs as an electrocatalytic process is hampered by poor conductivity and limited diffusion due to the pore size of MOFs. Table 5.11 shows some recent publications related to the use of MOFs as electrocatalytic process.

5.6 MOFS AS PHOTOCATALYST

MOFs photocatalyst got great interest among scientists owing to its attractive properties such as high surface area, easily tunable pore size, rich topology, good stability, easy synthesis methods, and low-cost process. MOFs-based photocatalysts could be used for environmental remediation, water splitting, oxygen evolution reactions, hydrogen evolution reactions, photocatalytic degradation, carbon dioxide reduction, etc. [118]. Several approaches have been done to enhance the photocatalytic abilities of MOFs, including ligand functionalization, metal-metal linker

TABLE 5.11
MOFs Functions as Electrocatalysts

Catalytic reactions	MOFs Catalyst	Significance	Ref.
Methanol oxidation reaction	Fe/Ni-MOFs	Low overpotentials, lower Tafel slope, high electrochemical stability, faster catalytic kinetics and higher conductivity	[112]
Methanol oxidation reaction	NiO-MOF	Best performance with highest current density, lower over-potential values and higher stability	[113]
Oxygen reduction	Mn-MOF-based@AC	Best performance, PGM-free catalysts	[114]
CH$_3$OH oxidation process	NiNH$_2$BDC MOF	Electrocatalytic activity	[115]
Methanol oxidation process	Pt/Ni(OH)$_2$/rGO ternary hybrids	Unprecedented performances	[116]
Methanol electrooxidation	Nanohybrid NiCr-LDH (layered double hydroxide)	Higher electrocatalytic activity	[117]

modification, dye sensitization and semiconductor coupling [119]. Based on the conventional metal oxide-based photocatalyst, the electron excited from valence band to conduction band under the illumination condition and leave a hole in the valence band. Excited electrons come to the surface of metal oxides, and similarly, electron transfer ensues in MOFs. In the MOFs, the electron in the highest occupied molecular orbital (HOMO) excited to lowest unoccupied molecular orbital (LUMO). The excited electron transferred to O$_2$ and creates $O_2^{-\cdot}$. The hole in HOMO oxidized the surface hydroxide group and water and produce $OH\cdot$. Due to the presence of those radicals' organic pollutants could be degraded.

There are several MOFs were showed semiconductor-like behavior, such as MOF-5, MIL-125, and UiO-66. A large number of choices of metal ions, clusters, and organic stunts in MOFs have endowed leads to the vast number of MOFs with tunable photocatalyst to harvest a wide range of the solar spectrum. The stable MOFs could be formed between hard or soft Lewis base and metals in which the high-valence metal ions form a stable bond with hard Lewis base (carboxylates). The divalent metal ions form stable MOFs with soft Lewis bases (Azolates) [119]. The details of MOF photocatalysts are summarized in the Table 5.12.

5.7 CONCLUSION AND OUTLOOKS

The study on MOF materials with intrinsic catalytic activity has become a hot topic and reached great progress over the past few years. MOFs have a higher surface area, porosity, great topology variations, defects associated with catalytic properties, and tuneable diversity of properties. This chapter summarizes MOFs' recent

TABLE 5.12

Some Recent Reports Related to MOFs as Photocatalyst

Catalytic reactions	MOFs catalyst	Significance	Ref.
Hydrogenation of olefins	Ti-MOFs@Pt@DM-LZU1	Pt facilitates the charge separation, DM-LZU1 shell leads to enrichment of reactant, high photocatalytic activity	[120]
Methylene blue, rhodamine B, and methyl orange degradation	TiO_2/Cu-MOF/PPy	Enhanced photocatalytic activities under visible light, high charge separation, carrier mobility	[121]
Degradation of rhodamine-b	La-MOFs	Enhance photocatalyst	[122]
Organic pollutant degradation and hydrogen evolution	ZIF-8@ZIF-67 MOFs	Higher light absorption, higher charge separation capabilities, and stability	[123]
Degradation of rhodamine B and methyl orange	$[Zn(L)(H_2O)]\cdot H_2O$ in which H_2L is 4-(pyridine-4-yl) phthalic acid	Higher catalytic activity and reusability	[124]
Methylene blue degradation	$[Cd_2(dda)(bibp)_2]\cdot H_2O$ (YAU-10, H_4dda = 3,3,5,5-azodibenzoic acid, bibp = 4,4'-di(1H-imidazol-1-yl)-1,1'-biphenyl)	Remarkable high, visible-light catalytic performance	[125]

progress as heterogeneous catalysts, including post-treatments on MOF, nanomaterials in MOFs, and catalytic applications of MOFs. The broad spectrum of catalytic processes associated with MOFs and the potential for applications is endless. Further, the great understanding of topology, defects associated with variation in properties, tunable pore size, incorporation of nanomaterials, and associated studies via characterization and molecular modeling leads to the discovery towards the novel applications of MOFs.

REFERENCES

[1] Lammert, M., et al., *Cerium-based metal organic frameworks with UiO-66 architecture: Synthesis, properties and redox catalytic activity.* Chemical Communications, 2015. **51**(63): p. 12578–12581.

[2] Bavykina, A., et al., *Metal-organic frameworks in heterogeneous catalysis: Recent progress, new trends, and future perspectives.* Chemical Reviews, 2020. **120**(16): p. 8468–8535.

[3] Safaei, M., et al., *A review on metal-organic frameworks: Synthesis and applications.* TrAC Trends in Analytical Chemistry, 2019. **118**: p. 401–425.

[4] Fleker, O., et al., *Preparation and properties of metal organic framework/activated carbon composite materials.* Langmuir, 2016. **32**(19): p. 4935–4944.

[5] Vakili, R., et al., *PtNi bimetallic structure supported on UiO-67 metal-organic framework (MOF) during CO oxidation.* Journal of Catalysis, 2020. **391**: p. 522–529.

[6] Altass, H.M., et al., *Enhanced catalytic activity for CO oxidation by highly active Pd nanoparticles supported on reduced graphene oxide/copper metal organic framework.* Journal of the Taiwan Institute of Chemical Engineers, 2021. **128**: p. 194–208.

[7] Lozano, L.A., et al., *Metal–organic framework-based sustainable nanocatalysts for CO oxidation.* Journal of Nanomaterials, 2020. **10**(1): p. 165.

[8] Lee, H.K., et al., *Applying a nanoparticle@MOF interface to activate an unconventional regioselectivity of an inert reaction at ambient conditions.* Journal of the American Chemical Society, 2020. **142**(26): p. 11521–11527.

[9] Lin, A., et al., *Palladium nanoparticles supported on ce-metal–organic framework for efficient CO oxidation and low-temperature CO_2 capture.* ACS Applied Materials & Interfaces, 2017. **9**(21): p. 17961–17968.

[10] Nagarkar, S.S. and Ghosh, S.K. *Reversible structural transformations in a Co (II)-based 2D dynamic metal-organic framework showing selective solvent uptake.* Journal of Chemical Sciences 2015. **127**(4): p. 627–633.

[11] Kim, T., et al., *Low-temperature CO oxidation using a metal organic framework with unsaturated Co_2+ sites.* Polyhedron, 2015. **90**: p. 18–22.

[12] Kökçam-Demir, Ü., et al., *Coordinatively unsaturated metal sites (open metal sites) in metal–organic frameworks: Design and applications.* Chemical Society Reviews, 2020. **49**(9): p. 2751–2798.

[13] Zhu, L., et al., *Metal–organic frameworks for heterogeneous basic catalysis.* Chemical Reviews, 2017. **117**(12): p. 8129–8176.

[14] Kim, H.K., et al., *A chemical route to activation of open metal sites in the copper-based metal–organic framework materials HKUST-1 and Cu-MOF-2.* Journal of the American Chemical Society, 2015. **137**(31): p. 10009–10015.

[15] Almáši, M., V. Zeleňák, and A. Zeleňáková, *Magnetic and structural studies into the effect of solvent exchange process in metal-organic framework MOF-76 (Gd).* Acta Physica Polonica A, 2017. **131**: p. 991–993.

[16] Espín, J., et al., *Photothermal activation of metal–organic frameworks using a UV–Vis Light Source.* ACS Applied Materials & Interfaces, 2018. **10**(11): p. 9555–9562.

[17] Zhang, J., et al., *Thermally activated construction of open metal sites on a Zn-organic framework: An effective strategy to enhance Lewis acid properties and catalytic performance for CO_2 cycloaddition reactions.* Applied Surface Science, 2022. **572**: p. 151408.

[18] Liang, X., et al., *The activation of Co-MOF-74 with open metal sites and their corresponding CO/N2 adsorptive separation performance.* Microporous and Mesoporous Materials, 2021. **320**: p. 111109.

[19] Villajos, J.A., et al., *Increasing exposed metal site accessibility in a Co-MOF-74 material with induced structure-defects.* Frontiers in Materials, 2019. **6**: p. 230.

[20] Moustafa, M., et al., *Thermal Decomposition and Kinetic Analysis of a Mil-88a Metal-organic Framework.* Research square, 2021.

[21] Yang, J., et al., *Effects of activation temperature and densification on adsorption performance of MOF MIL-100 (Cr).* Journal of Chemical Engineering Data, 2019. **64**(12): p. 5814–5823.

[22] Tanabe, K.K., C.A. Allen, and S.M. Cohen, *Photochemical activation of a metal–organic framework to reveal functionality.* Angewandte Chemie, International Edition, 2010. **49**(50): p. 9730–9733.

[23] Wang, J., et al., *Engineering effective structural defects of metal–organic frameworks to enhance their catalytic performances.* Journal of Materials Chemistry A, 2020. **8**(8): p. 4464–4472.

[24] Yoskamtorn, T., et al., *Responses of defect-rich Zr-based metal–organic frameworks toward NH3 adsorption.* Journal of the American Chemical Society, 2021. **143**(8): p. 3205–3218.

[25] Fang, Z., et al., *Structural complexity in metal–organic frameworks: Simultaneous modification of open metal sites and hierarchical porosity by systematic doping with defective linkers.* Journal of the American Chemical Society, 2014. **136**(27): p. 9627–9636.

[26] Wu, H., et al., *Unusual and highly tunable missing-linker defects in zirconium metal–organic framework UiO-66 and their important effects on gas adsorption.* Journal of the American Chemical Society, 2013. **135**(28): p. 10525–10532.

[27] Liang, W., et al., *Linking defects, hierarchical porosity generation and desalination performance in metal–organic frameworks.* Chemical Science, 2018. **9**(14): p. 3508–3516.

[28] Ma, X., et al., *Switching on the photocatalysis of metal–organic frameworks by engineering structural defects.* Angewandte Chemie, International Edition, 2019. **58**(35): p. 12175–12179.

[29] Kalaj, M. and S.M. Cohen, *Postsynthetic modification: An enabling technology for the advancement of metal–organic frameworks.* ACS Central Science, 2020. **6**(7): p. 1046–1057.

[30] Dalapati, R., S. Nandi, and S. Biswas, *Post-synthetic modification of a metal–organic framework with a chemodosimeter for the rapid detection of lethal cyanide via dual emission.* Dalton Transactions, 2020. **49**(25): p. 8684–8692.

[31] Rui, Z., J.B. James, and Y.S. Lin, *Highly CO_2 perm-selective metal-organic framework membranes through CO_2 annealing post-treatment.* Journal of Membrane Science, 2018. **555**: p. 97–104.

[32] Dalapati, R. and S. Biswas, *Post-synthetic modification of a metal-organic framework with fluorescent-tag for dual naked-eye sensing in aqueous medium.* Sensors and Actuators B: Chemical, 2017. **239**: p. 759–767.

[33] Zamani, S., A. Abbasi, and M. Masteri-Farahani, *Post-synthetic modification of porous [Cu3(BTC)2] (BTC = benzene-1,3,5-tricarboxylate) metal organic framework with molybdenum and vanadium complexes for the epoxidation of olefins and allyl alcohols.* Reaction Kinetics, Mechanisms and Catalysis, 2021. **132**(1): p. 235–250.

[34] Liu, F., et al., *Dilute sulfonic acid post functionalized metal organic framework as a heterogeneous acid catalyst for esterification to produce biodiesel.* Fuel, 2020. **266**: p. 117149.

[35.] Cai, X., et al., *Nano-sized metal-organic frameworks: Synthesis and applications.* Coordination Chemistry Reviews, 2020. **417**: p. 213366.

[36] Mian, M.R., et al., *Precise control of Cu nanoparticle size and catalytic activity through pore templating in Zr metal–organic frameworks.* Chemistry of Materials, 2020. **32**(7): p. 3078–3086.

[37] Wang, S., et al., *Encapsulation of metal oxide nanoparticles inside metal-organic frameworks via surfactant-assisted nanoconfined space.* Nanotechnology, 2020. **31**(25): p. 255604.

[38] Chen, Y.-Z., et al., *Location determination of metal nanoparticles relative to a metal-organic framework.* Nature Communications, 2019. **10**(1): p. 3462.

[39] Chen, D., et al., *Boosting catalysis of Pd nanoparticles in MOFs by pore wall engineering: The roles of electron transfer and adsorption energy.* Advanced Materials, 2020. **32**(30): p. 2000041.

[40] Wu, R., et al., *Highly dispersed Au nanoparticles immobilized on Zr-based metal–organic frameworks as heterostructured catalyst for CO oxidation.* Journal of Materials Chemistry A, 2013. **1**(45): p. 14294–14299.

[41] Nabi, S., et al., *Au-nanoparticle loaded nickel-copper bimetallic MOF: An excellent catalyst for chemical degradation of Rhodamine B.* Inorganic Chemistry Communications, 2020. **117**: p. 107949.

[42] Liu, H., et al., *Metal–organic framework supported gold nanoparticles as a highly active heterogeneous catalyst for aerobic oxidation of alcohols.* The Journal of Physical Chemistry C, 2010. **114**(31): p. 13362–13369.

[43] Guo, S., et al., *Isolated atomic catalysts encapsulated in MOF for ultrafast water pollutant treatment.* Nano Research, 2021. **14**(5): p. 1287–1293.

[44] Liu, G.-F., et al., *Aluminum metal–organic framework–silver nanoparticle composites for catalytic reduction of nitrophenols.* ACS Applied Nano Materials, 2020. **3**(11): p. 11426–11433.

[45] Chang, S., et al., *Fe₃O₄ nanoparticles coated with Ag-nanoparticle-embedded metal–organic framework MIL-100(Fe) for the catalytic reduction of 4-nitrophenol.* ACS Applied Nano Materials, 2020. **3**(3): p. 2302–2309.

[46] Tombesi, A. and C. Pettinari, *Metal organic frameworks as heterogeneous catalysts in olefin epoxidation and carbon dioxide cycloaddition.* Inorganics, 2021. **9**(11): p. 81.

[47] Jia, G., et al., *Metal-organic frameworks as heterogeneous catalysts for electrocatalytic oxidative carbonylation of methanol to dimethyl carbonate.* Electrochemistry Communications, 2013. **34**: p. 211–214.

[48] Zhao, Y., et al., *CO catalytic oxidation by a metal organic framework containing high density of reactive copper sites.* Chemical communications (Cambridge, England), 2011. **47**: p. 6377–6379.

[49] Ye, J.-Y. and C.-J. Liu, *Cu₃(BTC)₂: CO oxidation over MOF based catalysts.* Chemical Communications, 2011. **47**(7): p. 2167–2169.

[50] Kim, J.Y., et al., *In situ-generated metal oxide catalyst during CO oxidation reaction transformed from redox-active metal-organic framework-supported palladium nanoparticles.* Nanoscale Research Letters, 2012. **7**(1): p. 461.

[51] Zhang, C., et al., *Metal organic framework-derived Co₃O₄ microcubes and their catalytic applications in CO oxidation.* New Journal of Chemistry, 2017. **41**(4): p. 1631–1636.

[52] Zhang, X., et al., *Thermal activation of CuBTC MOF for CO oxidation: The effect of activation atmosphere.* Journal of Catalysis, 2017. **7**(4): p. 106.

[53] Carson, F., et al., *Ruthenium complexation in an aluminium metal-organic framework and its application in alcohol oxidation catalysis.* Chemistry—A European Journal, 2012. **18**(48): p. 15337–15344.

[54] Babahydari, A.K., et al., *Heterogeneous oxidation of alcohols with hydrogen peroxide catalyzed by polyoxometalate metal–organic framework.* Journal of the Iranian Chemical Society, 2016. **13**(8): p. 1463–1470.

[55] Kim, B.R., et al., *Robust aerobic alcohol oxidation catalyst derived from metal–organic frameworks.* Catalysis Letters, 2016. **146**(4): p. 734–743.

[56] Peng, L., et al., *Oxidation of benzyl alcohol over metal organic frameworks M-BTC (M = Co, Cu, Fe).* New Journal of Chemistry, 2017. **41**(8): p. 2891–2894.

[57] Taher, A., D.W. Kim, and I.-M. Lee, *Highly efficient metal organic framework (MOF)-based copper catalysts for the base-free aerobic oxidation of various alcohols.* RSC Advances, 2017. **7**(29): p. 17806–17812.

[58] Jumpathong, W., et al., *Exploitation of missing linker in Zr-based metal-organic framework as the catalyst support for selective oxidation of benzyl alcohol.* APL Materials, 2019. **7**(11): p. 111109.

[59] Wang, X., et al., *Vanadium catalyst on isostructural transition metal, lanthanide, and actinide based metal–organic frameworks for alcohol oxidation.* Journal of the American Chemical Society, 2019. **141**(20): p. 8306–8314.

[60] Chen, Y., et al., *Insights into the structure–activity relationship in aerobic alcohol oxidation over a metal-organic-framework-supported Molybdenum(VI) catalyst.* Journal of the American Chemical Society, 2021. **143**(11): p. 4302–4310.

[61] Sen, R., et al., *Heterogeneous catalytic epoxidation of olefins over hydrothermally synthesized lanthanide containing framework compounds.* European Journal of Inorganic Chemistry, 2011. **2011**: p. 241–248.

[62] Beier, M.J., et al., *Aerobic epoxidation of olefins catalyzed by the cobalt-based metal–organic framework STA-12(Co).* Journal name: Chemistry–A European Journal, 2012. **18**(3): p. 887–898.

[63] Sun, J., et al., *Epoxidation of styrene over Fe(Cr)-MIL-101 metal–organic frameworks.* RSC Advances, 2014. **4**(72): p. 38048–38054.

[64] Tang, J., et al., *Efficient molybdenum(vi) modified Zr-MOF catalysts for epoxidation of olefins.* RSC Advances, 2014. **4**(81): p. 42977–42982.

[65] Brown, J.W., et al., *Epoxidation of alkenes with molecular oxygen catalyzed by a manganese porphyrin-based metal–organic framework.* Catalysis Communications, 2015. **59**: p. 50–54.

[66] Kardanpour, R., et al., *Efficient alkene epoxidation catalyzed by molybdenyl acetylacetonate supported on aminated UiO-66 metal–organic framework.* Journal of Solid State Chemistry, 2015. **226**: p. 262–272.

[67] Ahn, S., et al., *Stable metal–organic framework-supported niobium catalysts.* Inorganic Chemistry, 2016. **55**(22): p. 11954–11961.

[68] Hui, J., et al., *Multicomponent metal–organic framework derivatives for optimizing the selective catalytic performance of styrene epoxidation reaction.* Nanoscale, 2018. **10**(18): p. 8772–8778.

[69] Stubbs, A.W., et al., *Selective catalytic olefin epoxidation with MnII-exchanged MOF-5.* ACS Catalysis, 2018. **8**(1): p. 596–601.

[70] Berijani, K., A. Morsali, and J.T. Hupp, *An effective strategy for creating asymmetric MOFs for chirality induction: A chiral Zr-based MOF for enantioselective epoxidation.* Catalysis Science & Technology, 2019. **9**(13): p. 3388–3397.

[71] Ganina, O.G., et al., *Cu-MOF-catalyzed carboxylation of alkynes and epoxides.* Russian Journal of Organic Chemistry, 2019. **55**(12): p. 1813–1820.

[72] Huang, K., et al., *One-pot synthesis of bimetal MOFs as highly efficient catalysts for selective oxidation of styrene.* Journal of Chemical Sciences, 2020. **132**(1): p. 1–10.

[73] Wang, F., et al., *A highly efficient heterogeneous catalyst of bimetal-organic frameworks for the epoxidation of olefin with H_2O_2.* Molecules, 2020. **25**(10): p. 2389.

[74] Kolilananthan, S., A. Manjceevan, and S. Rasalingam, *Extraction of mahua (Madhuca indica) and veralu (Elaeocarpus serratus) oils: A study on their potential use in biodiesel production.* ASRS - FAS 2018, 2018. **Faculty of Applied Science, South Eastern University of Sri Lanka.**

[75] Cong, W.-J., et al., *Metal–organic framework-based functional catalytic materials for biodiesel production: A review.* Green Chemistry, 2021. **23**(7): p. 2595–2618.

[76] Mishra, V.K. and R. Goswami, *A review of production, properties and advantages of biodiesel.* Biofuels, 2018. **9**(2): p. 273–289.

[77] Chen, Y.-Z., et al., *Palladium nanoparticles stabilized with N-doped porous carbons derived from metal–organic frameworks for selective catalysis in biofuel upgrade: The role of catalyst wettability.* Green Chemistry, 2016. **18**(5): p. 1212–1217.

[78] Ibrahim, A.A., et al., *Palladium nanoparticles supported on a metal–organic framework-partially reduced graphene oxide hybrid for the catalytic hydrodeoxygenation of vanillin as a model for biofuel upgrade reactions.* ChemCatChem, 2017. 9(3): p. 469–480.

[79] Li, D.-D., et al., *Amphiphilic cellulose supported PdNi alloy nanoparticles towards biofuel upgrade under mild conditions.* Catalysis Communications, 2019. **122**: p. 43–46.

[80] Zhai, Y., et al., *Bimetal Co₈Ni₂ catalyst supported on chitin-derived N-containing carbon for upgrade of biofuels.* Applied Surface Science, 2020. **506**: p. 144681.

[81] Jrad, A., et al., *Structural engineering of Zr-based metal-organic framework catalysts for optimized biofuel additives production.* Chemical Engineering Journal, 2020. **382**: p. 122793.

[82] Zhou, S., et al., *Scale-up biopolymer-chelated fabrication of cobalt nanoparticles encapsulated in N-enriched graphene shells for biofuel upgrade with formic acid.* Green Chemistry, 2019. **21**(17): p. 4732–4747.

[83] Zhong, W., et al., *Base-free oxidation of alcohols to esters at room temperature and atmospheric conditions using nanoscale co-based catalysts.* ACS Catalysis, 2015. **5**(3): p. 1850–1856.

[84] Villoria-del-Álamo, B., et al., *Zr-MOF-808 as catalyst for amide esterification.* Chemistry—A European Journal, 2021. **27**(14): p. 4588–4598.

[85] Zhang, Q., et al., *Metal–organic framework-derived nanoporous titanium dioxide–heteropoly acid composites and its application in esterification.* Green Processing and Synthesis, 2021. **10**(1): p. 284–294.

[86] Cirujano, F., A. Corma, and F.L. i Xamena, *Conversion of levulinic acid into chemicals: Synthesis of biomass derived levulinate esters over Zr-containing MOFs.* Chemical Engineering Science, 2015. **124**: p. 52–60.

[87] de la Iglesia, Ó., et al., *Metal-organic framework MIL-101(Cr) based mixed matrix membranes for esterification of ethanol and acetic acid in a membrane reactor.* Renewable Energy, 2016. **88**: p. 12–19.

[88] Larasati, I., et al., *Synthesis of metal-organic frameworks based on Zr4+ and benzene 1, 3, 5-tricarboxylate linker as heterogeneous catalyst in the esterification reaction of palmitic acid.* in *IOP conference series: Materials science and engineering.* 2017. IOP Publishing.

[89] Dou, Y., et al., *Highly efficient catalytic esterification in an –SO₃H-functionalized Cr(III)-MOF.* Industrial & Engineering Chemistry Research, 2018. **57**(25): p. 8388–8395.

[90] Han, M., et al., *Immobilization of thiol-functionalized ionic liquids onto the surface of MIL-101(Cr) frameworks by SCr coordination bond for biodiesel production.* Colloids and Surfaces A: Physicochemical and Engineering Aspects, 2018. **553**: p. 593–600.

[91] Xu, Z., et al., *Acidic ionic liquid based UiO-67 type MOFs: A stable and efficient heterogeneous catalyst for esterification.* RSC Advances, 2018. **8**(18): p. 10009–10016.

[92] Zhang, Q., et al., *Heteropoly acid-encapsulated metal–organic framework as a stable and highly efficient nanocatalyst for esterification reaction.* RSC Advances, 2019. **9**(29): p. 16357–16365.

[93] Zhang, M., et al., *Metal-organic framework loaded manganese oxides as efficient catalysts for low-temperature selective catalytic reduction of NO with NH3.* Frontiers of Chemical Science and Engineering, 2017. **11**(4): p. 594–602.

[94] Zhao, L., et al., *Improved NO reduction by using metal–organic framework derived MnOx–ZnO.* RSC Advances, 2020. **10**(53): p. 31780–31787.

[95] Smolders, S., et al., *Selective catalytic reduction of NO by cerium-based metal–organic frameworks.* Catalysis Science & Technology, 2020. **10**(2): p. 337–341.

[96] Shi, Y., et al., *Synthesis and characterization of bimetallic Cu-Al-BTC MOFs as an efficient catalyst for selective catalysis reduction of NO with CO.* Ferroelectrics, 2020. **565**(1): p. 58–65.

[97] Cai, Z., et al., *Direct NO reduction by a biomimetic iron(II) pyrazolate MOF.* Angewandte Chemie, International Edition, 2021. **60**(39): p. 21221–21225.

[98] Shi, Y., et al., *A new type bimetallic NiMn-MOF-74 as an efficient low-temperatures catalyst for selective catalytic reduction of NO by CO.* Chemical Engineering and Processing - Process Intensification, 2021. **159**: p. 108232.

[99] Cui, W.-G., et al., *Metal-organic framework-based heterogeneous catalysts for the conversion of C_1 chemistry: CO, CO_2 and CH_4.* Coordination Chemistry Reviews, 2019. **387**: p. 79–120.

[100] Lippi, R., et al., *Highly active catalyst for CO_2 methanation derived from a metal organic framework template.* Journal of Materials Chemistry A, 2017. **5**(25): p. 12990–12997.

[101] Chen, X., et al., *MOF-templated preparation of highly dispersed Co/Al_2O_3 composite as the photothermal catalyst with high solar-to-fuel efficiency for CO_2 methanation.* ACS Applied Materials & Interfaces, 2020. **12**(35): p. 39304–39317.

[102] Lippi, R., et al., *Unveiling the structural transitions during activation of a CO_2 methanation catalyst $Ru0/ZrO_2$ synthesised from a MOF precursor.* Catalysis Today, 2021. **368**: p. 66–77.

[103] Zhen, W., et al., *Enhancing catalytic activity and stability for CO_2 methanation on Ni@MOF-5 via control of active species dispersion.* Chemical Communications, 2015. **51**(9): p. 1728–1731.

[104] Xu, W., et al., *Plasma-assisted Ru/Zr-MOF catalyst for hydrogenation of CO_2 to methane.* Plasma Science and Technology, 2019. **21**(4): p. 044004.

[105] Khan, I.S., et al., *Bimetallic metal-organic framework mediated synthesis of Ni-Co catalysts for the dry reforming of methane.* Catalysts, 2020. **10**(5): p. 592.

[106] Dhankhar, S.S., et al., *Chemical fixation of CO_2 under solvent and co-catalyst-free conditions using a highly porous two-fold interpenetrated Cu(II)-metal–organic framework.* Crystal Growth & Design, 2021. **21**(2): p. 1233–1241.

[107] Li, P.-Z., et al., *Highly effective carbon fixation via catalytic conversion of CO_2 by an acylamide-containing metal–organic framework.* Chemistry of Materials, 2017. **29**(21): p. 9256–9261.

[108] Musa, S., et al., *Application of POMOF composites for CO_2 fixation into cyclic carbonates.* in *IOP Conference Series: Earth and Environmental Science.* 2021. IOP Publishing.

[109] Luo, Z., et al., *A stable Zn-based metal–organic framework as an efficient catalyst for carbon dioxide cycloaddition and alcoholysis at mild conditions.* Catalysis Letters, 2020. **150**(5): p. 1408–1417.

[110] Alhumaimess, M.S., *Metal–organic frameworks and their catalytic applications.* Journal of Saudi Chemical Society, 2020. **24**(6): p. 461–473.

[111] Tang, C., H.-F. Wang, and Q. Zhang, *Multiscale principles to boost reactivity in gas-involving energy electrocatalysis.* Accounts of Chemical Research, 2018. **51**(4): p. 881–889.

[112] Zheng, F., et al., *Fe/Ni bimetal organic framework as efficient oxygen evolution catalyst with low overpotential.* Journal of Colloid and Interface Science, 2019. **555**: p. 541–547.

[113] Hanif, S., et al., *Electrocatalytic study of NiO-MOF with activated carbon composites for methanol oxidation reaction.* Scientific Reports, 2021. **11**(1): p. 17192.

[114] Gonen, S., et al., *Metal organic frameworks as a catalyst for oxygen reduction: An unexpected outcome of a highly active Mn-MOF-based catalyst incorporated in activated carbon.* Nanoscale, 2018. **10**(20): p. 9634–9641.

[115] Yaqoob, L., et al., *Electrocatalytic performance of NiNH2BDC MOF based composites with rGO for methanol oxidation reaction.* Scientific Reports, 2021. **11**(1): p. 13402.

[116] Huang, W., et al., *Highly active and durable methanol oxidation electrocatalyst based on the synergy of platinum–nickel hydroxide–graphene.* Nature Communications, 2015. **6**(1): p. 10035.

[117] Gamil, S., et al., *Nanohybrid layered double hydroxide materials as efficient catalysts for methanol electrooxidation.* RSC Advances, 2019. **9**(24): p. 13503–13514.

[118] Qian, Y., F. Zhang, and H. Pang, *A review of MOFs and their composites-based photocatalysts: Synthesis and applications.* Advanced Functional Materials, 2021. **31**(37): p. 2104231.

[119] Wang, Q., et al., *Recent advances in MOF-based photocatalysis: Environmental remediation under visible light.* Inorganic Chemistry Frontiers, 2020. **7**(2): p. 300–339.

[120] Sun, D. and D.-P. Kim, *Hydrophobic MOFs@Metal Nanoparticles@COFs for interfacially confined photocatalysis with high efficiency.* ACS Applied Materials & Interfaces, 2020. **12**(18): p. 20589–20595.

[121] Liu, G., et al., *TiO2/Cu-MOF/PPy composite as a novel photocatalyst for decomposition of organic dyes.* Journal of Materials Science: Materials in Electronics, 2021. **32**(4): p. 4097–4109.

[122] Buhori, A., A. Zulys, and J. Gunlazuardi. *Synthesis of lanthanum metal-organic frameworks (La-MOFs) as degradation photocatalyst of Rhodamine-B.* in *AIP Conference Proceedings.* 2020. AIP Publishing LLC.

[123] Huang, Z., et al., *Stable core–shell ZIF-8@ZIF-67 MOFs photocatalyst for highly efficient degradation of organic pollutant and hydrogen evolution.* Journal of Materials Research, 2021. **36**(3): p. 602–614.

[124] Dong, J.-P., et al., *Synthesis of a novel 2D zinc(ii) metal–organic framework for photocatalytic degradation of organic dyes in water.* Dalton Transactions, 2019. **48**(47): p. 17626–17632.

[125] Zhang, F.-F., et al., *Visible light-assisted photocatalytic degradation of methylene blue in water by highly chemically stable Cd-coordination polymers at room temperature.* New Journal of Chemistry, 2021. **45**(42): p. 19660–19665.

6 Metal-Organic Frameworks as Chemical Sensors for Detection of Environmental Pollutants

Nidhi Goel

CONTENTS

DOI: 10.1201/9781003252061-6

6.1   INTRODUCTION

Due to global urbanization and industrial establishment, the contamination in environment has become a thoughtful issue for the human beings as well as our ecosystem [1]. According to GAHP (Global Alliance on Health and Pollution), pollution is responsible for 21% of all deaths globally as it causes several incurable diseases such as cancer, immunologic abnormalities, organ damage and malformation, and the conditions in coming years will be worse. There is various causes of pollution, but the chemical sources such as heavy metal ions, radioactive ions, organic pollutants, toxic gases and volatile organic compounds are highly toxic and key sources of environment contamination [2,3]. Although, disparate analytical techniques such as gas chromatography (GC), high-performance liquid chromatography (HPLC), capillary electrophoresis (CE), ultraviolet spectrophotometry (UV), Fourier-transform infrared spectroscopy (FTIR), near infrared (NIR) spectroscopy, surface-enhanced Raman spectroscopy (SERS) have been used for the identification of environmental pollutants [4–8], but there are numerous disadvantages like the requirement of proficient manpower, sophisticated and expensive instruments, lack of sensitivity, long test times and high cost. Hence, it is necessary to design novel sensing methodologies to detect toxic and dangerous pollutants in an accurate way. Metal-organic frameworks (MOFs) are an extensive family of crystalline and porous materials, and exhibit exclusive applications in diverse field. Owing to their versatile structures, great surface area, high stability, porosity as well as controlled functionality [9,10], fluorescent MOFs have become the promising material for treating an extensive range of pollutants (Figure 6.1) [11].

They have been fabricated by the use of d-, f- block and main group metal ions/ metal clusters as well as organic linkers [12]. Fluorescent MOFs can be categorised into transition metal-MOFs, lanthanide-MOFs (Ln-MOFs), heterometal-organic frameworks (H-MOFs), and main group-MOFs [12]. Lately, several transition metal ions have been noted to formulate the fluorescent MOFs because they have the tendency to readily bound with organic linkers and exhibit a variety of coordination numbers as well as geometries, which help in enhancing the luminescent properties of created MOF [13]. Ln-MOFs have also attracted huge consideration due to the exclusive luminescent properties of lanthanide ions as well as their inherent porosity [14]. The mixed Ln-MOFs and co-doped Ln-MOFs produce the different luminescence properties [15,16]. The luminescence properties in Ln-MOFs are influenced from both Ln^{3+} and organic linkers, where the organic linkers act as a sensitizer for lanthanide ions [17]. Therefore, Ln-MOFs are favorable for chemical sensing [18]. In the preparation of H-MOFs, heterometallic ions are used to obtain the alluring structures. The hetero metals in MOFs are also responsible for the change of energy

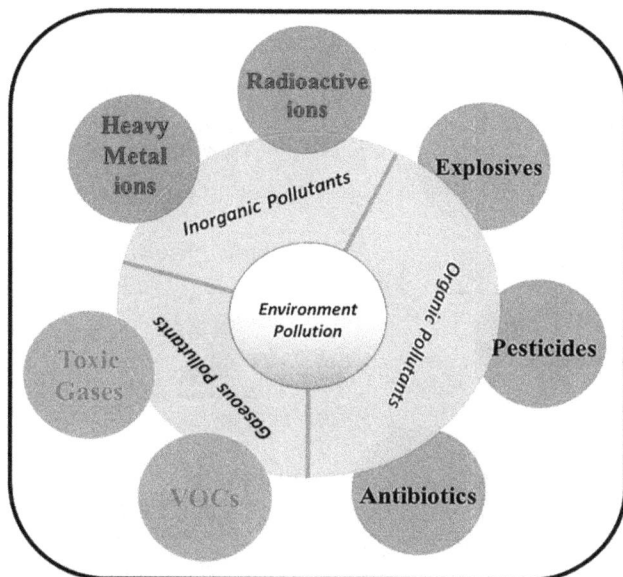

FIGURE 6.1   Schematic presentation of environment pollution.

levels of MOFs, which appreciate the luminescence properties of H-MOFs [19]. Hence, H-MOFs have a prominent place in the sensing of wide range of pollutants. Main group metal ions are also used for the preparation of fluorescent MOFs, but their numbers are less compared to transition metal-MOFs, Ln-MOFs and H-MOFs [20]. Customarily, MOFs achieve the luminescence properties due to the presence of conjugated organic moieties, metal ions/clusters, and the guests enclosed by MOFs [21]. Different mechanisms are responsible for the origin of photoluminescence such as photoinduced electron transfer (PET), intramolecular charge transfer (ICT), and Förster energy transfer (FRET) (Figure 6.2) [22].

In PET, the fluorochrome is associated to a receptor through the spacer, which has the capability to connect with analytes and generate the luminescent sensing, while in the case of ICT, MOFs having electron withdrawing and electron donating groups interact with analytes, which develop the shifts in luminescent intensity. FRET, also known as resonance energy transfer (RET) is an optical process, where the excess energy is transferred from donor to acceptor molecule, and show the excellent luminescence [23–25]. All of these mechanisms may govern the fabrication of MOF-based chemical sensors for detection of environmental pollutants [26]. The present chapter provides extensive vision for the readers and introduces the current research status of MOF-based chemical sensors for environmental pollutants such as inorganic, organic and gaseous pollutants.

6.2   MOF-BASED SENSORS FOR INORGANIC POLLUTANTS

Various inorganic pollutants (IPs) compile day by day and persist in an environment for a long time. Expanding industrialization, immediately developing agriculture

(a)

(b)   (c)

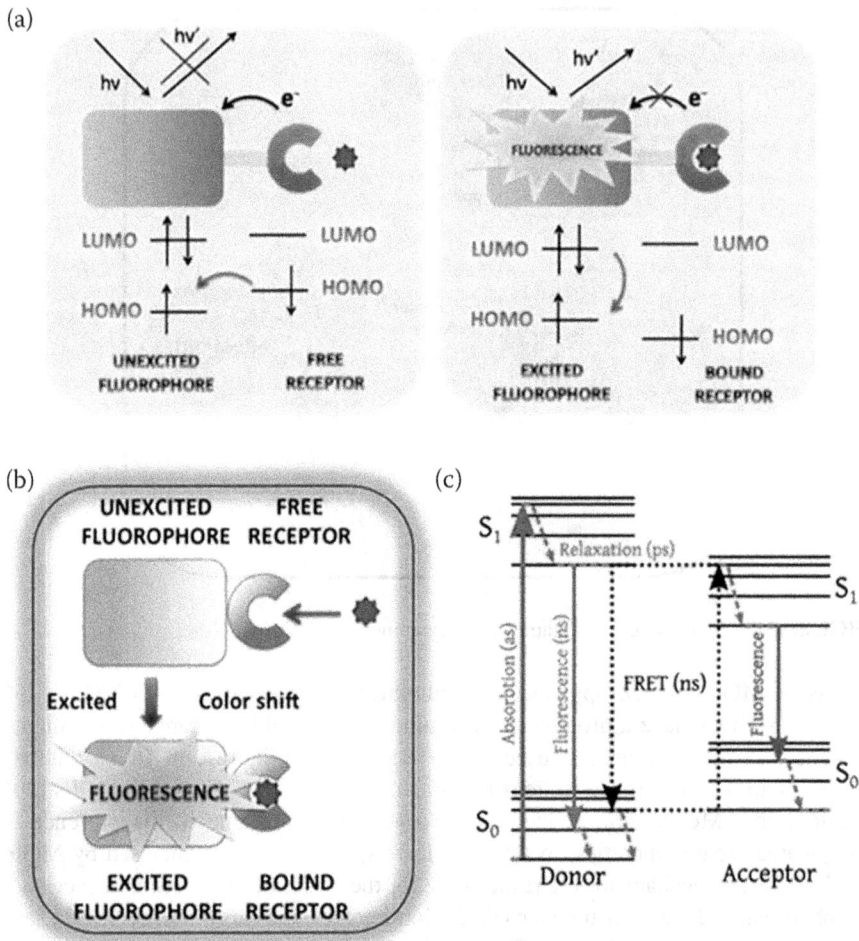

FIGURE 6.2 The general concept of the principle of (a) PET fluorescence sensor, (b) ICT fluorescence sensors, (c) Forster Resonance Energy Transfer (Jablonski diagram). (Reproduced with permission (Creative Commons Attribution (CC BY-NC) License) from Duan et al., 2015 [22(a)]).

and erroneous waste disposal are the main cause of IPs that expedite a severe danger to human health and ecosystem [27]. Hence, sensitive detection of IPs at a very small concentration is required for the protection of environment.

6.2.1 HEAVY METAL IONS

6.2.1.1 Chromium

Because of anthropogenic practice, chromium is highly toxic for the ecosystem [28]. Lang et. al. synthesized Zn-MOF namely; $\{[Zn_2(tpeb)_2(2,3-ndc)_2] \cdot H_2O\}_n$ under solvothermal condition. The reported MOF revealed the selective quenching

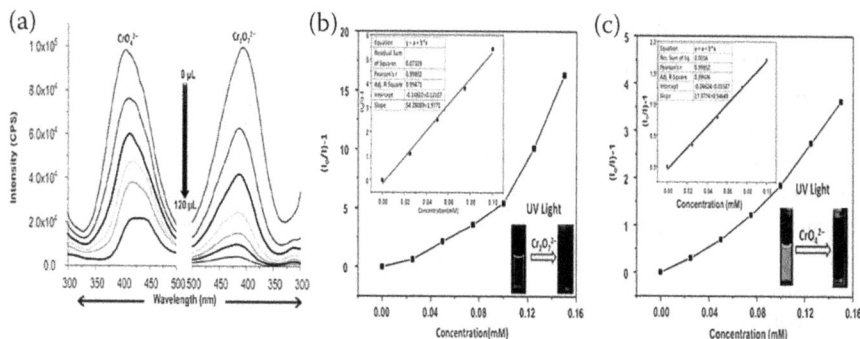

FIGURE 6.3 (a) Emission spectra of Cd-MOF upon incremental addition of CrO_4^{2-} (left) and $Cr_2O_7^{2-}$ (right) solution (2.5 mM) in water, SV plots for Cd-MOF upon incremental addition of 2.5 mM aqueous solutions of (b) $Cr_2O_7^{2-}$ and (c) CrO_4^{2-} (inset shows the visible color change of solution under UV light) [32]. Copyright permission 2020. Inorganic Chemistry.

for Cr^{3+}, $Cr_2O_7^{2-}$, and CrO_4^{2-} ions in aqueous media with detection limits 0.88 ppb, 2.623 ppb and 1.734 ppb, respectively [29]. Li et al. reported 2D Ln-MOFs (Eu^{3+} and Tb^{3+}), which act as excellent and recyclable chemical sensor towards selective sensing of Cr^{3+}, CrO_4^{2-} and $Cr_2O_7^{2-}$ in an aqueous solution [30]. Wang et al. synthesized Cd-MOF by solvothermal method, which displayed turn-on luminescence responses toward Cr^{3+} (K_{sv}: 2.70×10^4 M^{-1}) with detection limit 1.84 µM [31]. Neogi et al. developed a twofold interpenetrated Cd-MOF. The synthesized MOF showed rapid and sensitive response for noxious CrO_4^{2-} and $Cr_2O_7^{2-}$ anions with quenching constants 1.73×10^4 M^{-1} (LOD: 280 ppb) and 5.42×10^4 M^{-1} (LOD: 320 ppb), respectively (Figure 6.3) [32].

Mondal and co-workers prepared a two-dimensional Zn-MOF by the layer diffusion method. The luminescence study showed that the resulted MOF was highly sensible for Cr^{3+} ions in existence of other metal ions with the limit of detection 0.46 µM [33].

6.2.1.2 Copper

Heavy metal Cu^{2+} ions show adverse effects on human health as well as environment [34]. Su et al. reported Cd-MOF-74, which was used to identify the Cu^{2+} ions from water as well as imitated biological fluids that consist of Na^+, K^+, Mg^{2+}, Fe^{2+}, Co^{2+}, Ni^{2+}, and Zn^{2+} ions. Moreover, Cd-MOF-74 displayed 189.5 mg g^{-1} adsorption capacity for Cu^{2+} ions within 10 minutes [35]. Safari et al.,described a Zn-MOF (Zn-TMU-24) for the sensing of Cu^{2+} ions in real samples such as milk as well as milk powder with with the limit of detection 0.014 nM [36]. Wang et al. prepared a fluorescent RhB@ZIF-8 by encapsulating Rhodamine B (RhB) into ZIF-8, which revealed highly selective response toward Cu^{2+} ions with great sensitivity (0.04 ppm, 1.91×10^{-7} M) in existence of several interfering ions. [37]. Wu and co-workers described a porous MOF (PCN222) by the reaction of $ZrOCl_2 \cdot 8H_2O$ and mixed organic linkers (benzoic acid as well as 5,10,15,20-Tetrakis(4-carboxyphenyl)porphyrin) for the selective and fast determination of Cu^{2+} ions with a limit of detection (50 nmol L^{-1}) [38].

6.2.1.3   Cadmium

It is a lethal heavy metal for most of the organisms including human beings [39]. Yan et al. inserted Eu^{3+} ion into voids of UiO-66(Zr)-$(COOH)_2$ to obtain a potent and fluorescent Ln-MOF for Cd^{2+} ion with fast detection time (LOD: 0.06 mm) [40]. Ma et al. constructed UiO-66-NH_2@PANI sensor for the decisive revelation of massive Cd^{2+} ions with a 0.3 µg L^{-1} limit of detection in presence of co-existing ions [41]. Li and group fabricated UiO-66-NH2@MWCNTs composites under hydrothermal condition for the sensing of Cd^{2+} ions in meat samples (LOD: 0.2 µg/L) [42]. He et al. synthesized Fe_3O_4/MOF/L-cysteine for "turn-off" detection of Cd^{2+} ions in waste water with the detection limit 0.94 ng mL^{-1} (Figure 6.4a) [43]. Reynolds and co-workers revealed a Zr-porphyrin-based MOF (NU-902) that showed significant fluorescence quenching for Cd^{2+} ion at levels of 0.3 ppb [44].

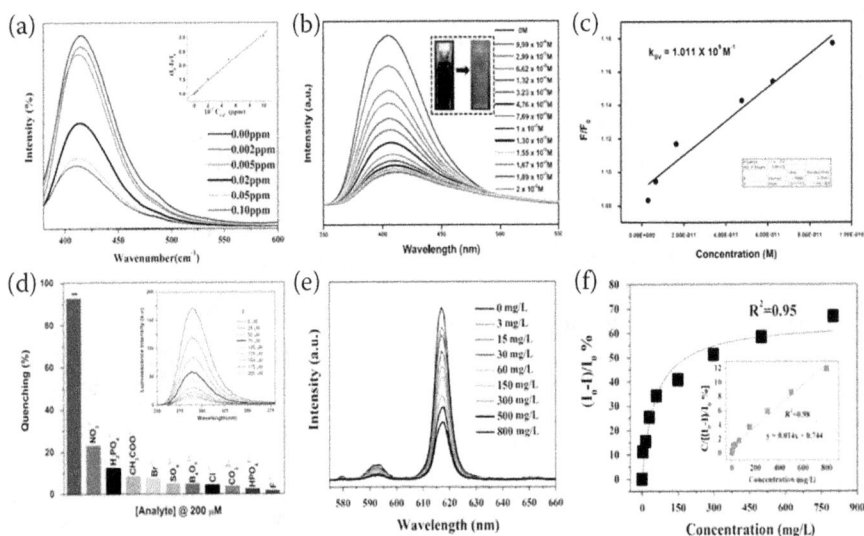

FIGURE 6.4   (a) Fluorescence emission spectra of the Fe_3O_4/MOF/L-cysteine sensor exposed to various concentrations of Cd(II): 0, 0.002, 0.005, 0.02, 0.05, and 0.1 ppm from top to bottom. The inset is the linear plot of the fluorescence intensity *versus* the concentration of Cd(II) [43]. Copyright permission 2018 (Creative Commons Attribution-NonCommercial 3.0 Unported Licence.). RSC Advances (b) Graph shows the quenching of fluorescence by adding Hg(II) ion. Photograph in the inset is the color change under UV light. (c) Percentage of fluorescence quenching for different metal ions in an aqueous medium at room temperature [47]. Copyright permission 2019. Inorganic Chemistry. (d) Percentage of luminescence quenching with respect of emission at 290 nm of Cd-MOF with 200 µM of different anions (Inset shows the emission spectra of Cd-MOF dispersed in water upon incremental addition of water solution of I^- ions (λex = 226 nm) [Ref. 62]. Copyright permission 2018. Journal of Photochemistry and Photobiology A: Chemistry. (e) Luminescence spectra of compound Eu-MOF in varied concentrations of UO_2^{2+}. (f) The correlation between the quenching ratio versus concentration simulated by a Langmuir model. The inset is the correlation between C/[(I0− I)/I0]% versus UO_2^{2+} concentration [Ref. 68]. Copyright permission 2019. Talanta.

6.2.1.4 Mercury

Mercury (Hg^{2+}) is the most noxious and perilous metal ion. Several natural (volcanoes, degradation of minerals or evaporation from soils) and human activities (industrial processes, coal burning, medical products and their disposal) are the origin of mercury contamination in environment that is the cause of various incurable diseases [45]. Several MOFs have been used for the detection of Hg^{2+} ion. Recently, Qian and co-workers synthesized UiO-66-PSM, which was able to quantitatively detect Hg^{2+} in real water samples [46]. Mandal et al. synthesized a Zn-MOF, namely, $\{[Zn(4,4'-AP)(5-AIA)]. (DMF)_{0.5}\}_n$, which has ability to detect femtogram concentration of Hg^{2+} ion with K_{sv} value of 1.011×10^9 M^{-1} without any interference of other metal ions in water (Figure 6.4b-c) [47]. Ghosh et al. developed a stable and butyne functionalized MOF (UiO-66@Butyne) for the identification of Hg^{2+} ion in aqueous solution (LOD: 10.9 nm) [48]. Morsali et al. designed a novel fluorescent probe by using pillar ligands that exhibited great efficiency for the detection of Hg^{2+} (LOD: 0.1 ppm) in existence of others [49]. Sabouni and colleagues prepared NH_2-Cd-BDC that showed the capability towards the excellent finding of Hg^{2+} ions in an aqueous media with a limit of detection 0.58 µM [50]. Wang and co-researchers demonstrated a luminescent MOF (Cd–EDDA), which displayed high sensitivity towards Hg^{2+} (LOD: 2 nM) for the first time with quick response (~15 s) [51].

6.2.1.5 Lead and Arsenic

On increasing the use of lead (Pb^{2+}) during the 20th century, risk for environment is continuously elevated [52]. Liu et al. determined a 2D Ln-MOF; $[Tb(L)(H_2O)_5]_n$ (L^- = 3,5-dicarboxyphenol anion), which displayed the strong ability for the detection of Pb^{2+} ion. This was the first MOF to detect Pb^{2+} ions at very low concentration (LOD: 10^{-7} M) [53]. Wu et al. synthesized of a series of [Ag(bpy)(DA-X)] (X = -H, -OH, -NH_2) coordination polymers. The DA (dicarboxylic acid) ligand with -NH_2 (electron-donating group) showed the excellent photoluminescence property for the selective identification of Pb^{2+} ions with limit of detection 4.9 µm [54]. Shirsat et al. investigated Au@CuBTC through solvothermal method that performed the higher selectivity toward Pb^{2+} ions with a limit of detection 1 nM/L, which was below the detection limit recommended by the USEPA drinking water regulations (0.03 µM/L) [55]. Similarly, Wang et al. revealed a novel composite (MOFL-TpBD) by solvothermal methods for the senstive identification of Pb^{2+} ions with a limit of detection 0.32 µgL^{-1} [56]. Arsenic is also highly toxic and displayed cancer, neurological disorder, and damage the immune system [57]. Li et al. synthesized isostructural Al-MOFs; BUT-18 and BUT-19 for the selective identification of arsenics in water (roxarsone and nitarsone). Because of water stability as well as great porosity, BUT-18 and BUT-19 displayed the quenching efficiencies above 98% with a detection limit of 15.7 and 13.5 ppb towards ROX (roxarsone) and 32.2 and 13.3 ppb towards NIT (nitarsone), respectively [58]. Ghosh et al. reported a luminescent cationic Zn-MOF (iMOF-4C), which displayed excellent "turn-on" response for $HAsO_4^{2-}$ and $HAsO_3^{2-}$ with an enhancement coefficient of 1.98×10^4 M^{-1} and 3.56×10^3 M^{-1} respectively [59].

6.2.2 RADIOACTIVE IONS

6.2.2.1 Iodide

It is rapidly combined with other organic materials and easily spread in water, air and soil that can be serious for human beings and environment. Zhao et al. reported two H-MOFs [Ln$_2$Zn(L)$_3$(H$_2$O)$_4$](NO$_3$)$_2$·12H$_2$O}$_n$ (Ln = Eu, Tb; L = 4,4'-dicarboxylate-2,2'-dipyridine) for the selective identification of I$^-$ ions in water. Out of these H-MOFs, Tb-Zn MOF showed an extremely fast response (10 s) towards I$^-$ ions with very low detection limit (0.001 ppm) [60]. Zhang and co-researchers constructed a Zn-MOF through [Zn$_2$(COO)$_4$] clusters and terphenyl-3,3'',5,5''-tetracarboxylic acid through solvothermal method that displayed great adsorption capacity for I$_2$ (up to 216 wt %) due to large porosity and charge-transfer interactions [61]. Mahata and colleagues synthesized a 3D Cd-MOF that exhibited excellent luminescence property. It was also observed that Cd-MOF showed great selectivity for I$^-$ ions in water with K$_{SV}$ value of 1.8×104 M^{-1} (LOD = 80 ppb) in the presence of other anions (Figure 6.4d) [62]. Shahid et al. prepared four Co/Mn-MOFs via solvothermal method, and investigated their adsorption/release performance towards I$_2$ in vapor as well as solution phase. Among all the reported MOFs, [Co$_{1.5}$(PhCOO)$_3$(bpy)$_{0.5}$]$_n$ exhibited 551 and 131.05 mg/g sorption capacity in vapor as well as a solution state, respectively [63].

6.2.2.2 Uranium

Elevated concentration of uranium in environment causes serious health issues [64]. Therefore, it is necessary to detect a trace level of uranium to protect the ecosystem. A luminescent In-MOF (In$_2$(OH)$_2$(H$_2$TTHA)(H$_2$O)$_2$)$_n$ was reported by the Xing group [65]. This MOF was synthesized hydrothermally by the use of indium chloride and 1,3,5-triazine-2,4,6-triamine hexaacetic acid that showed the excellent selectivity towards UO$_2^{2+}$ (K$_{sv}$ = 4.8×104 M^{-1}). Sun et al. constructed a Tb-MOF-76 via Tb^{3+} ion and 1,3,5-benzene tri-carboxylate ligand, which has the strong ability for sensitive selection of UO$_2^{2+}$ ions in existence of various Cr^{3+}, Ni^{2+}, Co^{2+}, Zn^{2+}, Cs$^+$, Sr^{2+} and Pb^{2+} ions. It also displayed great adsorption capacity (298 mg g^{-1}) at a low pH (~3.0) [66]. Pan et al. synthesized a Zn-MOF (Zn$_2$(PMA)(NPYC)H$_2$O)$_2$·2H$_2$O) for the potent detection of UO$_2^{2+}$ ions (NPYC = N-pyridin-4-ylpyridine-4-carboxamide; PMA = pyromellitic acid). The synthesized MOF showed the limit of detection (1.2×10^{-8} M) that is below the WHO maximum pollution standards for potable water (6.3×10^{-8} M) [67]. Wang and co-workers reported a Eu-MOF, [Eu$_2$(MTBC) (OH)$_2$(DMF)$_3$(H$_2$O)$_4$]·2DMF·7H$_2$O for the selective identification of UO$_2^{2+}$ ions with 309.2 µg/L detection limit (K$_{sv}$ = 3631.5 M^{-1}) (Figure 6.4e-f) [68].

6.2.2.3 Thorium

A high level of thorium causes cancer (lung, pancreas, bone), liver diseases and also damage the environment. Hence, its detection is a must to protect the ecosystem. Wang et al. synthesized [Eu$_2$(FDC)$_3$(DMA)$_2$]·4H$_2$O MOF (H$_2$FDC = 2,5-furan dicarboxylic acid) that exhibited excellent selective detection for radioactive Th^{4+} ions (K$_{SV}$ = 6.68×10^4) with detection limit of 3.49×10^{-5} mol L^{-1} that was found to be very close to the WHO standard [69].

6.2.2.4 Others

Other radioactive ions (Sr^{2+}, Cs^+, TcO_4^-, and ReO_4^-) are also perilous to the atmosphere. Thallapally et al. synthesized MIL-101-SO_3H to selective sensing of Cs^+ and Sr^{2+} ions from water in the existence of other interfering ions. It has also been observed that the reported MOF also detects Cs^+ and Sr^{2+} ions at various pH levels [70]. Song et al. prepared UiO-66-NH–SO_3H, which showed the elimination capacities for Sr^{2+} and Cs^+ are 173.9 and 233.4 mg g^{-1}, respectively, which may be due to the –SO_3H functional groups [71]. Desai et al. fabricated a series of Cd-MOFs whose pores were engaged with ClO_4^-, NO_3^-, $HCOO^-$, SO_4^{2-}, NO_3^- anions. Only Cd-MOF with ClO_4^- was quenched with $^{99}TcO_4^-$ ($K_{SV} = 0.68 \times 104$ m^{-1}), which was ascribed to the strong interactions between Cd-MOFs and guest anions [72]. Zhou et al. reported Ag-MOF with 1,2,4,5-tetra(pyridin-4-yl)benzene organic linker, which showed an excellent fluorescence quenching for $ReO4^-$ ion (86%) [73].

6.3 MOF-BASED SENSORS FOR ORGANIC POLLUTANTS

Organic pollutants are powerful toxic pollutants and cause critical health disorders, and are also risky to the ecosystem. Hence, the fast and selective detection of organic pollutants is very crucial for environmental safety as well as the health of human beings.

6.3.1 Antibiotics

One of the most notable classes of environmental pollutants is posed by pharmaceutical antibiotics, which are widely used in human medicines and agriculture. However, people and animals struggle to digest and absorb antibiotics, which could lead to the development of antibiotic-resistant bacteria and antibiotic-resistance genes that have negative consequences for the ecosystem. Therefore, the ultrasensitive detection of them must be a first priority for the world [74–77]. In this aspect, Fan et al. prepared zinc-based fluorescence materials for the sensitive selection and identification of antibiotics. His group revealed six new Zn-MOFs by using various organic linkers such as 5-(2-carboxylphenoxy)isophthalic acid (L), 1,4-bis(imidazol-l-ylmethyl) benzene, 4,4′-bis(benzoimidazo-1-ly) biphenyl, 4,4′-bis(imidazolyl)diphenyl ether, 4,4′-bis(imidazolyl)biphenyl, 3,5′-bis(1-imidazoly)pyridine and 1,3-bis(l-imidazoly)toluene). They also demonstrated that an efficient luminescent sensor could also be developed with these synthesized MOFs to detect tetracycline antibiotic. Among all the reported six MOFs, [Zn_3(L)$_2$(1,4-bis(imidazol-l-ylmethyl)benzene)$_3$]$_n$ displayed greater luminescence intensity, and it was the potential chemical sensor for tetracycline antibiotic with the detection limit 0.15 M (Figure 6.5) [78].

Among the resulting MOFs, [Zn_3(5-(2-carboxylphenoxy)isophthalic acid)$_2$(1,4-bis(imidazol-l-ylmethyl) benzene)$_3$]$_n$ showed potential and selective behavior towards tetracycline with 1.99×105 M^{-1} K_{SV} value and 0.15 μM limit of detection. In comparison to other literature, the results obtained in the present study displayed better detection limitations.

(a)

(b)

(c)

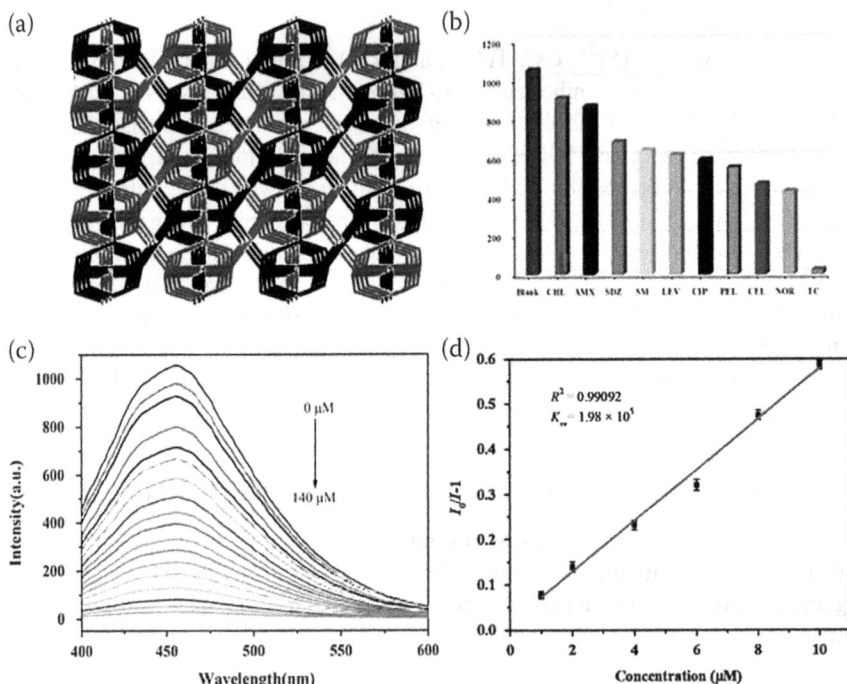

Intensity(a.u.)

0 μM

140 μM

Wavelength(nm)

(d)

$R^2 = 0.99092$
$K_{sv} = 1.98 \times 10^5$

$I_0/I-1$

Concentration (μM)

FIGURE 6.5 (a) Schematic representation of the twofold interpenetrating networks of $[Zn_3(L)_2(1,4\text{-bis(imidazol-1-ylmethyl)benzene})_3]_n$. (b) Fluorescence intensities of 1 introduced into different antibiotics dissolved in aqueous solution (375 nm). (c) Emission spectra in aqueous solutions of different tetracycline concentrations. (d) The linear relationships at low concentration in aqueous solutions of different tetracycline concentrations [Ref. 78]. Copyright permission 2020. Journal of Pharmaceutical and Biomedical Analysis.

Wang and colleagues described a 3D Eu-MOF with (2,5-bis(2H-tetrazol-5-yl) terephthalic acid). They also concluded that the MOF prepared by them showed strong fluorescence, and was able to detect metronidazole as well as dimetridazole antibiotics in calf serum and lake water with low limit of detection (13.0 ppm 13.4 ppm, respectively). The findings anticipated its utility in the clinical diagnosis and treatment of antibiotics toxicity [79]. Fu et al. synthesized an Eu-MOF by employing 2,6-naphthalenedicarboxylic acid linker, which showed the excellent fluorescent properties and selective detection of enrofloxacin, norfloxacin and ciprofloxacin over others. Moreover, the limit of detection for enrofloxacin, norfloxacin and ciprofloxacin estimated were 0.12 μmol/L, 0.53 μmol/L, and 0.75 μmol/L, respectively, which were lower than previous literature. They also discovered that photoinduced electron transfer and competing absorption of UV light could be the cause of quenching [80]. Jin et al. constructed a 3D Tb-MOF through the 5-(4′-carboxyphenoxy) isophthalic acid organic linker via the solvothermal method, which was used as extinguished luminescent sensor for selective identification of sulfamethazine, a sulfa drug ($K_{sv} = 1.28 \times 10^3$ M^{-1} and LOD = 1.43 ppm). They also explained the credible mechanism for sensing through computational analyzes [81]. Zheng and co-workers demonstrated a porous Tb-MOF with urea functionalized organic linker. Moreover, this complex has been effectively employed to determine antibiotics in real-world water samples [82]. Pan et al. synthesized a novel Zn-MOF through solvothermal method by using 5-chloro-8-hydroxyquinoline and 2, 5-dihydroxy-terephthalic acid. Additionally, the developed MOF could effectively detect nitrofurantoin and nitrofurazone antibiotics with low limits of detection 9.2×10^{-7} M and 7.2×10^{-7} M, respectively [83]. Long et al. demonstrated a zinc-MOF; namely, $\{[Zn(1,4\text{-benzenedicarboxylate}) (4,4'\text{-bis} (\text{imidazolyl})\text{diphenyl ether})]\cdot H_2O\}_n$ for the sensitive and selective finding of antibiotics in water. The obtained results displayed that Zn-MOF exhibited sensitive behavior towards tetracycline with a limit of detection 3.72×10^{-7} M (165 ppb) through the transfer of resonance energy [84]. Zhu and co-workers revealed an anionic MOF, $\{[Zn_3(OH)(bmipia)(H_2O)_3]_4\cdot[Zn(H_2O)_{6.5}]_2\}_n$ from zinc metal ions and 5-[N, N-bis(5-methylisophthalic acid)amion] isophthalic acid for the hypersensitive detection of ofloxacin antibiotic (LOD of 0.19 ppm) in aqueous phases [85]. Li et al. constructed two isomeric In(III)-MOFs for the efficient detection of ciprofloxacin antibiotic with 50 ppb ($K_{sv} = 6.0 \times 10^4$ M^{-1}) limits of detection [86]. Hu et al. prepared a Tb-MOF by using the tripodal organic linker; 1,3,5-tris(4-carbonylphenyloxy) benzene via the solvothermal method. The synthesized Tb-MOF showed selective sensing for nitrofurazone and nitrofurantoin in aqueous solutions [87], while Niu and co-researchers prepared a Cd-MOF by using two different organic linkers, such as 1,4-bis(2-methyl-imidazol-1-yl)butane and 1,2-phenylenediacetic acid for the selective sensing of ceftriaxone sodium antibiotic with 3.51×10^5 mol^{-1} K_{sv} and 55 ppb detection of limit values [88]. Bu and co-workers constructed a heterometallic MOF formulated as $[NaEu_2(4,4',4''\text{-s-triazine-} 1,3,5\text{-triyltri-m-aminobenzoic acid})_2(DMF)_3]\cdot OH$ for the sensitive and selective identification of oridazole antibiotics [89].

6.3.2 PESTICIDES

Pesticides have negative effects not only in the area of application but also in distant water supplies and other fields, pastures, and human settlements [90]. In this respect, Zhang and co-workers reported a novel $[Mg_2(APDA)_2(H_2O)_3]\cdot 5DMA\cdot 5H_2O$ MOF by using amino-decorated 4,4'-(4-aminopyridine-3,5-diyl)dibenzoic acid organic linker through the solvothermal process. The resulting MOF showed the fluorescence emission quenching towards 2,6-dichloro-4-nitroaniline among other selected pesticides with K_{sv} 7.50×10^4 M^{-1} (LOD = 150 ppb) through electron-transfer and energy-transfer processes [91]. Zhao et al. designed and constructed a thiophene-based MOF by using $Cd(NO_3)_2$ 1,1'-(1,4-butanediyl)bis(imidazole), tetrakis(4-pyridyloxy-methylene)methane and benzo-(1,2;4,5)-bis(thiophene-2'-carboxylic acid) under solvothermal process. The synthesized MOF was able to trap 2,4-dichlorophenol in the presence of other pesticides [92]. Xu and co-workers prepared a water-stable Zr-MOF using $ZrCl_4$ and 1,2,4,5-tetrakis(4-carboxyphenyl)benzene for the detection of parathion-methyl in water with limit of detection 0.438 nM [93]. Wang and co-researchers synthesized a Zn-MOF, namely, $[Zn_2(4\text{-(tetrazol-5-yl)phenyl-} 4,2':6',4''\text{-terpyridine})_2(\text{terephthalic acid})]\cdot 2H_2O$ through the solvothermal reaction, which produced the blue luminescence, and performed a potential chemosensor for 2,6-dichloro-4-nitroaniline with a low limit of detection 0.39 ppm ($K_{sv} = 2.36 \times 10^4 \text{ M}^{-1}$). The quenching in fluorescence intensity was obtained through host-guest interactions between an amino group of 2,6-dichloro-4-nitroaniline and free Lewis's base sites of pyridine and tetrazole rings as well as uncoordinated carboxylate group of terephthalic acid, present in Zn-MOF [94]. Yuan et al. described a new chemosensor based on Tb^{3+}-functionalized Zr-MOF. The synthesized MOF showed the selective and sensing detection toward the thiabendazole pesticide in oranges with a 0.271 µM limit of detection [95]. Jiao et al. constructed four new Ln-MOFs, coordinated as $[Ln_3(HDDB)(DDB)(H_2O)_6] \cdot NMP \cdot 3H_2O$ by using 3,5-di(2′,4′-dicarboxylphenyl) benzoic acid. Out of the resulting four MOFs, Tb-MOF exhibit exemplary sensitivity and selectivity towards 2,6-dichloro-4-nitroaniline pesticide with a small limit of detection 1.4×10^{-7} M (∼86.1 ppb) in water (Figure 6.6) [96] and in existence in extracts of grapes, carrots, and nectarines, which indicated that the Tb-MOF showed a potential luminescent sensor.

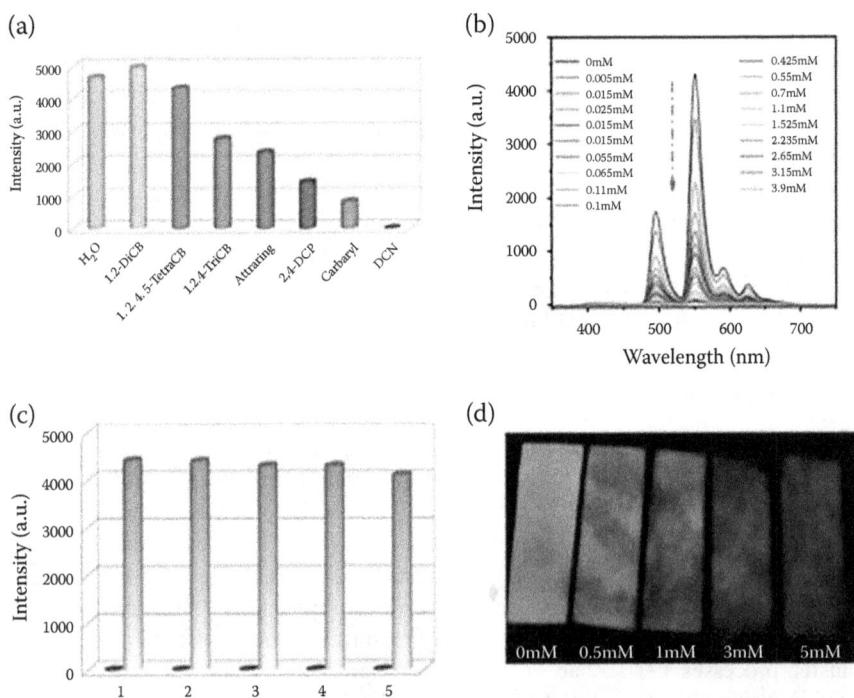

FIGURE 6.6 (a) Fluorescence intensity of Tb-MOF dispersed in various pesticides at 551 nm. (b) Fluorescence spectra of Tb-MOF with addition of DCN. (c) The luminescence intensity of Tb-MOF for detecting DCN after five cycles. (d) The color of the test paper with addition of different amounts of DCN (mM) under ultraviolet light (254 nm) irradiation [Ref. 96]. Copyright permission 2021. Crystal Growth & Design.

6.3.3   EXPLOSIVES

Nitro-aromatic analytes or explosives are the primary constituent of various unexploded land mines worldwide [97,98]. Apart from this, they can cause various diseases, including liver or kidney damage, and bladder tumor. Hence, their fast detection is quite essential for the environment and human beings. Here, I have discussed various categories of nitroaromatic explosives such as nitrophenols, nitrobenzenes, and methyl nitrobenzenes.

6.3.3.1   Nitrophenols

Our group constructed a series of mixed ligand-based Cd-MOFs. Among these synthesized MOFs, only MOF-AA, which is a three-dimensional coordination framework with a honeycomb topology, showed the interesting as well as good luminescent property, and exhibited notable applications for the sensitive as well as selective identification of 4-nitrophenol with quenching constant 5.07×10^5 M^{-1} in the presence of other aromatic analytes [99]. Liu and co-workers synthesized Ln-MOFs, $[(Ln_2(L)_2(H_2O)_2) \cdot 5H_2O \cdot 6DMAC]_n$, through $Ln(NO_3)_3.6H_2O$ (Ln = Eu/Tb) and 4,4'-(((5-carboxy-1,3-phenylene)bis(azanediyl))bis(carbonyl)) dibenzoic acid, which exhibited an attractive luminescence sensing properties for 4-nitrophenol, with a high K_{sv} value (5,352 M^{-1}) and a low limit of detection (7.6×10^{-5} M) [100]. Cao et al. reported an anionic Zn-MOF (FJI-C8) by $Zn(NO_3)_2$ and 2,4,6-tris(3,5-dicarboxylphenylamino)-1,3,5-triazine through a solvothermal process. They also demonstrated the emission properties of FJI-C8, which displayed the excellent selective performance for 2,4-dinitrophenol (K_{sv} = 5.11×10^4 M^{-1}; LOD = 0.002866 mM) with fast detection time (less than 30 s). The sensitive detection of FJI-C8 for 2,4-dinitrophenol among other nitro-explosives may be due to energy transfer mechanism [101]. Batten and co-workers disclosed a 3D Cd-MOF, $\{(CH_3)_2NH_2)_6[Cd_5(L)_4] \cdot H_2O \cdot 3DMF\}_n$ through terphenyl-3,3'',5,5''-tetracarboxylic acid rigid organic linker for the selective fluoroscent sensing towards 2,4-dinitrophenol, 2,4,6-trinitrophenol. Computational analysis introduced the possible mechanism for a decrease in fluorescence intensity of synthesized MOF in the presence of other nitro-aromatic analytes [102]. Pan et al. constructed a luminescent 3D MOF, $[Zn_2(TCPE)(tta)_2] \cdot 2DMF \cdot 4H_2O \cdot 2Me_2NH_2^+$ (TCPE = 1,1,2,2-tetra(4-carboxylphenyl) ethylene, 1H-tta = 1H-tetrazole) for the detection of nitrophenol explosives with limits of detection 29.45 nM (Ksv = 3.75×10^4 M^{-1}) and 36.15 nM (Ksv = 3.06×10^4 M^{-1}) for 2,6-dinitrophenol and 2,4,6-trinitrophenol, respectively [103]. Similarly, our group also synthesized a 2D luminescent TPA-MOF $\{[Tb(L_1) (L_2)_{0.5}(NO_3)(DMF)] \cdot DMF\}_n$, by 1,10-phenanthroline and 3,3',5,5'-azobenzene-tetracarboxylate linkers. The TPA-MOF, which was π-electron rich, acted as a sensitive and selective probe for electron-deficient 2,4,6-trinitrophenol with a quenching constant 5.67×10^5 M^{-1} through both photo-induced electrons and energy transfer mechanisms (Figure 6.7) [104].

6.3.3.2   Nitrobenzenes

Wang et al. fabricated four Zn-MOFs by the use of $ZnSO_4 \cdot 7H_2O$, 5-(2-carboxybenzyloxy) isophthalic acid and N-donor linkers through hydrothermal

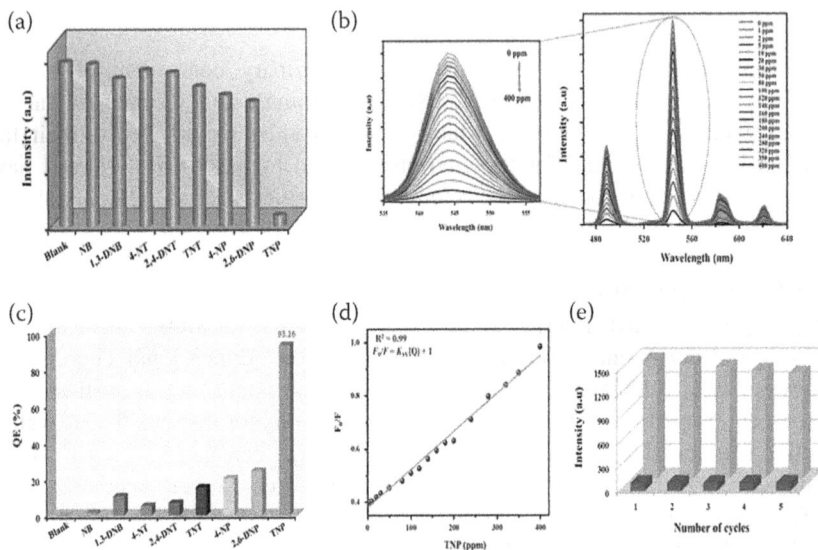

FIGURE 6.7 (a) Reduction in fluorescence intensity of TPA-MOF ($^5D_4 \rightarrow {}^7F_5$ transition 544 nm) after addition of different NAC analytes, (b) fluorescence titration of TPA-MOF by gradual addition of an aqueous solution of TNP (magnified image shows reduction in fluorescence intensity ($^5D_4/{}^7F_5$ transition, 544 nm)), and (c) QE of fluorescence intensity for TPA-MOF upon the addition of several NAC analytes, (d) Stern–Volmer plot of F0/F vs. TNP (ppm) concentration in aqueous solution for TPA-MOF, (e) quenching and reproducibility test for TPA-MOF (initial fluorescence intensity, green; intensity after quenching (400 ppm TNP), blue) [Ref. 104]. Copyright permission 2018. RSC Advances.

method. The luminescent studies described that all Zn-MOFs showed highly selective detection towards nitrobenzene in aqueous solution [105]. Kim et al. explored a Zr-MOF (UiO-66-NH$_2$) for the detection of electron deficient nitrobenzene in water with 95% quenching efficiency and 0.9 ppm limit of detection through electron transfer mechanism [106]. Sama group synthesized three Cu-MOFs [Cu(4,4'-DP)Cl]$_n$, [Cu(4,4'-DP)$_{0.5}$Cl]$_n$, and [Cu(4,4'-TMDP)Cl]$_n$ (4,4'-DP = 4,4'-dipyridyl, 4,4'-TMDP = 4,4'-trimethylenedipyridyl) under solvothermal conditions. Among three Cu-MOFs, [Cu(4,4'-DP)$_{0.5}$Cl]$_n$ acted as a potent fluorescent sensor for sensitive detection of meta-dinitrobenzene with K$_{SV}$ = 5.73 × 10^5 M^{-1} and limit of detection of 1.23 × 10^{-7} M [107].

6.3.3.3 Methyl Nitrobenzene

Guo and group constructed H$_2$N-Fe-MIL-88B@OMC composites through the hydrothermal method for the efficient detection of p-nitrotoluene with limit of detection of 8 μM [108]. Li et al., manufactures a fluorescent and microporous [Zn2(4,4'-biphenyldicarboxylate)2(1,2-bipyridylethene)] MOF for the fast detection of 2,4-dinitrotoluene by a redox fluorescence quenching mechanism [109]. Zheng et al. explored 3D Mg-MOF for selective sensing of 2,4,6-Trinitrotoluene through a photo-induced electron transfer mechanism [110].

6.4 MOF-BASED SENSORS FOR GASEOUS POLLUTANTS

Gaseous pollutants lead to serious health problems and are hazardous to the environment. As a result, rapid identification of these contaminants is critical for both the environment and human beings.

6.4.1 VOLATILE ORGANIC COMPOUNDS

Many alkanes, alkenes, toluene, xylene, terpenes, halogenated hydrocarbons, alcohols, aldehydes, and ketones are the examples of volatile organic molecules (VOCs). These VOCs have the potential to cause serious human health such as dizziness; unconsciousness; shortness of breath; damage to liver, kidney, and central nervous system; and harm environmental contamination due to their pervasiveness [111].

6.4.1.1 Aliphatic, Aromatic and Halogenated Hydrocarbons

Long et al. prepared a Fe_2(2,5-dioxido-1,4-benzenedicarboxylate) MOF, which performed exceptionally well in the separation of ethylene and propylene over others due to the presence of functional groups, strong sorbent-sorbate interactions, high surface areas and open metal sites [112]. Zhang et al. synthesized ultra-microporous MOFs, $Cu[M(pdt)_2]$, where M is Cu, Ni and pdt is pyrazine-2,3-dithiol. Because of the binding affinity between the active metal sites and guest, as well as hydrogen bonding and π–π interactions between acetylene and pdt, the produced MOFs showed high stability and selectivity against acetylene over multiple gases [113]. Li and colleagues demonstrated $[Zr_6(\mu_3-O)_4(\mu_3-OH)_4(4,4'-(benzene-1,3-diyl)dibenzoate)_6]$ (BUT-66) and $[Zr_6(\mu_3-O)4(\mu_3-OH)_4(4,4'-(naphthalene-2,7-diyl)dibenzoate)_6]$ (BUT-67) MOFs. BUT-66 displayed much greater benzene adsorption capacity than the other synthesized MOF (1.75 mmol cm^{-3}). It also has a great capacity for catching many parts per million of benzene (ca. 0.28 mmol cm^{-3}) in humid air (RH 50%) due to presence of uniform hydrophobic micropores [114]. To make luminescent MOF, Zhao et al. utilized a tetraphenylethene derivative with an aggregation-induced emission property as an organic linker, which worked as an excellent luminescent sensor for aromatic hydrocarbons. By changing the confirmation of the aggregation-induced emission linker, numerous aromatic hydrocarbons caused different emission peaks [115]. Khavasi et al. synthesized UiO-67(I)$_2$ MOF under solvothermal condition. Furthermore, it showed a strong fluorescence response when exposed to halogenated aromatic hydrocarbons [116]. Lang and co-workers designed and synthesized a ($[Cu_4I_4(Py_3P)_2]_n$) MOF with a Cu_4I_4 cluster as the node and tris(2-pyridyl)phosphine as the organic linker. The resulted MOF showed large pores to accommodate the guest molecules. Furthermore, the authors revealed the guest-lock-induced light-up mechanism, and this process was used for the detection of chlorobenzene [117]. Dong et al. synthesized a porous Cu^I-MOF ($H_2O⊂Cu2(L)_2I_2$) by using 1-benzimidazolyl-3,5-bis (4-pyridyl)benzene] organic linker that was successfully used to detect tiny polar aliphatic VOCs including halocarbons [118]. Han and co-researchers investigated a chemically stable {$Cd_3(L)(bbib).6DMF$}$_n$ MOF (NBU-18) through $Cd(NO_3)_2.4H_2O$, hexa[4-(carboxyphenyl)-oxamethyl]-3-oxapentane acid and 1,4-bis(1H-benzo[d]-

imidazol-1-yl)benzene) ligands under a solvothermal condition. The NBU-18 showed a diffusion-controlled turn-off mechanism to selectively detect CCl_4 vapour in the existence of other volatile chloromethanes [119].

6.4.1.2 Aldehydes and Ketones

Yan et al. constructed a ZnO@Uio-MOFs through a solvothermal reaction. The resulting MOF was tailored to a dual-emitting material by changing the excitation wavelength after the Eu^{3+} fluorescent core was introduced, which displayed reusable and selective fluorescence response to aldehyde gases with low detection limit (42 ppb for formaldehyde, 58 ppb for acetaldehyde and 66 ppb for acraldehyde). This potential ZnO-doped Uio-MOF will open a new door for practical vehicle detecting application [120]. Li et al. prepared a {Zn(3,3'-dpdc)bpp} where bpp was 1,3-bis(4-pyridyl)propane and 3,3'-dpdc was 3,3'-diphenyldicarboxylic acid. By using a "turn-off" effect, the synthesized MOF was employed as a fluorescence probe to detect benzaldehyde [121,122]. Biswas and co-workers fabricated a [Al (OH)(BDC−N_2H_3)]·0.85DEF·1.0H_2O for the selective detection of formaldehyde. The creation of the hydrazone moiety inhibits the photoinduced electron transfer process, resulting in the fluorescence "turn-on" characteristic of the reaction-based probe, which showed a 0.25 ppm detection limit for formaldehyde. The probe was also discovered to be capable of detecting endogenous formaldehyde in cancer cells [123]. Lazarides et al. reported [Ca_6(bpdc-(NH_2)$_2$)$_5$(μ_3-CO_2)$_2$(H_2O)$_{2.5}$ (DMF)$_{0.5}$]·0.5H_2O·2.5DMF and [Sr_4(bpdc-(NH_2)$_4$)(μ_2-DMF)$_2$(DMF)$_{1/3}$]·2/3 (DMF) MOFs for the sensing of acetaldehyde vapors. The amino group that was present in 2,2'-diamino-[1,1'-biphenyl]-4,4'-dicarboxylic acid organic linker reacted with the aldehyde group that showed the capability to alter the fluorescence properties [124]. Wang et al. synthesized Ln-MOFs by using 5-(3',5'-dicarboxylphenyl) nicotinic acid through a solvothermal method. The luminous selective sensing tests revealed that Eu-MOF and Tb MOF may be used as recyclable ketone sensors in water systems, with low detection limits (0.45 μM and 0.24 μM, respectively) [125]. Feng et al. demonstrated a 3D microporous Cd-MOF through 2-(4-carboxyphenoxy) terephthalic acid and 1,10-phenanthroline ligands, and this synthesized MOF displayed the selective and sensitive detection for ketones (acetone/2-butanone) in an aqueous solution [126]. Goel et al. prepared a {[Tb(L_1)(L_2)$_{0.5}$(NO_3)(DMF)].DMF}$_n$ MOF (where L_1 and L_2 were 1,10-phenanthroline and 3,3',5,5'-azobenzene-tetracarboxylic acid), which exhibited an eminent fluorescence quenching effect by the progressive addition of acetone in a dispersed aqueous medium of synthesized MOF, and its fluorescence nearly vanished at an acetone content of 18% (v/v). The graph of fluorescence intensity ($^5D_4 \rightarrow {}^7F_5$, 544 nm) against acetone content (vol%) displayed a first-order exponential decay, which indicated that the acetone quenching behavior (98.4% QE; QE ¼ quenching efficiency) of TPA-MOF was controlled by a diffusion process [127–130]. The reason behind the mechanism for the quenching of acetone may be explained by the existence of binding interactions of azo (−N=N−) group of 3,3',5,5'-azobenzene-tetracarboxylic acid with the analytes (Figure 6.8) [104]. Lang et al. revealed a fluorescent, recyclable and reusable Zn-MOF

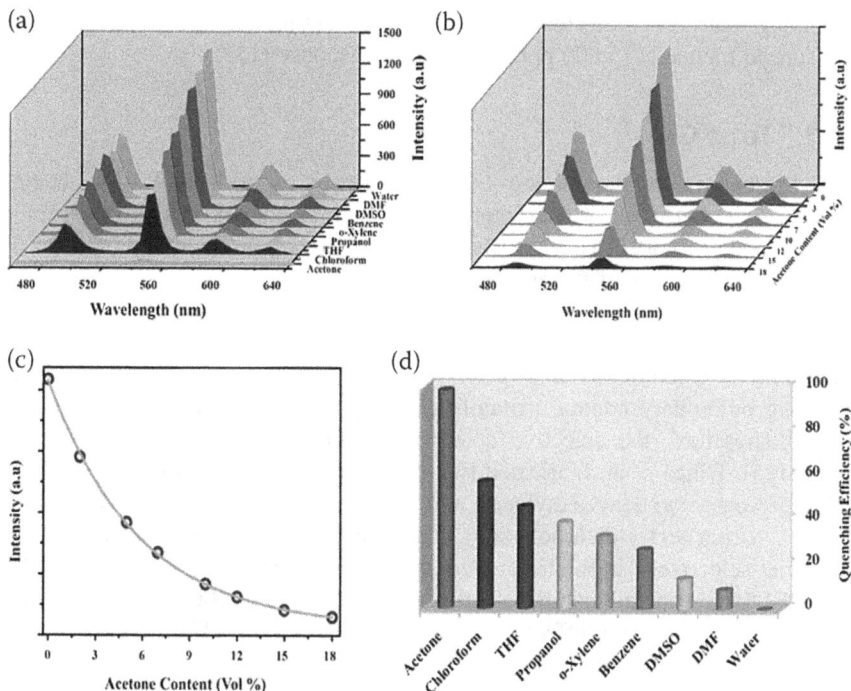

FIGURE 6.8 The photoluminescence spectra of TPA-MOF: (a) intensity of TPA-MOF dispersed in H_2O and various organic solvent molecules, (b) spectra of TPA-MOF in presence of variable amounts of acetone, (c) emission intensities ($^5D_4 \rightarrow {}^7F_5$ transition, 544 nm transition) as a function of acetone content, (d) quenching efficiency of $^5D_4 \rightarrow {}^7F_5$ transition fluorescence intensity for TPA-MOF dispersed in selected solvents at room temperature. [Ref. 104]. Copyright permission 2018. RSC Advances.

by the use of 3,5-bis-(2-(pyridin-4-yl)vinyl)pyridine and 1,3,5-benzenetricarboxylic acid organic ligands. The synthesized MOF displayed high sensitivity towards p-benzoquinone with a low limit of detection (1.0×10^{-6} mol L^{-1}) [131].

6.4.1.3   Alcohols

Alcohols are generally used as solvents in commercial manufacturing, and their considerable use increases the environmental pollutant. Desai and co-researchers investigated two fluorescent MOFs by using In(II) and Mn(II) metal salts as well as [L-tris(4-(1H-imidazol-1-yl)phenyl)amine] organic ligands. These discussed MOFs showed sensitive and selective detection of toxic and volatile allyl alcohol, which may be due to presence of distinctive π-electrons of allyl alcohol [132]. Roeffaers et al. produced ZIF-8 and ZIF-93, which had been grown on fiber optic–based totally floor plasmon resonance sensors for the sensing of methanol with a limit of detection of 2.5 ppm in the gaseous phase [133]. Smet et al. revealed the MOF-polymer composites for capacitive sensor applications. NH_2-MIL-53(Al) MOF-containing Matrimid® films (ca. 2.5%) exhibited a double capacitive response upon exposure to methanol (5,000 ppm) with faster response [134].

6.4.2 TOXIC GASES

Toxic gases are unfavorable to the surroundings as well as human beings. Therefore, the detection of unfavorable gases such as ammonia, hydrogen sulfide, sulfur dioxide, nitrogen oxides, carbon oxides, and carbon disulfide is a must for monitoring the quality of the environment.

6.4.2.1 Ammonia

Ammonia is a colorless and pernicious gas that causes various diseases, including pulmonary edema, irritability in respiratory tract, and skin of human beings. Therefore, the selective detection of ammonia is required in various fields [135]. Dincá et al. fabricated two fluorescent MOFs, $Zn_2(TCPE)$ (TCPE = tetrakis-(4-carboxyphenyl)ethylene) and $Mg(H_2DHBDC)$ (H_2DHBDC^{2-} =2,5-dihydroxybenzene-1,4-dicarboxylate). $Zn_2(TCPE)$ showed the ability to sense ammonia selectively at high temperatures (350°C). On the other hand, Mg (H_2DHBDC) displayed notable sensing for ammonia at 100°C. However, the mechanism of the above effects continues to be unknown [136]. Dincá and co-researchers also reported the an electrically conductive two-dimensional $Cu_3(HITP)_2$ MOF (HITP = 2,3,6,7,10,11-hexaiminotriphenylene), which acted as reversible chemiresistive sensor for ammonia vapor [137]. Marti-Gastaldo et al. synthesized ultrathin Cu_3-$(HHTP)_2$ films (HHTP = 2,3,6,7,10,11-hexahydroxytriphenylene) through a bottom-up approach. Additionally, the computational analysis revealed that guest molecules (ammonia and water) exhibited stronger interaction with open metal sites [138]. Yan et al. demonstrated (Eu^{3+}@Ga(OH)bpydc) MOF for the sensitive selection of ammonia in the presence of other gases with a 2.4 ppm detection limit. Furthermore, the resulting MOF was able to determine the biological metabolite of ammonia in the human urinary urea [139]. Salama and co-researchers described rare earth (RE) MOFs (RE-fcu-MOF). Out of which, NDC-Y-fcu-MOF (NDC = 1, 4-naphthalene dicarboxylic acid acid) as a thin film sensing layer on a capacitive interdigitated electrodes sensor showed the selective detection of ammonia with a limit of detection of approximately 100 ppb at room temperature. The excellent property of sensing towards ammonia may be due to the presence of interaction between lone pairs of electrons of ammonia and active metal sites as well as the hydrogen bonding [140].

6.4.2.2 Hydrogen Sulfide

Hydrogen sulfide is a noxious gas and evolves in diverse enterprise operations [141,142]. Upon exposure, it can bind with Fe^{2+} in mitochondria and damage the central nervous system. It also produces the lethal diseases such as cancer and Alzheimer's disease [108]. Ghosh et al. reported an azide ($-N_3$) group functionalized MOF, which exhibited a fluorescence turn-on response towards hydrogen sulfide through strong interactions between azide and hydrogen sulfide (Figure 6.9) [143].

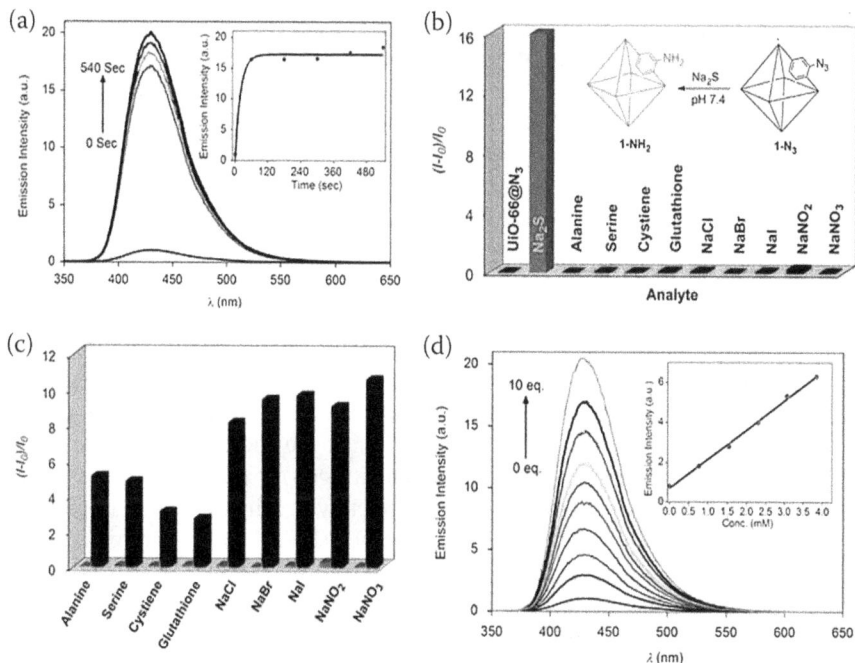

FIGURE 6.9 Fast and selective turn-on response of 1-N$_3$ towards Na$_2$S. (a) Fluorescence response of 1-N$_3$ towards addition of Na$_2$S after 0, 60, 180, 300, 420 and 540 seconds (Inset: Time dependence of emission intensity at 436 nm). (b) Relative fluorescence response of 1-N$_3$ towards various analytes (10 equivalents/azide group) after 540 seconds of analyte addition (Inset: Reduction of 1-N$_3$ to 1-NH$_2$ upon addition of Na$_2$S under physiological pH). (c) Turn-on response of probe 1-N$_3$ at 436 nm in presence of respective analyte (blue), followed by addition of Na$_2$S in solution containing analyte (red). (d) Fluorescence response of 1-N$_3$ with increasing concentrations of Na$_2$S [Ref. 143]. Copyright permission 2015. Scientific Reports.

Tang et al. described {CuL-[AlOH]$_2$}$_n$ (H$_6$L = meso-tetrakis-(4-carboxylphenyl) porphyrin), which was a porphyrin-derived MOF. On exposure to hydrogen sulfide, the fluorescence intensity of the organic linker was increased by eliminating the Cu^{2+} ions from the porphyrin centers. The synthesized MOF showed the light-on sensing mechanism (LOD 16 nM). The authors also reported the selective detection of hydrogen sulfide in living cells [144]. Qian et al. used MIL-100(In)@Eu^{3+}/Cu^{2+} film that displayed fluorescence turn-on sensing towards hydrogen sulfide with a limit of detection of 0.535 ppm. Hydrogen sulfide in gaseous form showed a strong affinity with Cu^{2+} ions, which greatly improved Eu^{3+} fluorescence [145]. Wang et al. reported Cu–HIA (HIA = 5-hydroxyisophthalic acid) MOF, which highlighted the selective and ultrasensitive sensing of hydrogen sulfide with a limit of detection 0.21 μM [146].

6.4.2.3 Sulfur Dioxide

Excessive sulfur dioxide emissions cause extreme pollution of the environment, and affect the human health [147,148]. Cao et al. synthesized MOF-5-NH$_2$ through zinc nitrate and 2-amino-1,4-benzene dicarboxylate at room temperature. The resulting luminescent probe showed the selective detection of sulfur dioxide derivatives (i.e., SO_3^{2-}) among various gases with a 0.168 ppm limit of detection [149]. Kalidindi et al. fabricated a Zr-NH_2-benzenedicarboxylate MOF (NH_2-UiO-66) that acted as an exquisite chemiresistive sensor for sulfur dioxide. It may be due to the formation of a charge transfer complex, which was formed through the interactions between functional groups of organic linkers and guest molecules [150].

6.4.2.4   Others

Nitrogen oxides are an extremely reactive family of toxic gases. Nitrogen dioxide is acidic as well as an extremely corrosive gas, and has a considerable adverse impact on the environment and human health. Schaate et al. reported a Zr-based MOF through calixarene. The calixarene cavities accommodate the nitrogen dioxide and the exhibited calixarene operated as a visual nitrogen dioxide sensor due to the formation of charge-transfer complex between nitrogen dioxide and the MOF [151]. Nitric oxide primarily affects the respiratory tract, with minor irritation to the eyes. Duan et al. reported Cu-TCA and Eu-TCA MOFs (H3TCA = tri-carboxytriphenyl amine). Cu–TCA exhibited a high selectivity for NO. Moreover, Cu-TCA also displayed the applications of bioimaging in living cells. At the same time, the synthesized Eu-TCA also operated as an exclusive fluorescent sensor for NO with a limit of detection 140 μm [152]. Dincá et al. synthesized a two-dimensional Cu_3(hexaiminobenzene)$_2$ MOF that acted as a chemiresistive sensor for carbon dioxide [153].

6.5   CONCLUSIONS

Metal-organic frameworks are a notable group of effective and functional materials with exclusive properties and various vital uses. Different metal nodes and multi-topic organic linkers in the MOFs increase the sensitivity and selectivity through the interactions with guest molecules. Therefore, enhanced performance can be achieved by careful selection of the organic linkers and the nature of the metal ion. Hence, the resulting MOFs with multiple binding sites and large surface area can increase the fluorescence-emission by decreasing non-radiative relaxation. A lot of luminescent MOFs have been described in the literature for the detection of pernicious environmental pollutants through different sensing mechanisms, and this is believed to be one of the most promising applications of these materials. Herein, the prevailing situation of MOF-derived chemical sensors for the detection of various inorganic, organic, and gaseous pollutants have been discussed.

ACKNOWLEDGMENT

N.G. gratefully acknowledges the financial support from UGC, New Delhi (Letter No. F.30-431/2018(BSR), M-14-55) and BHU (Letter No. R/Dev/D/IoE/Seed Grant/2020-23). Author also thanks to BHU for providing the research work facilities.

REFERENCES

[1] Howarth, A. J., Liu, Y., Hupp, J. T., & Farha, O. K. 2015. "Metal–Organic Frameworks for Applications in Remediation of Oxyanion/Cation-Contaminated Water". *CrystEngComm.* 17(38), 7245–7253. 10.1039/C5CE01428J

[2] Manisalidis, I., Stavropoulou, E., Stavropoulos, A., & Bezirtzoglou, E. 2020. "Environmental and Health Impacts of Air Pollution: A Review". *Frontiers in Public Health.* 8(2020), 1–14. 10.3389/fpubh.2020.00014

[3] Global Assessment of Soil Pollution: Report; FAO and UNEP, 2021. 10.4060/cb4894en

[4] Huelsmann, R., & Martendal, E. 2020. "A Simple and Effective Liquid-Liquid-Liquid Microextraction Method with Ultraviolet Spectrophotometric Detection for the Determination of Bisphenol A in Aqueous Matrices and Plastic Leachates". *Journal of the Brazilian Chemical Society.* 31(8), 1575–1584. 10.21577/0103-5053.20200043

[5] Lee, K.-J., & Dabrowski, K. 2002. "High-Performance Liquid Chromatographic Determination of Gossypol and Gossypolone Enantiomers in Fish Tissues Using Simultaneous Electrochemical and Ultraviolet Detectors". *Journal of Chromatography B.* 779(2), 313–319. 10.1016/S1570-0232(02)00402-6

[6] Mirghani, M. E. S., & Che Man, Y. B. 2003. "A New Method for Determining Gossypol in Cottonseed Oil by FTIR Spectroscopy". *Journal of the American Oil Chemists Society.* 80(7), 625–628. 10.1007/s11746-003-0749-2

[7] Moliner Martínez, Y., Muñoz-Ortuño, M., Herráez-Hernández, R., & Campíns-Falcó, P. 2014. "Rapid Analysis of Effluents Generated by the Dairy Industry for Fat Determination by Preconcentration in Nylon Membranes and Attenuated Total Reflectance Infrared Spectroscopy Measurement". *Talanta.* 119, 11–16. 10.1016/j.talanta.2013.10.032

[8] Emamian, S., Eshkeiti, A., Narakathu, B. B., Avuthu, S. G. R., & Atashbar, M. Z. 2015. "Gravure Printed Flexible Surface Enhanced Raman Spectroscopy (SERS) Substrate for Detection of 2,4-Dinitrotoluene (DNT) Vapor". *Sensors and Actuators B: Chemical.* 217, 129–135. 10.1016/j.snb.2014.10.069

[9] Zhou, H.-C., Long, J. R., & Yaghi, O. M. 2012. "Introduction to Metal–Organic Frameworks". *Chemical Reviews.* 112(2), 673–674. 10.1021/cr300014x

[10] Zhou, H.-C. "Joe", & Kitagawa, S. 2012. "Metal–Organic Frameworks (MOFs)". *Chemical Society Reviews.* 43(16), 5415–5418. 10.1039/C4CS90059F

[11] Shi, L., Li, N., Wang, D., Fan, M., Zhang, S., & Gong, Z. 2021. "Environmental Pollution Analysis Based on the Luminescent Metal Organic Frameworks: A Review". *Trends in Analytical Chemistry.* 134, 116131. 10.1016/j.trac.2020.116131

[12] Cui, Y., Yue, Y., Qian, G., & Chen, B. 2012. "Luminescent Functional Metal–Organic Frameworks". *Chemical Reviews.* 112(2), 1126–1162. 10.1021/cr200101d

[13] Kreno, L. E., Leong, K., Farha, O. K., Allendorf, M., Van Duyne, R. P., & Hupp, J. T. 2012. "Metal–Organic Framework Materials as Chemical Sensors". *Chemical Reviews.* 112(2), 1105–1125. 10.1021/cr200324t

[14] Jin, G., Liu, Z., Sun, H., & Tian, Z. 2016. "Pyrolytic Synthesis and Luminescence of Porous Lanthanide Eu-MOF". *Luminescence.* 31(1), 190–194. 10.1002/bio.2944

[15] Cui, Y., Chen, B., & Qian, G. 2014. "Lanthanide Metal-Organic Frameworks for Luminescent Sensing and Light-Emitting Applications". *Coordination Chemistry Reviews.* 273–274, 76–86. 10.1016/j.ccr.2013.10.023

[16] Rao, X., Huang, Q., Yang, X., Cui, Y., Yang, Y., Wu, C., Chen, B., & Qian, G. 2012. "Color Tunable and White Light Emitting Tb^{3+} and Eu^{3+} Doped Lanthanide Metal–Organic Framework Materials". *Journal of Materials Chemistry.* 22(7), 3210. 10.1039/c2jm14127b

[17] Usman, M., Haider, G., Mendiratta, S., Luo, T.-T., Chen, Y.-F., & Lu, K.-L. 2016. "Continuous Broadband Emission from a Metal–Organic Framework as a Human-Friendly White Light Source". *Journal of Materials Chemistry C*. 4(21), 4728–4732. 10.1039/C6TC00839A

[18] Meng, Q., Xin, X., Zhang, L., Dai, F., Wang, R., & Sun, D. 2015. "A Multifunctional Eu MOF as a Fluorescent PH Sensor and Exhibiting Highly Solvent-Dependent Adsorption and Degradation of Rhodamine B". *Journal of Materials Chemistry A*. 3(47), 24016–24021. 10.1039/C5TA04989J

[19] Zhao, X.-Q., Zhao, B., Shi, W., & Cheng, P. 2009. "Structures and Luminescent Properties of a Series of Ln–Ag Heterometallic Coordination Polymers". *CrystEngComm*. 11(7), 1261. 10.1039/b900430k

[20] Stylianou, K. C., Heck, R., Chong, S. Y., Bacsa, J., Jones, J. T. A., Khimyak, Y. Z., Bradshaw, D., & Rosseinsky, M. J. 2010. "A Guest-Responsive Fluorescent 3D Microporous Metal–Organic Framework Derived from a Long-Lifetime Pyrene Core". *Journal of the American Chemical Society*. 132(12), 4119–4130. 10.1021/ja906041f

[21] Zhang, Y., Yuan, S., Day, G., Wang, X., Yang, X., & Zhou, H.-C. 2018. "Luminescent Sensors Based on Metal-Organic Frameworks". *Coordination Chemistry Reviews*. 354, 28–45. 10.1016/j.ccr.2017.06.007

[22] (a) Jiao, Y., Zhu, B., Chen, J., & Duan, X. 2015. "Fluorescent Sensing of Fluoride in Cellular System". *Theranostics*. 5(2), 173–187. 10.7150/thno.9860; (b) Jung, H. S., Verwilst, P., Kim, W. Y., & Kim, J. S. 2016. "Fluorescent and Colorimetric Sensors for the Detection of Humidity or Water Content". *Chemical Society Reviews*. 45(5), 1242–1256. 10.1039/C5CS00494B.

[23] Sun, W., Li, M., Fan, J., & Peng, X. 2019. "Activity-Based Sensing and Theranostic Probes Based on Photoinduced Electron Transfer". *Accounts of Chemical Research*. 52(10), 2818–2831. 10.1021/acs.accounts.9b00340

[24] Sun, X., Liu, T., Sun, J., & Wang, X. 2020. "Synthesis and Application of Coumarin Fluorescence Probes". *RSC Advances*. 10(18), 10826–10847. 10.1039/C9RA10290F

[25] Dinda, D., Gupta, A., Shaw, B. K., Sadhu, S., & Saha, S. K. 2014. "Highly Selective Detection of Trinitrophenol by Luminescent Functionalized Reduced Graphene Oxide through FRET Mechanism". *ACS Applied Materials & Interfaces*. 6(13), 10722–10728. 10.1021/am5025676

[26] Yang, G., Jiang, X., Xu, H., Zhao, B. 2021. "Applications of MOFs as Luminescent Sensors for Environmental Pollutants". *Small*. 17(22), 2005327. 10.1002/smll.202005327

[27] Li, J., Wang, X., Zhao, G., Chen, C., Chai, Z., Alsaedi, A., Hayat, T., & Wang, X. 2018. "Metal–Organic Framework-Based Materials: Superior Adsorbents for the Capture of Toxic and Radioactive Metal Ions". *Chemical Society Reviews*. 47(7), 2322–2356. 10.1039/C7CS00543A

[28] Oliveira, H. 2012. "Chromium as an Environmental Pollutant: Insights on Induced Plant Toxicity". *Journal of Botany*. 2012, 1–8. 10.1155/2012/375843

[29] Gu, T.-Y., Dai, M., Young, D. J., Ren, Z.-G., & Lang, J.-P. 2017. "Luminescent Zn (II) Coordination Polymers for Highly Selective Sensing of Cr(III) and Cr(VI) in Water". *Inorganic Chemistry*. 56 (8), 4668–4678. 10.1021/acs.inorgchem.7b00311

[30] Sun, Z., Yang, M., Ma, Y., & Li, L. 2017. "Multi-Responsive Luminescent Sensors Based on Two-Dimensional Lanthanide–Metal Organic Frameworks for Highly Selective and Sensitive Detection of Cr(III) and Cr(VI) Ions and Benzaldehyde". *Crystal Growth & Design*. 17(8), 4326–4335. 10.1021/acs.cgd.7b00638

[31] Yu, Y., Wang, Y., Yan, H., Lu, J., Liu, H., Li, Y., Wang, S., Li, D., Dou, J., & Yang, L. 2020. "Multiresponsive Luminescent Sensitivities of a 3D Cd-CP with Visual Turn-on and Ratiometric Sensing toward Al^{3+} and Cr^{3+} as Well as Turn-off Sensing toward Fe^{3+}". *Inorganic Chemistry*. 59(6), 3828–3837. 10.1021/acs.inorgchem.9b03496

[32] Singh, M., Senthilkumar, S., Rajput, S., & Neogi, S. 2020. "Pore-Functionalized and Hydrolytically Robust Cd(II)-Metal–Organic Framework for Highly Selective, Multicyclic CO_2 Adsorption and Fast-Responsive Luminescent Monitoring of Fe (III) and Cr(VI) Ions with Notable Sensitivity and Reusability". *Inorganic Chemistry*. 59(5), 3012–3025. 10.1021/acs.inorgchem.9b03368

[33] Daga, P., Manna, P., Majee, P., Singha, D. K., Hui, S., Ghosh, A. K., Mahata, P., & Mondal, S. K. 2021. "Response of a Zn(II)-Based Metal–Organic Coordination Polymer towards Trivalent Metal Ions (Al^{3+}, Fe^{3+} and Cr^{3+}) Probed by Spectroscopic Methods". *Dalton Transactions*. 50(21), 7388–7399. 10.1039/D1 DT00729G

[34] Sun, T., Fan, R., Xiao, R., Xing, T., Qin, M., Liu, Y., Hao, S., Chen, W., & Yang, Y. 2020. "Anionic Ln-MOF with Tunable Emission for Heavy Metal Ion Capture and l-Cysteine Sensing in Serum". *Journal of Materials Chemistry A*. 8(11), 5587–5594. 10.1039/c9ta13932j

[35] Zheng, T.-T., Zhao, J., Fang, Z.-W., Li, M.-T., Sun, C.-Y., Li, X., Wang, X.-L., Su, Z.-M. 2017. "A Luminescent Metal Organic Framework with High Sensitivity for Detecting and Removing Copper Ions from Simulated Biological Fluids". *Dalton Transactions*. 46(8), 2456–2461. 10.1039/C6DT04630D

[36] Ehzari, H., Amiri, M., Safari, M., & Samimi, M. 2020. "Zn-Based Metal-Organic Frameworks and p-Aminobenzoic Acid for Electrochemical Sensing of Copper Ions in Milk and Milk Powder Samples". *International Journal of Environmental Analytical Chemistry*. 2020, 1–14. 10.1080/03067319.2020.1784410

[37] Du, T., Wang, J., Zhang, T., Zhang, L., Yang, C., Yue, T., Sun, J., Li, T., Zhou, M., & Wang, J. 2020. "An Integrating Platform of Ratiometric Fluorescent Adsorbent for Unconventional Real-Time Removing and Monitoring of Copper Ions". *ACS Applied Materials & Interfaces*. 12(11), 13189–13199. 10.1021/acsami.9b23098

[38] Xu, Z., Meng, Q., Cao, Q., Xiao, Y., Liu, H., Han, G., Wei, S., Yan, J., & Wu, L. 2020. "Selective Sensing of Copper Ions by Mesoporous Porphyrinic Metal–Organic Framework Nanoovals". *Analytical Chemistry*. 92(2), 2201–2206. 10.1021/acs.analchem.9b04900

[39] Saini, S., & Dhania, G. 2020. "Cadmium as an Environmental Pollutant: Ecotoxicological Effects, Health Hazards, and Bioremediation Approaches for Its Detoxification from Contaminated Sites". In *Bioremediation of Industrial Waste for Environmental Safety*. 357–387. Springer. 10.1007/978-981-13-3426-9_15

[40] Hao, J.-N., & Yan, B. 2015. "A Water-Stable Lanthanide-Functionalized MOF as a Highly Selective and Sensitive Fluorescent Probe for Cd^{2+}". *Chemical Communications*. 51(36), 7737–7740. 10.1039/C5CC01430A

[41] Wang, Y., Wang, L., Huang, W., Zhang, T., Hu, X., Perman, J. A., & Ma, S. 2017. "A Metal–Organic Framework and Conducting Polymer Based Electrochemical Sensor for High Performance Cadmium Ion Detection". *Journal of Materials Chemistry A*. 5(18), 8385–8393. 10.1039/C7TA01066D

[42] Wang, X., Xu, Y., Li, Y., Li, Y., Li, Z., Zhang, W., Zou, X., Shi, J., Huang, X., & Liu, C. 2021. "Rapid Detection of Cadmium Ions in Meat by a Multi-Walled Carbon Nanotubes Enhanced Metal-Organic Framework Modified Electrochemical Sensor". *Food Chemistry*. 357, 129762. 10.1016/j.foodchem.2021.129762

[43] Fan, L., Deng, M., Lin, C., Xu, C., Liu, Y., Shi, Z., Wang, Y., Xu, Z., Li, L., & He, M. 2018. "A Multifunctional Composite Fe_3O_4 /MOF/ 1 -Cysteine for Removal, Magnetic Solid Phase Extraction and Fluorescence Sensing of Cd(II)". *RSC Advances*. 8(19), 10561–10572. 10.1039/C8RA00070K

[44] Hibbard, H. A. J., Burnley, M. J., Rubin, H. N., Miera, J. A., & Reynolds, M. M. 2020. "Porphyrin-Based Metal-Organic Framework and Polyvinylchloride Composites for Fluorescence Sensing of Divalent Cadmium Ions in Water". *Inorganic Chemistry Communications*. 115, 107861. 10.1016/j.inoche.2020.107861

[45] Centers for Disease Control and Prevention, A.f.T.S.a.D.R.A. 1999. "Toxicological profile for mercury". https://www.atsdr.cdc.gov/toxprofiles/tp46.pdf.

[46] Zhang, X., Xia, T., Jiang, K., Cui, Y., Yang, Y., & Qian, G. 2017. "Highly Sensitive and Selective Detection of Mercury (II) Based on a Zirconium Metal-Organic Framework in Aqueous Media". *Journal of Solid State Chemistry*. 253, 277–281. 10.1016/j.jssc.2017.06.008

[47] Pankajakshan, A., Kuznetsov, D., & Mandal, S. 2019. "Ultrasensitive Detection of Hg(II) Ions in Aqueous Medium Using Zinc-Based Metal–Organic Framework". *Inorganic Chemistry*. 58(2), 1377–1381. 10.1021/acs.inorgchem.8b02898

[48] Samanta, P., Desai, A. V., Sharma, S., Chandra, P., & Ghosh, S. K. 2018. "Selective Recognition of Hg^{2+} Ion in Water by a Functionalized Metal–Organic Framework (MOF) Based Chemodosimeter". *Inorganic Chemistry*. 57(5), 2360–2364. 10.1021/acs.inorgchem.7b02426

[49] Esrafili, L., Gharib, M., & Morsali, A. 2019. "Selective Detection and Removal of Mercury Ions by Dual-Functionalized Metal–Organic Frameworks: Design-for-Purpose". *New Journal of Chemistry*. 43(46), 18079–18091. 10.1039/C9NJ03951A

[50] El Taher, B. J., Sabouni, R., & Ghommem, M. 2020. "Luminescent Metal Organic Framework for Selective Detection of Mercury in Aqueous Media: Microwave-Based Synthesis and Evaluation". *Colloids and Surfaces A: Physicochemical and Engineering Aspects*. 607, 125477. 10.1016/j.colsurfa.2020.125477

[51] Wu, P., Liu, Y., Liu, Y., Wang, J., Li, Y., Liu, W., & Wang, J. 2015. "Cadmium-Based Metal–Organic Framework as a Highly Selective and Sensitive Ratiometric Luminescent Sensor for Mercury(II)". *Inorganic Chemistry*. 54(23), 11046–11048. 10.1021/acs.inorgchem.5b01758

[52] Wani, A. L., Ara, A., & Usmani, J. A. 2015. "Lead Toxicity: A Review". *Interdisciplinary Toxicology*. 8(2), 55–64. 10.1515/intox-2015-0009

[53] Ji, G., Liu, J., Gao, X., Sun, W., Wang, J., Zhao, S., & Liu, Z. 2017. "A Luminescent Lanthanide MOF for Selectively and Ultra-High Sensitively Detecting Pb^{2+} Ions in Aqueous Solution". *Journal of Materials Chemistry A*. 5(21), 10200–10205. 10.1039/C7TA02439H

[54] Van Nguyen, C., Matsagar, B. M., Ahamad, T., Alshehri, S. M., Chiang, W.-H., & Wu, K. C.-W. 2020. "Unraveling the Highly Selective Nature of Silver-Based Metal–Organic Complexes for the Detection of Metal Ions: The Synergistic Effect of Dicarboxylic Acid Linkers". *Journal of Materials Chemistry C*. 8(15), 5051–5057. 10.1039/C9TC07078H

[55] Bodkhe, G. A., Hedau, B. S., Deshmukh, M. A., Patil, H. K., Shirsat, S. M., Phase, D. M., Pandey, K. K., & Shirsat, M. D. 2020. "Detection of Pb(II): Au Nanoparticle Incorporated CuBTC MOFs". *Frontiers in Chemistry*. 803(8), 1–11. 10.3389/fchem.2020.00803

[56] Li, W.-T., Hu, Z.-J., Meng, J., Zhang, X., Gao, W., Chen, M.-L., & Wang, J.-H. 2021. "Zn-Based Metal Organic Framework-Covalent Organic Framework Composites for Trace Lead Extraction and Fluorescence Detection of TNP". *Journal of Hazardous Materials*. 411, 125021. 10.1016/j.jhazmat.2020.125021

[57] Hong, Y.-S., Song, K.-H., & Chung, J.-Y. 2014. "Health Effects of Chronic Arsenic Exposure". *Journal of Preventive Medicine and Public Health.* 47(5), 245–252. 10.3961/jpmph.14.035

[58] Lv, J., Wang, B., Xie, Y., Wang, P., Shu, L., Su, X., & Li, J.-R. 2019. "Selective Detection of Two Representative Organic Arsenic Compounds in Aqueous Medium with Metal–Organic Frameworks". *Environmental Science: Nano.* 6(9), 2759–2766. 10.1039/C9EN00316A

[59] Dutta, S., Let, S., Shirolkar, M. M., Desai, A. V., Samanta, P., Fajal, S., More, Y. D., & Ghosh, S. K. 2021. "A Luminescent Cationic MOF for Bimodal Recognition of Chromium and Arsenic Based Oxo-Anions in Water". *Dalton Transactions.* 50(29), 10133–10141. 10.1039/D1DT01097B

[60] Shi, P.-F., Hu, H.-C., Zhang, Z.-Y., Xiong, G., & Zhao, B. 2015. "Heterometal–Organic Frameworks as Highly Sensitive and Highly Selective Luminescent Probes to Detect I – Ions in Aqueous Solutions". *Chemical Communications.* 51(19), 3985–3988 10.1039/C4CC09081K

[61] Yao, R.-X., Cui, X., Jia, X.-X., Zhang, F.-Q., & Zhang, X.-M. 2016. "A Luminescent Zinc(II) Metal–Organic Framework (MOF) with Conjugated π-Electron Ligand for High Iodine Capture and Nitro-Explosive Detection". *Inorganic Chemistry.* 55(18), 9270–9275. 10.1021/acs.inorgchem.6b01312

[62] Singha, D. K., Majee, P., Mondal, S. K., & Mahata, P. 2018. "A Luminescent Cadmium Based MOF as Selective and Sensitive Iodide Sensor in Aqueous Medium". *Journal of Photochemistry and Photobiology A: Chemistry.* 356, 389–396. 10.1016/j.jphotochem.2018.01.024

[63] Qasem, K. M. A., Khan, S., Ahamad, M. N., Saleh, H. A. M., Ahmad, M., & Shahid, M. 2021. "Radioactive Iodine Capture by Metal Organic Frameworks in Liquid and Vapour Phases: An Experimental, Kinetic and Mechanistic Study". *Journal of Environmental Chemical Engineering.* 9(6), 106720. 10.1016/j.jece.2021.106720

[64] Ma, M., Wang, R., Xu, L., Xu, M., & Liu, S. 2020. "Emerging Health Risks and Underlying Toxicological Mechanisms of Uranium Contamination: Lessons from the Past Two Decades". *Environment International.* 145, 106107. 10.1016/j.envint.2020.106107

[65] Du, N., Song, J., Li, S., Chi, Y.-X., Bai, F.-Y., & Xing, Y.-H. 2016. "A Highly Stable 3D Luminescent Indium–Polycarboxylic Framework for the Turn-off Detection of UO_2^{2+}, Ru^{3+}, and Biomolecule Thiamines". *ACS Applied Materials & Interfaces.* 8(42), 28718–28726. 10.1021/acsami.6b09456

[66] Yang, W., Bai, Z.-Q., Shi, W.-Q., Yuan, L.-Y., Tian, T., Chai, Z.-F., Wang, H., & Sun, Z.-M. 2013. "MOF-76: From a Luminescent Probe to Highly Efficient U[VI] Sorption Material". *Chemical Communications.* 49(88), 10415–10417. 10.1039/C3CC44983A

[67] Qin, X., Yang, W., Yang, Y., Gu, D., Guo, D., & Pan, Q. 2020. "A Zinc Metal–Organic Framework for Concurrent Adsorption and Detection of Uranium". *Inorganic Chemistry.* 59 (14), 9857–9865. 10.1021/acs.inorgchem.0c01072

[68] Liu, W., Wang, Y., Song, L., Silver, M. A., Xie, J., Zhang, L., Chen, L., Diwu, J., Chai, Z., & Wang, S. 2019. "Efficient and Selective Sensing of Cu^{2+} and UO_2^{2+} by a Europium Metal-Organic Framework". *Talanta.* 196, 515–522. 10.1016/j.talanta.2018.12.088

[69] Song, L., Liu, W., Wang, Y., Chen, L., Wang, X.-F., Wang, S. 2019. "A Hydrolytically Stable Europium–Organic Framework for the Selective Detection of Radioactive Th^{4+} in Aqueous Solution". *CrystEngComm.* 21(22), 3471–3477. 10.1039/C9CE00241C

[70] Aguila, B., Banerjee, D., Nie, Z., Shin, Y., Ma, S., & Thallapally, P. K. 2016. "Selective Removal of Cesium and Strontium Using Porous Frameworks from High Level Nuclear Waste". *Chemical Communications.* 52(35), 5940–5942. 10.1039/C6CC00843G

[71] Wu, J., Zhang, Y., Zhou, J., Cao, R., Wang, C., Li, J., & Song, Y. 2021. "Efficient Removal of Sr^{2+} and Cs^+ from Aqueous Solutions Using a Sulfonic Acid-Functionalized Zr-Based Metal–Organic Framework". *Journal of Radioanalytical and Nuclear Chemistry.* 328(3), 769–783. 10.1007/s10967-020-07477-y

[72] Desai, A. V., Sharma, S., Roy, A., & Ghosh, S. K. 2019. "Probing the Role of Anions in Influencing the Structure, Stability, and Properties in Neutral N-Donor Linker Based Metal–Organic Frameworks". *Crystal Growth & Design.* 19(12), 7046–7054. 10.1021/acs.cgd.9b00873

[73] Li, C.-P., Zhou, H., Chen, J., Wang, J.-J., Du, M., Zhou, W. 2020. "A Highly Efficient Coordination Polymer for Selective Trapping and Sensing of Perrhenate/Pertechnetate". *ACS Applied Materials & Interfaces.* 12(13), 15246–15254. 10.1021/acsami.0c00775

[74] Zhu, X.-D., Zhang, K., Wang, Y., Long, W.-W., Sa, R.-J., Liu, T.-F., Lü, J. 2018. "Fluorescent Metal–Organic Framework (MOF) as a Highly Sensitive and Quickly Responsive Chemical Sensor for the Detection of Antibiotics in Simulated Wastewater". *Inorganic Chemistry.* 57(3), 1060–1065. 10.1021/acs.inorgchem.7b02471

[75] Martinez, J. L. 2009. "Environmental Pollution by Antibiotics and by Antibiotic Resistance Determinants". *Environmental Pollution.* 157(11), 2893–2902. 10.1016/j.envpol.2009.05.051

[76] Hirsch, R., Ternes, T., Haberer, K., & Kratz, K.-L. 1999. "Occurrence of Antibiotics in the Aquatic Environment". *Science of The Total Environment.* 225(1–2), 109–118. 10.1016/S0048-9697(98)00337-4

[77] Halling-Sørensen, B., Nors Nielsen, S., Lanzky, P. F., Ingerslev, F., Holten Lützhøft, H. C., & Jørgensen, S. E. 1998. "Occurrence, Fate and Effects of Pharmaceutical Substances in the Environment- A Review". *Chemosphere.* 36(2), 357–393. 10.1016/S0045-6535(97)00354-8

[78] Fan, C., Zhang, X., Li, N., Xu, C., Wu, R., Zhu, B., Zhang, G., Bi, S., & Fan, Y. 2020. "Zn-MOFs Based Luminescent Sensors for Selective and Highly Sensitive Detection of Fe^{3+} and Tetracycline Antibiotic". *Journal of Pharmaceutical and Biomedical Analysis.* 188, 113444. 10.1016/j.jpba.2020.113444

[79] Wang, G.-D., Li, Y.-Z., Shi, W.-J., Zhang, B., Hou, L., & Wang, Y.-Y. 2021. "A Robust Cluster-Based Eu-MOF as Multi-Functional Fluorescence Sensor for Detection of Antibiotics and Pesticides in Water". *Sensors and Actuators B: Chemical.* 331, 129377. 10.1016/j.snb.2020.129377

[80] Wang, C.-Y., Wang, C.-C., Zhang, X.-W., Ren, X.-Y., Yu, B., Wang, P., Zhao, Z.-X., & Fu, H. 2022. "A New Eu-MOF for Ratiometrically Fluorescent Detection toward Quinolone Antibiotics and Selective Detection toward Tetracycline Antibiotics". *Chinese Chemical Letters.* 33(3), 1353–1357. 10.1016/j.cclet.2021.08.095

[81] Li, G., Wang, T., Zhou, S., Wang, J., Lv, H., Han, M., Singh, D. P., Kumar, A., & Jin, J. 2021. "New Highly Luminescent 3D Tb(III)-MOF as Selective Sensor for Antibiotics". *Inorganic Chemistry Communications.* 130, 108756. 10.1016/j.inoche.2021.108756

[82] Lei, M., Ge, F., Gao, X., Shi, Z., & Zheng, H. 2021. "A Water-Stable Tb-MOF As a Rapid, Accurate, and Highly Sensitive Ratiometric Luminescent Sensor for the Discriminative Sensing of Antibiotics and D_2O in H_2O". *Inorganic Chemistry.* 60(14), 10513–10521. 10.1021/acs.inorgchem.1c01145

[83] Yang, Y., Ren, G., Yang, W., Qin, X., Gu, D., Liang, Z., Guo, D.-Y., & Qinhe, P. 2021. "A New MOF-Based Fluorescent Sensor for the Detection of Nitrofuran Antibiotics". *Polyhedron*. 194, 114923. 10.1016/j.poly.2020.114923

[84] Cui, L., Zhu, B., Huang, K., Gan, Y., Li, Y., & Long, J. 2020. "Synthese, Structure of Three Zn-MOFs and Potential Sensor Material for Tetracycline Antibiotic in Water: {[Zn(Bdc)(4,4'-Bidpe)]·H_2O}$_n$". *Journal of Solid State Chemistry*. 290, 121526. 10.1016/j.jssc.2020.121526

[85] Li, C.-P., Long, W.-W., Lei, Z., Guo, L., Xie, M.-J., Lü, J., & Zhu, X.-D. 2020. "Anionic Metal–Organic Framework as a Unique Turn-on Fluorescent Chemical Sensor for Ultra-Sensitive Detection of Antibiotics". *Chemical Communications*. 56(82), 12403–12406. 10.1039/D0CC05175F

[86] Zhong, W.-B., Li, R.-X., Lv, J., He, T., Xu, M.-M., Wang, B., Xie, L.-H., & Li, J.-R. 2020. "Two Isomeric In(III)-MOFs: Unexpected Stability Difference and Selective Fluorescence Detection of Fluoroquinolone Antibiotics in Water". *Inorganic Chemistry Frontiers*. 7(5), 1161–1171. 10.1039/C9QI01490J

[87] Zhang, J., Gao, L., Wang, Y., Zhai, L., Niu, X., & Hu, T. 2019. "A Bifunctional 3D Tb-Based Metal–Organic Framework for Sensing and Removal of Antibiotics in Aqueous Medium". *CrystEngComm*. 21(47), 7286–7292. 10.1039/C9CE01303B

[88] Xing, P., Wu, D., Chen, J., Song, J., Mao, C., Gao, Y., & Niu, H. 2019. "A Cd-MOF as a Fluorescent Probe for Highly Selective, Sensitive and Stable Detection of Antibiotics in Water". *The Analyst*. 144 (8), 2656–2661. 10.1039/C8AN02442A

[89] Han, M.-L., Wen, G.-X., Dong, W.-W., Zhou, Z.-H., Wu, Y.-P., Zhao, J., Li, D.-S., Ma, L.-F., & Bu, X. 2017. "A Heterometallic Sodium–Europium-Cluster-Based Metal–Organic Framework as a Versatile and Water-Stable Chemosensor for Antibiotics and Explosives". *Journal of Materials Chemistry C*. 5(33), 8469–8474. 10.1039/C7TC02885G

[90] Hassaan, M. A., & El Nemr, A. 2020. "Pesticides Pollution: Classifications, Human Health Impact, Extraction and Treatment Techniques". *The Egyptian Journal of Aquatic Research*. 46(3), 207–220. 10.1016/j.ejar.2020.08.007

[91] Xu, N., Zhang, Q., Hou, B., Cheng, Q., & Zhang, G. 2018. "A Novel Magnesium Metal–Organic Framework as a Multiresponsive Luminescent Sensor for Fe(III) Ions, Pesticides, and Antibiotics with High Selectivity and Sensitivity". *Inorganic Chemistry*. 57(21), 13330–13340. 10.1021/acs.inorgchem.8b01903

[92] Zhao, Y., Xu, X., Qiu, L., Kang, X., Wen, L., & Zhang, B. 2017. "Metal–Organic Frameworks Constructed from a New Thiophene-Functionalized Dicarboxylate: Luminescence Sensing and Pesticide Removal". *ACS Applied Materials & Interfaces*. 9(17), 15164–15175. 10.1021/acsami.6b11797

[93] He, K., Li, Z., Wang, L., Fu, Y., Quan, H., Li, Y., Wang, X., Gunasekaran, S., & Xu, X. 2019. "A Water-Stable Luminescent Metal–Organic Framework for Rapid and Visible Sensing of Organophosphorus Pesticides". *ACS Applied Materials & Interfaces*. 11(29), 26250–26260. 10.1021/acsami.9b06151

[94] Guo, X.-Y., Dong, Z.-P., Zhao, F., Liu, Z.-L., & Wang, Y.-Q. 2019. "Zinc (II)–Organic Framework as a Multi-Responsive Photoluminescence Sensor for Efficient and Recyclable Detection of Pesticide 2,6-Dichloro-4-Nitroaniline, Fe(III) and Cr(VI)". *New Journal of Chemistry*. 43(5), 2353–2361. 10.1039/C8NJ05647A

[95] Peng, X.-X., Bao, G.-M., Zhong, Y.-F., Zhang, L., Zeng, K.-B., He, J.-X., Xiao, W., Xia, Y.-F., Fan, Q., & Yuan, H.-Q. 2021. "Highly Sensitive and Rapid Detection of Thiabendazole Residues in Oranges Based on a Luminescent Tb^{3+}-Functionalized MOF". *Food Chemistry*. 343, 128504. 10.1016/j.foodchem.2020.128504

[96] Wang, X.-Q., Ma, X., Feng, D., Tang, J., Wu, D., Yang, J., & Jiao, J. 2021. "Four Novel Lanthanide(III) Metal–Organic Frameworks: Tunable Light Emission and Multiresponsive Luminescence Sensors for Vitamin B_6 and Pesticides". *Crystal Growth & Design*. 21(5), 2889–2897. 10.1021/acs.cgd.1c00080

[97] Germain, M. E., & Knapp, M. J. 2009. "Optical Explosives Detection: From Color Changes to Fluorescence Turn-On". *Chemical Society Reviews*. 38(9), 2543. 10.1039/b809631g

[98] Goel, N., & Kumar, N. 2020. "Study of Four New Cd(II) Metal-Organic Frameworks: Syntheses, Structures, and Highly Selective Sensing for 4-Nitrophenol". *Inorganica Chimica Acta*. 503, 119352. 10.1016/j.ica.2019.119352

[99] Vallero, D. A. 2014. "Fundamentals of Air Pollution". 5th edn. Elsevier, Academic Press, Cambridge. 10.1016/C2012-0-01172-6

[100] Ma, J.-J., & Liu, W. 2019. "Effective Luminescence Sensing of Fe^{3+}, $Cr_2O_7^{2-}$, MnO^{4-} and 4-Nitrophenol by Lanthanide Metal–Organic Frameworks with a New Topology Type". *Dalton Transactions*. 48(32), 12287–12295. 10.1039/C9DT01907C

[101] Wang, X.-S., Li, L., Yuan, D.-Q., Huang, Y.-B., & Cao, R. 2018. "Fast, Highly Selective and Sensitive Anionic Metal-Organic Framework with Nitrogen-Rich Sites Fluorescent Chemosensor for Nitro Explosives Detection". *Journal of Hazardous Materials*. 344, 283–290. 10.1016/j.jhazmat.2017.10.027

[102] Lu, L., Wu, J., Wang, J., Liu, J.-Q., Li, B.-H., Singh, A., Kumar, A., & Batten, S. R. 2017. "An Uncommon 3D 3,3,4,8-c Cd(II) Metal–Organic Framework for Highly Efficient Luminescent Sensing and Organic Dye Adsorption: Experimental and Theoretical Insight". *CrystEngComm*. 19(46), 7057–7067. 10.1039/C7CE01638G

[103] Zhang, X., Ren, G., Li, M., Yang, W., & Pan, Q. 2019. "Selective Detection of Aromatic Nitrophenols by a Metal–Organic Framework-Based Fluorescent Sensor". *Crystal Growth & Design*. 19(11), 6308–6314. 10.1021/acs.cgd.9b00793

[104] Goel, N., & Kumar, N. 2018. "A Dual-Functional Luminescent Tb(III) Metal–Organic Framework for the Selective Sensing of Acetone and TNP in Water". *RSC Advances*. 8(20), 10746–10755. 10.1039/C7RA13494K

[105] Yang, X., Ren, Y., Chai, H., Hou, X., Wang, Z., & Wang, J. 2021. "Highly Sensitive Detection of Nitrobenzene by a Series of Fluorescent 2D Zinc(II) Metal–Organic Frameworks with a Flexible Triangular Ligand". *RSC Advances*. 11(39), 23975–23984. 10.1039/D1RA03737D

[106] Vellingiri, K., Boukhvalov, D. W., Pandey, S. K., Deep, A., & Kim, K.-H. 2017. "Luminescent Metal-Organic Frameworks for the Detection of Nitrobenzene in Aqueous Media". *Sensors and Actuators B: Chemical*. 245, 305–313. 10.1016/j.snb.2017.01.126

[107] Ahamad, M. N., Shahid, M., Ahmad, M., & Sama, F. 2019. "Cu(II) MOFs Based on Bipyridyls: Topology, Magnetism, and Exploring Sensing Ability toward Multiple Nitroaromatic Explosives". *ACS Omega*. 4(4), 7738–7749. 10.1021/acsomega.9b00715

[108] Yuan, S., Bo, X., & Guo, L. 2018. "In-Situ Growth of Iron-Based Metal-Organic Framework Crystal on Ordered Mesoporous Carbon for Efficient Electrocatalysis of p -Nitrotoluene and Hydrazine". *Analytica Chimica Acta*. 1024, 73–83. 10.1016/j.aca.2018.03.064

[109] Lan, A., Li, K., Wu, H., Olson, D. H., Emge, T. J., Ki, W., Hong, M., & Li, J. 2009. "A Luminescent Microporous Metal-Organic Framework for the Fast and Reversible Detection of High Explosives". *Angewandte Chemie International Edition*. 48(13), 2334–2338. 10.1002/anie.200804853

[110] Hu, J.-S., Dong, S.-J., Wu, K., Zhang, X.-L., Jiang, J., Yuan, J., & Zheng, M.-D. 2019. "An Ultrastable Magnesium-Organic Framework as Multi-Responsive Luminescent Sensor for Detecting Trinitrotoluene and Metal Ions with High Selectivity and Sensitivity". *Sensors and Actuators B: Chemical*. 283, 255–261. 10.1016/j.snb.2018.12.022

[111] Li, H.-Y., Zhao, S.-N., Zang, S.-Q., & Li, J. 2020. "Functional Metal–Organic Frameworks as Effective Sensors of Gases and Volatile Compounds". *Chemical Society Reviews*. 49(17), 6364–6401. 10.1039/C9CS00778D

[112] Bloch, E. D., Queen, W. L., Krishna, R., Zadrozny, J. M., Brown, C. M., & Long, J. R. 2012. "Hydrocarbon Separations in a Metal-Organic Framework with Open Iron(II) Coordination Sites". *Science*. 335(6076), 1606–1610. 10.1126/science.1217544

[113] Peng, Y.-L., Pham, T., Li, P., Wang, T., Chen, Y., Chen, K.-J., Forrest, K. A., Space, B., Cheng, P., & Zaworotko, M. J. 2018. "Robust Ultramicroporous Metal-Organic Frameworks with Benchmark Affinity for Acetylene". *Angewandte Chemie International Edition*. 57(34), 10971–10975. 10.1002/anie.201806732

[114] Xie, L.-H., Liu, X.-M., He, T., & Li, J.-R. 2018. "Metal-Organic Frameworks for the Capture of Trace Aromatic Volatile Organic Compounds". *Chem*. 4(8), 1911–1927. 10.1016/j.chempr.2018.05.017

[115] Zhang, M., Feng, G., Song, Z., Zhou, Y.-P., Chao, H.-Y., Yuan, D., Tan, T. T. Y., Guo, Z., Hu, Z., & Tang, B. Z. 2014. "Two-Dimensional Metal–Organic Framework with Wide Channels and Responsive Turn-On Fluorescence for the Chemical Sensing of Volatile Organic Compounds". *Journal of the American Chemical Society*. 136(20), 7241–7244. 10.1021/ja502643p

[116] Norouzi, F., & Khavasi, H. R. 2020. "Iodine Decorated-UiO-67 MOF as a Fluorescent Sensor for the Detection of Halogenated Aromatic Hydrocarbons". *New Journal of Chemistry*. 44(21), 8937–8943. 10.1039/D0NJ01149E

[117] Liu, C.-Y., Chen, X.-R., Chen, H.-X., Niu, Z., Hirao, H., Braunstein, P., & Lang, J.-P. 2020. "Ultrafast Luminescent Light-Up Guest Detection Based on the Lock of the Host Molecular Vibration". *Journal of the American Chemical Society*. 142(14), 6690–6697. 10.1021/jacs.0c00368

[118] Yu, Y., Ma, J.-P., Zhao, C.-W., Yang, J., Zhang, X.-M., Liu, Q.-K., & Dong, Y.-B. 2015. "Copper(I) Metal–Organic Framework: Visual Sensor for Detecting Small Polar Aliphatic Volatile Organic Compounds". *Inorganic Chemistry*. 54(24), 11590–11592. 10.1021/acs.inorgchem.5b02150

[119] Yi, F. Y., Wang, S. C., Gu, M., Zheng, J. Q., & Han, L. 2018. "Highly Selective Luminescent Sensor for CCl₄ Vapor and Pollutional Anions/Cations Based on a Multi-Responsive MOF". *Journal of Materials Chemistry C*. 6(8), 2010–2018. 10.1039/c7tc05707e

[120] Xu, X.-Y., & Yan, B. 2017. "Eu(III)-Functionalized ZnO@MOF Heterostructures: Integration of Pre-Concentration and Efficient Charge Transfer for the Fabrication of a Ppb-Level Sensing Platform for Volatile Aldehyde Gases in Vehicles". *Journal of Materials Chemistry A*. 5(5), 2215–2223. 10.1039/C6TA10019H

[121] Fan, T.-T., Li, J.-J., Qu, X.-L., Han, H.-L., & Li, X. 2015. "Metal(II)–Organic Frameworks with 3,3′-Diphenyldicarboxylate and 1,3-Bis(4-Pyridyl)Propane: Preparation, Crystal Structures and Luminescence". *CrystEngComm*. 17(48), 9443–9451. 10.1039/C5CE01772F

[122] Li, J., Yao, S.-L., Liu, S.-J., & Chen, Y.-Q. 2021. "Fluorescent Sensors for Aldehydes Based on Luminescent Metal–Organic Frameworks". *Dalton Transactions*. 50(21), 7166–7175. 10.1039/D1DT00890K

[123] Nandi, S., Sharma, E., Trivedi, V., & Biswas, S. 2018. "Metal–Organic Framework Showing Selective and Sensitive Detection of Exogenous and Endogenous Formaldehyde". *Inorganic Chemistry*. 57(24), 15149–15157. 10.1021/acs.inorgchem.8b02411

[124] Diamantis, S. A., Pournara, A. D., Hatzidimitriou, A. G., Manos, M. J., Papaefstathiou, G. S., & Lazarides, T. 2018. "Two New Alkaline Earth Metal Organic Frameworks with the Diamino Derivative of Biphenyl-4,4′-Dicarboxylate as Bridging Ligand: Structures, Fluorescence and Quenching by Gas Phase Aldehydes". *Polyhedron*. 153, 173–180. 10.1016/j.poly.2018.07.010

[125] Zhao, F., Guo, X.-Y., Dong, Z.-P., Liu, Z.-L., & Wang, Y.-Q. 2018. "3D Ln(III)-MOFs: Slow Magnetic Relaxation and Highly Sensitive Luminescence Detection of Fe^{3+} and Ketones". *Dalton Transactions*. 47(27), 8972–8982. 10.1039/C8DT01034J

[126] Li, S., Lu, L., Zhu, M., Yuan, C., & Feng, S. 2018. "A Bifunctional Chemosensor for Detection of Volatile Ketone or Hexavalent Chromate Anions in Aqueous Solution Based on a Cd(II) Metal–Organic Framework". *Sensors and Actuators B: Chemical*. 258, 970–980. 10.1016/j.snb.2017.11.142

[127] Dang, S., Min, X., Yang, W., Yi, F.-Y., You, H., & Sun, Z.-M. 2013. "Lanthanide Metal-Organic Frameworks Showing Luminescence in the Visible and Near-Infrared Regions with Potential for Acetone Sensing". *Chemistry-A European Journal*. 19(50), 17172–17179. 10.1002/chem.201301346

[128] Wang, X., Zhang, L., Yang, J., Liu, F., Dai, F., Wang, R., & Sun, D. 2015. "Lanthanide Metal–Organic Frameworks Containing a Novel Flexible Ligand for Luminescence Sensing of Small Organic Molecules and Selective Adsorption". *Journal of Materials Chemistry A*. 3(24), 12777–12785. 10.1039/C5TA00061K

[129] Zhou, J.-M., Shi, W., Li, H.-M., Li, H., & Cheng, P. 2014. "Experimental Studies and Mechanism Analysis of High-Sensitivity Luminescent Sensing of Pollutional Small Molecules and Ions in Ln_4O_4 Cluster Based Microporous Metal–Organic Frameworks". *The Journal of Physical Chemistry C*. 118(1), 416–426. 10.1021/jp4097502

[130] Sharma, A., Kim, D., Park, J.-H., Rakshit, S., Seong, J., Jeong, G. H., Kwon, O.-H., & Lah, M. S. 2019. "Mechanistic Insight into the Sensing of Nitroaromatic Compounds by Metal-Organic Frameworks". *Communications Chemistry*. 2(1), 39. 10.1038/s42004-019-0135-2

[131] Shi, Y.-X., Hu, F.-L., Zhang, W.-H., & Lang, J.-P. 2015. "A Unique Zn(II)-Based MOF Fluorescent Probe for the Dual Detection of Nitroaromatics and Ketones in Water". *CrystEngComm*. 17(48), 9404–9412. 10.1039/C5CE02000J

[132] Desai, A. V., Sharma, S., Roy, A., Shirolkar, M. M., & Ghosh, S. K. 2020. "Specific Recognition of Toxic Allyl Alcohol by Pore-Functionalized Metal–Organic Frameworks". *Molecular Systems Design & Engineering*. 5(2), 469–476. 10.1039/C9ME00133F

[133] Vandezande, W., Janssen, K. P. F., Delport, F., Ameloot, R., De Vos, D. E., Lammertyn, J., & Roeffaers, M. B. J. 2017. "Parts per Million Detection of Alcohol Vapors via Metal Organic Framework Functionalized Surface Plasmon Resonance Sensors". *Analytical Chemistry*. 89(8), 4480–4487. 10.1021/acs.analchem.6b04510

[134] Sachdeva, S., Soccol, D., Gravesteijn, D. J., Kapteijn, F., Sudhölter, E. J. R., Gascon, J., & de Smet, L. C. P. M. 2016. "Polymer–Metal Organic Framework Composite Films as Affinity Layer for Capacitive Sensor Devices". *ACS Sensors*. 1(10), 1188–1192. 10.1021/acssensors.6b00295

[135] Woellner, M., Hausdorf, S., Klein, N., Mueller, P., Smith, M. W., & Kaskel, S. 2018. "Adsorption and Detection of Hazardous Trace Gases by Metal-Organic Frameworks". *Advanced Materials*. 30(37), 1704679. 10.1002/adma.201704679

[136] Shustova, N. B., Cozzolino, A. F., Reineke, S., Baldo, M., & Dincă, M. 2013. "Selective Turn-On Ammonia Sensing Enabled by High-Temperature Fluorescence in Metal–Organic Frameworks with Open Metal Sites". *Journal of the American Chemical Society*. 135(36), 13326–13329. 10.1021/ja407778a

[137] Campbell, M. G., Sheberla, D., Liu, S. F., Swager, T. M., & Dincă, M. 2015. "Cu$_3$ (Hexaiminotriphenylene)$_2$: An Electrically Conductive 2D Metal-Organic Framework for Chemiresistive Sensing". *Angewandte Chemie International Edition*. 54(14), 4349–4352. 10.1002/anie.201411854

[138] Rubio-Giménez, V., Almora-Barrios, N., Escorcia-Ariza, G., Galbiati, M., Sessolo, M., Tatay, S., & Martí-Gastaldo, C. 2018. "Origin of the Chemiresistive Response of Ultrathin Films of Conductive Metal–Organic Frameworks". *Angewandte Chemie International Edition*. 57(46), 15086–15090. 10.1002/anie.201808242

[139] Hao, J.-N., & Yan, B. 2016. "Simultaneous Determination of Indoor Ammonia Pollution and Its Biological Metabolite in the Human Body with a Recyclable Nanocrystalline Lanthanide-Functionalized MOF". *Nanoscale*. 8(5), 2881–2886. 10.1039/C5NR06066D

[140] Assen, A. H., Yassine, O., Shekhah, O., Eddaoudi, M., & Salama, K. N. 2017. "MOFs for the Sensitive Detection of Ammonia: Deployment of Fcu-MOF Thin Films as Effective Chemical Capacitive Sensors". *ACS Sensors*. 2(9), 1294–1301. 10.1021/acssensors.7b00304

[141] Vikrant, K., Kumar, V., Ok, Y. S., Kim, K.-H., & Deep, A. 2018. "Metal-Organic Framework (MOF)-Based Advanced Sensing Platforms for the Detection of Hydrogen Sulfide". *TrAC Trends in Analytical Chemistry*. 105, 263–281. 10.1016/j.trac.2018.05.013

[142] Liu, D., Li, B., Wu, J., & Liu, Y. 2020. "Sorbents for Hydrogen Sulfide Capture from Biogas at Low Temperature: A Review". *Environmental Chemistry Letters*. 18(1), 113–128. 10.1007/s10311-019-00925-6

[143] Nagarkar, S. S., Saha, T., Desai, A. V., Talukdar, P., & Ghosh, S. K. 2015. "Metal-Organic Framework Based Highly Selective Fluorescence Turn-on Probe for Hydrogen Sulphide". *Scientific Reports*. 4(1), 7053. 10.1038/srep07053

[144] Ma, Y., Su, H., Kuang, X., Li, X., Zhang, T., & Tang, B. 2014. "Heterogeneous Nano Metal–Organic Framework Fluorescence Probe for Highly Selective and Sensitive Detection of Hydrogen Sulfide in Living Cells". *Analytical Chemistry*. 86(22), 11459–11463. 10.1021/ac503622n

[145] Zhang, J., Liu, F., Gan, J., Cui, Y., Li, B., Yang, Y., & Qian, G. 2019. "Metal-Organic Framework Film for Fluorescence Turn-on H$_2$S Gas Sensing and Anti-Counterfeiting Patterns". *Science China Materials*. 62(10), 1445–1453. 10.1007/s4 0843-019-9457-5

[146] Zhao, X., Zhang, L., Bai, J., Wu, P., Li, Y., Liang, L., Xie, L., & Wang, J. 2020. "A Copper-Based Metal-Organic Framework for Ratiometric Detection of Hydrogen Sulfide with High Sensitivity and Fast Response". *Spectrochimica Acta Part A: Molecular and Biomolecular Spectroscopy*. 243, 118794. 10.1016/j.saa.2020.11 8794

[147] Tchalala, M. R., Bhatt, P. M., Chappanda, K. N., Tavares, S. R., Adil, K., Belmabkhout, Y., Shkurenko, A., Cadiau, A., Heymans, N., & De Weireld, G., 2019. "Fluorinated MOF Platform for Selective Removal and Sensing of SO2 from Flue Gas and Air". *Nature Communications*. 10(1), 1328. 10.1038/s41467-019-09157-2

[148] Martínez-Ahumada, E., López-Olvera, A., Jancik, V., Sánchez-Bautista, J. E., González-Zamora, E., Martis, V., Williams, D. R., & Ibarra, I. A. 2020. "MOF Materials for the Capture of Highly Toxic H$_2$S and SO$_2$". *Organometallics*. 39(7), 883–915. 10.1021/acs.organomet.9b00735

[149] Wang, M., Guo, L., & Cao, D. 2018. "Amino-Functionalized Luminescent Metal–Organic Framework Test Paper for Rapid and Selective Sensing of SO$_2$ Gas and Its Derivatives by Luminescence Turn-On Effect". *Analytical Chemistry*, 90(5), 3608–3614. 10.1021/acs.analchem.8b00146

[150] DMello, M. E., Sundaram, N. G., Singh, A., Singh, A. K., & Kalidindi, S. B. 2019. "An Amine Functionalized Zirconium Metal–Organic Framework as an Effective Chemiresistive Sensor for Acidic Gases". *Chemical Communications*. 55(3), 349–352. 10.1039/C8CC06875E

[151] Schulz, M., Gehl, A., Schlenkrich, J., Schulze, H. A., Zimmermann, S., & Schaate, A. 2018. "A Calixarene-Based Metal-Organic Framework for Highly Selective NO_2 Detection". *Angewandte Chemie International Edition*. 57(39), 12961–12965. 10.1002/anie.201805355

[152] Wu, P., Wang, J., He, C., Zhang, X., Wang, Y., Liu, T., & Duan, C. 2012. "Luminescent Metal-Organic Frameworks for Selectively Sensing Nitric Oxide in an Aqueous Solution and in Living Cells". *Advanced Functional Materials*. 22(8), 1698–1703. 10.1002/adfm.201102157

[153] Stassen, I., Dou, J.-H., Hendon, C., & Dincă, M. 2019. "Chemiresistive Sensing of Ambient CO_2 by an Autogenously Hydrated Cu_3(Hexaiminobenzene)$_2$ Framework". *ACS Central Science*. 5(8), 1425–1431. 10.1021/acscentsci.9b00482

7 Recent Advancements in Metal-Organic Frameworks for Drug Delivery

Sachin Mishra, Fulden Ulucan-Karnak, and Cansu İlke Kuru

CONTENTS

7.1 INTRODUCTION

Nanoscale metal-organic frameworks (NMOFs), an extraordinary class based on porous hybrid materials assembled with metal ions and organic ligands as a linker via coordination chemistry, have been designed and functionalized for the wide potentials in biomedical applications [1–5], catalysis [6,7] and energy [8,9]. As a unique type of porous hybrid material, NMOFs have creative particle composition and size that can be easily attuned to optimize their functions, which is different from the other nanomaterials in the biosensing technology. In 2008, Asefa and coworkers [10] proposed biosensors based on NMOF. In recent years, NMOFs gained much more attention in the field of drug delivery and disease diagnosis. Compared with the traditional methodology for the design of nanomaterials, NMOFs offer tremendous features such as high porosity, large surface area as well as loading efficiency of drugs, ease to functionalized, biodegradability, and good biocompatibility that can improve the bioavailability and efficiency of the drug.

DOI: 10.1201/9781003252061-7

Generally, NMOFs demonstrated a high degree of robustness along with a bendable nature; the greater flexibility while incorporating metal clusters with organic ligands. The series of NMOF structures presented a high degree of versatility in their chemical composition with their high surface area, excellent surface modification for biomedicine applications, and large pore sizes that ease for casing theranostic agents. In particular, the synthetic design, tunable features such as stability, diverse morphology, biodegradability, porosity, and conductivity can be altered and applicable in various fields, such as molecular sensing and so on. In this chapter, we will highlight the structural features, synthesis, functionalization, and applications of NMOFs as bioimaging and theranostics agents in the biomedical field. As compared to other chapters, this chapter will solely focus on and discuss the bioimaging and theranostics applications of NMOFs. NMOFs have the ability to make multifunctional entities using two different synthesis processes called direct or post-synthesis alteration for effective applications in the health sector.

7.2   SURFACE DESIGN AND MODIFICATIONS OF NMOFS

Surface design and modification of nanomaterials vary the MOF material applications, such as bioimaging and medical therapy. Surface modifications occur on the surface of the pores of the material. The manipulation of MOFs' surface to fit into the specific parameters and attain the desired functionality is of dominant importance and regulates the performance of nano-MOFs. Graft functional MOFs' surface of the particles attained stimuli-responsive features. Without altering their framework and porosity, the surface functionalization of MOFs is also substantial for biomedical applications. There are two post-synthesis modifications for bioengineering the surface of MOFs [11–13]. As we know that MOFs are made up of metal clusters and organic linkers with coordination bonds, the first step in modification is to locate the active site onto the linker and ready to conjugate with the target molecule [14,15]. The second modification step is to handle the chelation process where the metal ions and target molecule bind together to form a bridge for the better functionalization process of MOFs [16,17]. This study reduces the application of MOFs for bioimaging, analytical recognition, and drug therapy. DNA/RNA oligonucleotides improve the target specificity and selectivity, and high affinity in terms of the surface of MOF nanoparticles (Figure 7.1). In this chapter, we discuss the design, functionalization, and application of MOFs.

7.3   NMOFS AS SMART NANO-BASED DRUG CARRIER SYSTEMS

NMOFs gained attention as smart drug carriers due to increased surface areas and tunable properties like pore size and volume, and chemical and electrical properties. All of these unique properties have raised NMOFs [18] in especially therapeutic treatments of various diseases as promising and novel drug delivery platforms. NMOFs have unique features of structure and large surface area. With the aid of these properties, pharmaceutical agents can be bonded with the outer surface, or encapsulated into the NMOFs. It is possible to prevent the drug molecule from outside conditions such as pH, temperature, etc. with this novel outer surface. Also, drug

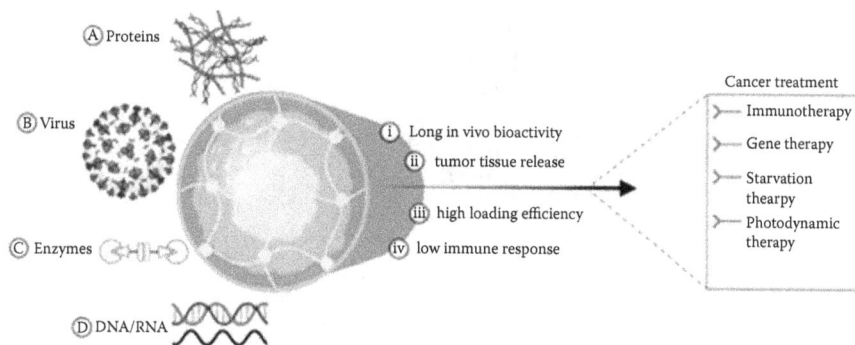

FIGURE 7.1 MOF-based biomineralization of proteins, enzymes, DNA/RNA, and virus, and their applications in cancer treatment.

carrier systems are helpful to enhance drug molecules' bioavailability and targeted releasing.

Enteral (into the gastrointestinal system) and parenteral (into the bloodstream) are the two most prevalent medication delivery methods (not into the gastrointestinal tract). Oral and rectal methods (into the gastrointestinal system) are two types of enteral routes. Sublingual (under the tongue) and injectable methods are two types of parenteral routes that differ depending on where they are administered: into a vein (intravenous), into a muscle (intramuscular), or beneath the skin (subcutaneous). Other parenteral methods include topical, which is applied to a bodily surface like the skin, nose, or eyes; pulmonary, which is applied to the lungs; and intraosseous, which is applied to the bone marrow [19,20].

When creating a medication dosage formulation, the desired route and volume of administration, the features and attributes of the planned absorption site, and the possible action of the drug on this site should all be considered [21]. In comparison to traditional systems, drug release systems have their unique administration pathway. The drug's journey into the targeted tissue or organ is referred to as this [22]. The medicine crosses over through the membrane to enter the circulatory system via diffusion or carrier-mediated membrane transporters mechanism. Drug-specific factors (solubility, pH, pKa, ionization, bioavailability), physicochemical variables, and patient-related factors are just a few of the variables that can affect the adsorption mechanism [23].

NMOF formulations for drugs could be performed to several environmental stimuli such as electricity, pressure, etc. These endogenous and exogenous stimuli provide conformational changes and lead the release profiles of the therapeutic agents [24–26]. Any parameter that affects the structural features of the NMOF nanocarrier can trigger a drug release (Figure 7.2). When designing NMOF-based nanocarrier systems that are for the delivery of pharmaceuticals, they should have a higher drug capacity [27].

For drug loading purposes, there are two techniques: the one-step method and the two-step method. The one-step method is the most convenient way of drug loading that could be performed during the synthesis (one-pot) step or utilizing drug

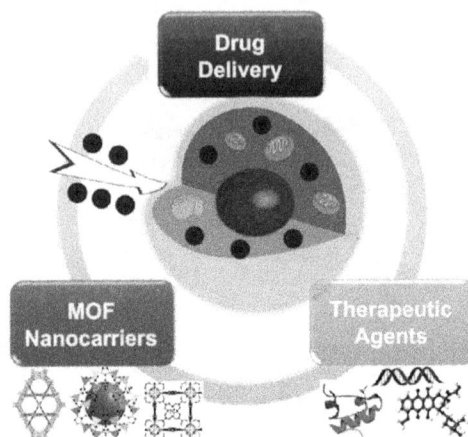

FIGURE 7.2 NMOFs as nanocarriers for drug delivery with the functionalization with therapeutic agents in biomedical applications (reproduced with permission from ref. Sun et al. 2020 [4]).

molecules as NMOF linkers. In the two-step method, drugs were loaded with the NMOFs in a drug solution or ground together with the drugs [4,28].

NMOFs' main limitations in drug delivery should be taken in consideration as toxicity, *in vivo* efficiency, and targeting. All of these parameters should be tested with many other studies. To translate these systems from laboratory to clinic, these limitations should be achieved by solutions. In these solution strategies, differentiation of synthesis pathways and advanced surface modification routes, use of endogenous bioligands should be examined. But they are still in proof-of-concept levels [29,30].

Bio-NMOFs are the novel and promising class of NMOFs, which are formulated with biological endogenous organic ligands. During the pandemic situation, encapsulated mRNA vaccines were developed against the SARS-CoV-2, but they still concern thrombosis and have possible side effects. It is suggested that if mRNA could be complexed with bio-NMOFs, mRNAs can be more stable and resistant to RNase-mediated degradation and improve targeted delivery of these in the cytosol with the aid of NMOFs' multi-responsive properties. Bio-NMOFs can solve the problems related with vaccine delivery via characteristics that can help biodistribution and protect the vaccine from degradation [31].

7.4 NMOFS AS MOLECULAR BIOIMAGING PROBES

NMOF-based nanomaterials were integrated to photoacoustic imaging (PAI), computed tomography imaging (CTI), photothermal imaging (PTI) and fluorescence imaging (FL). NMOF-based heterogeneous hybridization systems can be especially used in multimodal imaging methods that are essential for tumor diagnosis with defining an accurate position and realizing multimode image-guided therapy [32]. NMOFs can be designed with some intended properties [33]. For imaging applications, Mn, Fe, Zr, Hf, Gd and Tb are the typically used metal clusters in NMOFs [34].

NMOFs have a potential character as the nanocarrier to load fluorescent materials to provide NMOF-based fluorescent imaging probes. With these fluorescent-loaded NMOFs, in vivo imaging and living cell imaging can be realized. But of course, fluorescence intensity and efficient drug tracking should be considered [35]. NMOF properties can be differentiated regarding the intended use in bioimaging. To use NMOFs in optical imaging, they need to have fluorescence properties or be magnetic for MRI. As far as bioimaging is considered, the main limitations can be regarded as: (1) design of high resolution, stabile, and targeted lifetime imaging systems; (2) low biotoxicity and high biodegradability [36,37].

Bio-NMOFs can be also selected for imaging applications and ligands can be DNA, peptides, or biological molecules but there is a requirement of detailed studies to enlighten the interaction between bio-NMOFs and cells [2]. Besides DNA, amino acids, peptides, and saccharides have participated as bio-ligands to interact with metal ions to construct the novel bio-NMOFs for biomedical applications such as [Zn(curcumin)]n, called scCCNMOF-1 [38,39]. Recently, Yin and co-workers reported a type of 5-borobenzene 1,3-dicarboxylic acid (BBDC) [40] based bio-NMOF for MR image-guided tumor-targeted therapy. The structure of bio-NMOFs is a promising effect on good imaging of organisms with their controllable size and shape properties that allow endocytosis and help to increase their systemic circulation in blood [41].

7.5 NMOFS AS A THERANOSTICS

The theranostics concept has a meaning of a combination of diagnosis and treatment in a carrier; in this area, it can be NMOFs. This strategy is very useful for the clinicians to target tumors or other areas of interest accurately and provide certain treatment with the release of a required effective dose [42]. Integrating the imaging technologies and treatment strategies into one strategy for image-guided precise theranostics has become more interesting, especially in the cancer field. Zhu et al. developed a bio-based NMOF called Fe-DOX@Gd-NMOF (FDG) nanoparticles for multimodal image technique–guided compound antitumor therapy using chemo-phototherapy [43].

An ideal theranostic platform should have high cargo loading capacities, nanoscale sizes and targeted properties. NMOFs can co-deliver various bioactive agents to perform imaging-guided cancer theranostics [32,44,45]. NMOF-based theranostic platforms exhibit many impressive properties together that NMOF-based drug delivery systems and imaging probes have together. Also, they have challenges remaining as other NMOF-based systems [46]. Surface modification techniques can be implemented to improve the characteristics of the developed systems. Also, these technologies should be translated into clinics from the lab experiments, but the long-term toxicity, and biosafety problems should be solved immediately. It should be reminder that animal models and humans are not the same. Although the developed NMOF platforms exhibited great efficiency on animal models, their effects in humans must be researched. Further experimental studies are certainly needed to understand and validate NMOF-based systems and integrate them into clinical trials [47].

7.6   ADVANTAGES AND DISADVANTAGES OF MOFS IN DRUG DELIVERY SYSTEMS

Metal-organic frameworks (MOFs) are a special type of porous hybrid solid that combines metals and organic linkers. MOFs have several distinct advantages over conventional porous materials, including large surface area and porosity for high therapeutic drug loading [4,48]. Many medicinal compounds have been successfully delivered using MOF nanoparticles, owing to the particular properties of MOFs [4].

MOFs are more competitive as medication containers than other traditional pore materials [49,50]. Furthermore, interactions on medicinal molecules afforded by MOFs' distinctive open metal sites and possible organic functional groups may improve drug delivery efficiency. Various MOFs have been investigated as therapeutic agent carriers after being saturated in a composite containing drug molecules [48].

There are two types of organic linkers available: exogenous linkers and endogenous organic spacers. The most popular choice, exogenous linkers, are obtained or produced by ordinary substances without disturbing the body's function [51]. While using endogenous organic linkers is the best option for MOFs due to reuse in the body, the likelihood of unwanted effects is greatly decreased. As therapeutic agent carriers, the chemical instability of the composite is thought to be important with the product's interacts with the system without avoiding the endogenous buildup [48].

Furthermore, the degradable nature of the matrix's material has an impact on the efficiency with which drugs are released from the materials. Different levels of degradability are accomplished by the selection of the right metal ion, organic linker, and crystalline structure, which can range from a few hours to many weeks in the body's fluids [52,53]. However, a rigorous investigation of MOF stability is still needed to better understand its performance in organisms. Various therapeutic compounds have been effectively entrapped into MOFs using non-covalent or covalent techniques including highly difficult medicines and bioactive gas. In most cases, the absorbed medications were released in a controlled manner [48]. Furthermore, by using appropriate MOFs as carriers, the environment-triggered release might be achieved.

7.7   APPLICATIONS OF NANO METAL-ORGANIC FRAMEWORKS

7.7.1   MOF VACCINE

MOF-based vaccination platforms can be utilized to develop efficient vaccines for a variety of diseases, making MOFs more useful in biomedical fields [54]. Until now, MOF-based subunit vaccines have not been investigated. As a result, the creation of protein-encased MOF vaccines with simple synthesis processes and adequate size will pave the way for future MOF uses in biomedicine [55]. MOF-based vaccines have the potential to improve systemic immune response and induce a powerful immunological memory response. We believe that MOF-based vaccination platforms might be used to develop efficient vaccines for a variety of diseases, easing the application of MOFs in biomedical fields [56]. The MOF conjugated polymer delivery method is projected to be widely useful for peptides, nucleic acids, molecular immunopotentiators, chemotherapeutic medicines, and imaging contrast agents, among other therapeutic agents [52].

7.7.2 In-Vitro and In-Vivo Drug Delivery

Despite the obvious benefits, there are still a number of obstacles to overcome before clinical applications may be implemented. The first issue that must be addressed is the prevention of nanoscale MOF aggregation in bodily fluid. Although the strength of MOFs are defined in-vitro, and the regulating arrangement and shape of MOFs may give desirable steadiness in body fluid, in-vivo degradation mechanisms of MOFs must to be investigated to determine the true stability of MOFs in body fluids. Surface modification is another key feature of bioapplications to improve drug delivery efficiency. However, research on modified MOF nanoparticles is limited. Finally, examining the drug-loaded MOFs' in-vivo toxicity, efficiency, and pharmacokinetics are critical tasks that require thorough inquiry to estimate actual performance. Despite the fact that much more work needs to be executed before clinical trials, the MOFs have the right option in drug delivery functionalization [48].

7.8 CONCLUSION

Metal-organic frameworks (NMOFs) have developed as one of the most exciting archives of pore-based materials, with enormous applications possible in a wide range of fields. The bioanalytical and biomedical areas have benefited greatly from the development of the specific hybrid inorganic-organic NMOF-conjugated materials [57,58]. They have tremendous features for bio-applications, including low toxicity to live cells, inherent biodegradability, and the ability to synthesize with nanoscale diameters [59,60]. NMOFs have been reduced to nanometer diameters to generate nanoscale NMOFs in recent years, with initial biomedical applications in drug delivery and bioimaging focusing on their preliminary biomedical uses [58]. Finally, it is important to examine a NMOF-based drug delivery platform from operational design to trial evaluation within the contribution of expertize that would be the positive way in translational research technology [57].

REFERENCES

[1] Ma, X., Mathilde, L., and Serre, C., 2021. Metal-Organic Frameworks towards Bio-Medical Applications. *Materials Chemistry Frontiers*, 5, 15, 5573–5594. 10.1039/D1QM00784J

[2] Wang, H.-S., Wang, Y.-H., and Ding, Y., 2020. Development of Biological Metal-Organic Frameworks Designed for Biomedical Applications: From Bio-Sensing/Bio-Imaging to Disease Treatment. *Nanoscale Advances*, 2, 9, 3788–3797. 10.1039/D0NA00557F

[3] de Alencar Filho, J. M. T., Sampaio, P. A., de Carvalho, I. S., da Silva, A. R., Pereira, E. C. V., e Amariz, I. A., Nishimura, R. H. V., da Cruz Araújo, E. C., Rolim-Neto, P. J., and Rolim, L. A. 2021. Metal Organic Frameworks (NMOFs) with Therapeutic and Biomedical Applications: A Patent Review. *Expert Opinion on Therapeutic Patents*, 31, 10, 937–949. 10.1080/13543776.2021.1924149

[4] Sun, Y., Zheng, L. Yang Yu, and Li Xiaowei Yang Zunyi Yan He Cui Cheng Tan Weihong 2020. Metal-Organic Framework Nanocarriers for Drug Delivery in Biomedical Applications. *Nano-Micro Letters*, 12, 1, 103. 10.1007/s40820-020-00423-3.

[5] Keskin, S., and Kızılel, S., 2011. Biomedical Applications of Metal Organic Frameworks. *Industrial & Engineering Chemistry Research*, 50, 4, 1799–1812. 10.1021/ie101312k.

[6] Alhumaimess, M. S. 2020. Metal-Organic Frameworks and Their Catalytic Applications. *Journal of Saudi Chemical Society*, 24, 6, 461–473, 10.1016/j.jscs.2020.04.002

[7] Zhao, X., Feng, J., and Liu Jing Yang Guangming Wang Gui-Chang Cheng Peng 2018. An Efficient, Visible-Light-Driven, Hydrogen Evolution Catalyst NiS/ZnxCd1-xS Nanocrystal Derived from a Metal-Organic Framework. *Angewandte Chemie International Edition in English*, 57, 9790–9794. 10.1002/anie.201805425.

[8] Baumann, A. E., Burns, D. A., and Liu Binggian Toi V. 2019. Metal-Organic Framework Functionalization and Design Strategies for Advanced Electrochemical Energy Storage Devices. *Communications Chemistry*, 2, 86. 10.1038/s42004-019-0184-6

[9] Zhao, H. X., Zou, Q., and Sun Shao-Kai Yun Chunshui Zhang Xuejun Li Rui-Jun Fu Yan-Yan 2016. Theranostic Metal-Organic Framework Core-Shell Composites for Magnetic Resonance Imaging and Drug Delivery. *Chemical Science*, 7, 8, 5294–5301. 10.1039/C6SC01359G

[10] Asefa, T., and Shi, Y. L. 2008. Corrugated and Nanoporous Silica Microspheres: Synthesis by Controlled Etching, and Improving Their Chemical Adsorption and Application in Biosensing. *Journal of Materials Chemistry*, 18, 46, 5604–5614. 10.1039/B811034D

[11] Wang, S., Morris, W., and Liu Y. McGuirk M. Zhou Y. Hupp J. T., Farha O. K., Mirkin C. A., 2015. Surface-Specific Functionalization of Nanoscale Metal-Organic Frameworks. *Angewandte Chemie International Edition in English*, 54, 14738–14742. 10.1002/anie.201506888

[12] Morris, W., and Briley, W. E. Auyeung, E., and Cabezas, M. D., Mirkin, C. A., 2014. Nucleic Acid-Metal Organic Framework (MOF) Nanoparticle Conjugates. *Journal of the American Chemical Society*, 136, 7261–7264. 10.1021/ja503215w

[13] Deria, P., Bury, W., and Hod, I., Kung C. W., Karagiaridi O., et-al. 2015. MOF Functionalization via Solvent-Assisted Ligand Incorporation: Phosphonates vs Carboxylates. *Inorganic Chemistry*, 54, 2185–2192. 10.1021/ic502639v

[14] Zhu, W., Xiang, G. Shang J., and Guo J., Motevalli B., Durfee P., Agola J. o., Coker E. N., Brinker C. j., 2018. Versatile Surface Functionalization of Metal-Organic Frameworks through Direct Metal Coordination with a Phenolic Lipid Enables Diverse Applications. *Advanced Functional Materials*, 28, 1705274. 10.1002/adfm.201705274

[15] Yi, X.-C., Xi, F.-G. Oi Y., and Gao En-Qing 2015. Synthesis and Click Modification of an Azido-Functionalized Zr (iv) Metal-Organic Framework and a Catalytic Study. *RSC Advances*, 5, 893–900. 10.1039/C4RA09883H

[16] Wang, S., McGuirk, C. M. Ross M., B., and Wang S., Chen P., Xing Hang Liu Y., Mirkin C. A., 2017. General and Direct Method for Preparing Oligonucleotide-Functionalized Metal-Organic Framework Nanoparticles. *Journal of American Chemical Society*, 139, 9827–9830. 10.1021/jacs.7b05633

[17] Tamames-Tabar, C., García-Márquez, A., Blanco-Prieto, M. J., Serre, C., and Horcajada, P. 2014. MOFs in Pharmaceutical Technology. In *Bio- and Bioinspired Nanomaterials* (eds. D. Ruiz-Molina, F. Novio and C. Roscini). Wiley: 83–112. 10.1002/9783527675821.ch04

[18] Jeyaseelan, C., Jain, P., and Soin Deeya Gupta Deepshikha 2021. Metal Organic Frameworks: An Effective Application in Drug Delivery Systems. In *Inorganic and Nano-Metal Chemistry.* : 1–13. 10.1080/24701556.2021.1956966 Taylor & Francis

[19] Doggrell, Sheila A. 2017. *Introduction to pharmacology, and routes of drugs administration and absorption.*

[20] Verma, P., Thakur, A. S., Deshmukh, K., Jha, A. K., and Verma, S. 2010. Routes of Drug Administration. *International Journal of Pharmaceutical Studies and Research*, 1, 1, 54–59.

[21] Ruiz, M. E., and Scioli Montoto, S. 2018. Routes of Drug Administration. In Talavi Alan Dr. Quiroga Pablo A. M. Prof. *ADME Processes in Pharmaceutical Sciences.* Cham: Springer International Publishing: 97–133.

[22] Reza Rezaie, H., Esnaashary, M., Aref arjmand, A., and Öchsner, A. 2018. Classification of Drug Delivery Systems*SpringerBriefs in Applied Sciences and Technology book series (BRIEFSAPPLSCIENCES).* 9–25. https://doi.org/10.1007/978-981-10-0503-9_2

[23] Alagga, A. A., and Gupta, V. 2021. *Drug Absorption.* StatPearls.

[24] Yan, C., Jin, Y., and Zhao, C. 2021. Environment Responsive Metal-Organic Frameworks as Drug Delivery System for Tumor Therapy. *Nanoscale Research Letters*, 16, 1, 140. 10.1186/s11671-021-03597-w

[25] Wang, Y., Jianhua, Y., and When Nachuan Xiong Hongjie CaiShangdong He Qunye Hu Yaqin Peng Dongming et-al. 2020. Metal-Organic Frameworks for Stimuli-Responsive Drug Delivery. *Biomaterials*, 230, 119619. 10.1016/j.biomaterials.2019.119619

[26] Wei, M., Wan, Y., and Zhang, X. 2021. Metal-Organic Framework-Based Stimuli-Responsive Polymers. *Journal of Composites Science* 5, 4, 101. 10.3390/jcs5040101.

[27] Hashemzadeh, A., Drummen, G. P. C., Avan, A., Darroudi, M., Khazaei, M., Khajavian, R., Rangrazi, A., and Mirzaei, M. 2021. When Metal-Organic Framework Mediated Smart Drug Delivery Meets Gastrointestinal Cancers. *Journal of Materials Chemistry B*, 9, 19, 3967–3982. 10.1039/D1TB00155H

[28] He, S., Wu, L., Li, X., Sun, H., Xiong, T., Liu, J., Huang, C., Xu Huipeng Sun Huimin Chen Weidong Gref Xuandra Zhang Jiwen 2021. Metal-Organic Frameworks for Advanced Drug Delivery. *Acta Pharmaceutica Sinica B*, 11, 8, 2362–2395. 10.1016/j.apsb.2021.03.019

[29] Ma, X., Lepoitevin, M., and Serre, C. 2021. Metal–Organic Frameworks towards Bio-Medical Applications. *Materials Chemistry Frontiers*, 5, 5573–5594, 10.1039/D1QM00784J

[30] Shreedevi, A. K., Bhuvaneshwari, B., Bhunia, S., Jaiswal, M. K., Verma, K., Prateek, Khademhosseini, A., Gupta, R. K., and Gaharwar, A. K. 2021. Metal Organic Frameworks for Biomedical Applications. *WIREs Nanomedicine and Nanobiotechnology* 13, 2. 10.1002/wnan.1674

[31] Saeb, M. R., Rabiee, N., Mozafari, M., and Mostafavi, E. 2021. Metal-Organic Frameworks (MOFs)-Based Nanomaterials for Drug Delivery. *Materials*, 14, 13, 3652. 10.3390/ma14133652

[32] Lai, X., Jiang, H., and Wang, X. 2021. Biodegradable Metal Organic Frameworks for Multimodal Imaging and Targeting Theranostics. *Biosensors*, 11, 9, 299. 10.3390/bios11090299

[33] Pala, R., Pattnaik, S., Zeng, Y., Busi, S., Nauli, S. M., Mozafari Masoud , and Liu, G. 2020. Functional MOFs as Molecular Imaging Probes and Theranostics. In *Metal-Organic Frameworks for Biomedical Applications.* Elsevier: 425–443. 10.1016/B978-0-12-816984-1.00021-4

[34] Banerjee, S., Lollar, C. T., Xiao, Z., Fang, Y., and Zhou, H.-C. 2020. Biomedical Integration of Metal-Organic Frameworks. *Trends in Chemistry*, 2, 5, 467–479. 10.1016/j.trechm.2020.01.007

[35] Liu, M., Ren, X., Meng, X., and Li, H. 2021. Metal-Organic Frameworks-Based Fluorescent Nanocomposites for Bioimaging in Living Cells and in vivo. *Chinese Journal of Chemistry*, 39, 2, 473–487. 10.1002/cjoc.202000410

[36] Liu, Y., Jiang, T., and Liu, Z. 2022. Metal-Organic Frameworks for Bioimaging: Strategies and Challenges. *Nanotheranostics*, 6, 2, 143–160. 10.7150/ntno.63458

[37] Yang, C., Chen, K., Chen, M., Hu, X., Huan, S.-Y., Chen, L., Song, G., and Zhang, X.-B. 2019. Nanoscale Metal-Organic Framework Based Two-Photon Sensing Platform for Bioimaging in Live Tissue. *Analytical Chemistry*, 91, 4, 2727–2733. 10.1021/acs.analchem.8b04405

[38] Portol´es-Gil, N., Lanza, A., Aliaga-Alcalde, N., Ayll´on, J. A., Gemmi, M., Mugnaioli, E., L´opez-Periago, A. M., and Domingo, C. 2018. Crystalline Curcumin bioMOF Obtained by Precipitation in Supercritical CO2 and Structural Determination by Electron Diffraction Tomography, *ACS Sustainable Chemistry & Engineering*, 6, 12309–12319. 10.1021/acssuschemeng.8b02738.

[39] Li, Z., Peng, Y., Pang, X., and Tang, B. 2020. Potential Therapeutic Effects of Mg/HCOOH Metal Organic Framework on Relieving Osteoarthritis. *Chem Med Chem*, 15, 13–16. 10.1002/cmdc.201900546.

[40] Zhang, H., Shang, Y., Li, Y. H., Sun, S. K., and Yin, X. B. 2019. Smart Metal–Organic Framework-Based Nanoplatforms for Imaging-Guided Precise Chemotherapy. *ACS Applied Materials & Interfaces*, 11, 1886–1895. 10.1021/acsami.8b19048.

[41] Sun, B., Bilal, M., Jia, S., Jiang, Y., and Cui, J. 2019. Design and Bio-Applications of Biological Metal-Organic Frameworks. *Korean Journal of Chemical Engineering*, 36, 12, 1949–1964. 10.1007/s11814-019-0394-8

[42] Chedid, G., and Yassin, A. 2018. Recent Trends in Covalent and Metal Organic Frameworks for Biomedical Applications. *Nanomaterials*, 8, 11, 916. 10.3390/nano8110916

[43] Zhu, Y., Xin, N., Qiao, Z., Chen, S., Zeng, L., Zhang, Y., Wei, D., Sun, J., and Fan, H. 2020. Bioactive NMOFs Based Theranostic Agent for Highly Effective Combination of Multimodal Imaging and Chemo-Phototherapy. *Advanced Healthcare Materials*, 9, 2000205.

[44] Yang, J., and Yang, Y.-W. 2020. Metal-organic Framework-Based Cancer Theranostic Nanoplatforms. *View*, 1, 2. 10.1002/viw2.20

[45] Pandey, A., Dhas, N., Deshmukh, P., Caro, C., Patil, P., García-Martín, M. L., Padya, B., Nikam, A., Mehta, T., and Mutalik, S. 2020. Heterogeneous Surface Architectured Metal-Organic Frameworks for Cancer Therapy, Imaging, and Biosensing: A State-of-the-Art Review. *Coordination Chemistry Reviews*, 409 (May), 213212. 10.1016/j.ccr.2020.213212

[46] Lakshmi, B. A., and Kim, S. 2019. Current and Emerging Applications of Nanostructured Metal-Organic Frameworks in Cancer-Targeted Theranostics. *Materials Science and Engineering: C*, 105 (December), 110091. 10.1016/j.msec.2019.110091

[47] Sun, W., Li, S., Tang, G., Luo, Y., Ma, S., Sun, S., Ren, J., Gong, Y., and Xie, C. 2020. Recent Progress of Nanoscale Metal-Organic Frameworks in Cancer Theranostics and the Challenges of Their Clinical Application. *International Journal of Nanomedicine*, 14 (January), 10195–10207. 10.2147/IJN.S230524

[48] Sun, C. Y., Qin, C., Wang, X. L., and Su, Z. M. 2013. Metal-Organic Frameworks as Potential Drug Delivery Systems. *Expert Opinion on Drug Delivery*, 10, 1, 89–101.

[49] Simagina, A. A., Polynski, M. V., Vinogradov, A. V., and Pidko, E. A. 2018. Towards Rational Design of Metal-Organic Framework-Based Drug Delivery Systems. *Russian Chemical Reviews*, 87, 9, 831.

[50] Mallakpour, S., Nikkhoo, E., and Hussain, C. M. 2022. Application of MOF Materials as Drug Delivery Systems for Cancer Therapy aAnd Dermal Treatment. *Coordination Chemistry Reviews*, 451, 214262.

[51] Gutov, O. V., Hevia, M. G., Escudero-Adan, E. C., and Shafir, A. 2015. Metal-Organic Framework (MOF) Defects under Control: Insights into the Missing Linker Sites and Their Implication In The Reactivity of Zirconium-Based Frameworks. *Inorganic Chemistry*, 54, 17, 8396–8400.

[52] Li, X., Wang, X., Ito, A., and Tsuji, N. M. 2020. A Nanoscale Metal Organic Frameworks-Based Vaccine Synergises with PD-1 Blockade to Potentiate Antitumour Immunity. *Nature Communications*, 11, 1, 1–15.

[53] Horcajada, P., Gref, R., Baati, T., Allan, P. K., Maurin, G., Couvreur, P., and Serre, C. 2012. Metal–Organic Frameworks in Biomedicine. *Chemical Reviews*, 112, 2, 1232–1268. 10.1021/cr200256v.

[54] Liu, Y., Xu, Y., Tian, Y., Chen, C., Wang, C., and Jiang, X. 2014. Functional Nanomaterials Can Optimize the Efficacy of Vaccines. *Small*, 10, 22, 4505–4520.

[55] Zhang, Y., Liu, C., Wang, F., Liu, Z., Ren, J., and Qu, X. 2017. Metal-Organic-Framework-Supported Immunostimulatory Oligonucleotides for Enhanced Immune Response and Imaging. *Chemical Communications*, 53, 11, 1840–1843.

[56] Zhang, Y., Wang, F., Ju, E., Liu, Z., Chen, Z., Ren, J., and Qu, X. 2016. Metal-Organic-Framework-Based Vaccine Platforms for Enhanced Systemic Immune and Memory Response. *Advanced Functional Materials*, 26, 35, 6454–6461.

[57] Carrillo-Carrión, C. 2020. Nanoscale Metal-Organic Frameworks as Key Players in the Context of Drug Delivery: Evolution toward Theranostic Platforms. *Analytical and Bioanalytical Chemistry*, 412, 1, 37–54.

[58] Zhao, Y., Song, Z., Li, X., Sun, Q., Cheng, N., Lawes, S., and Sun, X. 2016. Metal Organic Frameworks for Energy Storage and Conversion, *Energy Storage Materials*, 2, 35–62, 10.1016/j.ensm.2015.11.005

[59] Yang, B., Shen, M., Liu, J., and Ren, F. 2017. Post-Synthetic Modification Nanoscale Metal-Organic Frameworks for Targeted Drug Delivery in Cancer Cells. *Pharmaceutical Research*, 34, 11, 2440–2450.

[60] Bal, Y. 2019. Nanomaterials for Drug Delivery: Recent Developments in Spectroscopic Characterization. Mohapatra S. S., Ranjan S., Thomas S., *Characterization and Biology of Nanomaterials for Drug Delivery*, 281–336.

8 Metal-Organic Frameworks for the Development of Biosensors

*Arpna Tamrakar, Kamlesh Kumar Nigam,
Bani Mahanti, Arun Kumar, and
Mrituanjay D. Pandey*

CONTENTS

8.1 INTRODUCTION

The metal-organic frameworks (MOFs) are crystalline polymeric materials of an organic linker and metal ions or metal clusters. MOFs offer unique possibilities owing to their porous nature and structural diversity. For a couple of decades, many research groups focused their attention on exploring the applications of MOFs in many areas, including biological (biosensing, drug storage and delivery, etc.) [1], energy, and environmental [2] fields. The applications of MOFs in the biosensing field have become one of the essential topics of research. With the continued advancement in this area, scientific communities have expanded the degree of various sensing methods and enormously improved qualitative and quantitative real-time identification.

For instance, many organic substances and biomacromolecules with low molecular weights are widely used in daily life, some of which have harmful effects on human health and the environment have grown substantially in recent years. Sometimes, these harmful compounds are present in preservatives of food, clinically used drugs, veterinary medicine, by-products, etc. The detection of such harmful substances is very important and poses difficulty for researchers, not only for qualitative but also from the quantitative identification point of view. Hence, several scientific communities and researchers have focused their attention on minimizing the effect of such compounds [3].

The development of MOF-based sensors is an exciting research area mainly because their tunable structure bears great active site and ultra-thin thickness, π-conjugated system, high surface area, high porosity, high tendency to form a crystalline compound, and even possibility for modification after synthesis of the frameworks [4]. These properties of MOF biosensors make them more valuable sensors and susceptible for sensing and capturing metal ions [5] in their applications [6] when compared to the other biosensors, which are mostly based on bulky organic and inorganic substances such as metal colloids [7], silica nanoparticles [8], fullerenes [9], and graphenes or graphene oxides [10], quantum dots [11], nanotubes, etc. [12].

The synthesis method, stability, shape, and size are the key factors for making the various MOF biosensors. At that, the applications of thebiosensors are centralized based on different types of MOFs used in the synthesis process. For example, the widely used synthetic methods for developing composite biosensors are either in-situ addition of biomolecules during the synthesis or post-synthesis incorporatingthe bioactive molecule to MOF [13]. Such MOF composite biosensors are useful in detecting biomolecules, antigen-antibody, enzymes, viruses, and bacteria [14] (Figure 8.1).

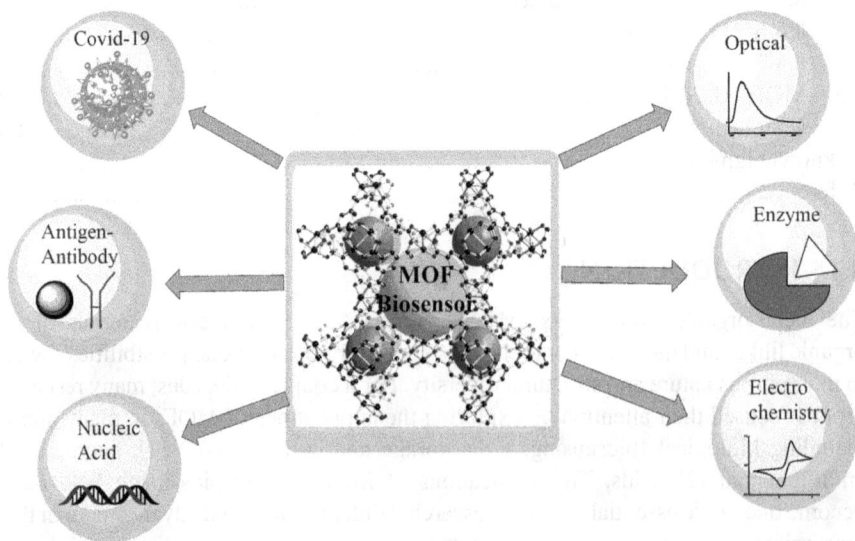

FIGURE 8.1 MOF as a biosensor.

Researchers have made extensive efforts to integrate MOF biocomposites in the design of diagnostic technologies for POC (point-of-care) tests [1,3,15–17]. Simultaneously, using bioconjugation, infiltration, and encapsulation strategies, the cells and viruses are integrated with MOFs to make the corresponding biocomposite probe systems [18–22]. For example, the strategically synthesized MOFs-ZIF-8 (zeolitic imidazole-based) in the use of encapsulation of biological kinds of stuff [23]. The MOFs-PCN-333(Al) (constructed from 4,4,4-s-triazine-2,4,6-triyl-tribenzoic acid ligands and Al_3 metal clusters) for trapping HRP (Horseradish peroxidase) in the detection of cholesterol and cholesterol oxide (ChOx) [24,25], etc. Recently, a virtual conference $MOF2020^{WEB}$ featuring MOFs, COFs (covalent organic frameworks), porous polymers, advanced porous materials, and their applications in many diverse areas was organized. A detailed highlight of a special collection of open-access articles is available [26].

This chapter reviews the development of various MOF-based biosensors, their functional properties, and their applications using some representative examples. This chapter will provide exclusively an update on MOF biosensing of biomolecules, biomarkers, drugs, and toxins with several biosensing mechanisms, which would be advantageous for the readers in many ways.

8.2 MOFS AS BIOSENSORS

At present, the chemistry of MOFs has reached the point where matrices such as composition, structure, functionality, and porosity can be designed precisely for specific applications, such as a developed biosensor provides diagnostic information by using its biological active elements. First of all, what is a biosensor? According to IUPAC, *biosensors are integrated receptor-transducer devices that provide selective quantitative or semi-quantitative analytical information using a biological recognition element* [27]. It has three important components: (i) signal receptor-transducer (electrical, thermal, oroptical); (ii) recognition of enzyme, antibody, DNA, etc.; and (iii) signal amplifier for data processing [28]. A schematic principle of operation of a biosensor is outlined in Figure 8.2 [29].

The MOF-based sensing materials demonstrate unique usefulness in biosensing. The researchers have published a few reviews on MOF biosensors with different perspectives. Their primary focus is on lanthanide-MOFs, nano-MOFs, functional-MOFs, biological-MOFs, and biomolecule@MOF composites based upon suitability, applicability, and continuously growing content. A few have sorted the MOF-sensor integrated with photoluminescent, optical, electrochemical, and colorimetric biomolecules. Several others included direct applications, such as detecting biomarkers or biomolecules. However, most of them describe a perspective to study mainly sensing markers viz. biomolecules, drugs, biomarkers, and toxins with their detection mechanisms (Figure 8.3). It is relevant to mention that the adequate concentrations of these markers in biological systems are critical for assessing potential health issues.

The advancement and development of structural understanding emerged MOFs to the level where they continuously break the higher surface record, both gravimetrically and volumetrically. For example, the worldwide highest specific surface area and surpassing all known crystalline framework materials is DUT-60,

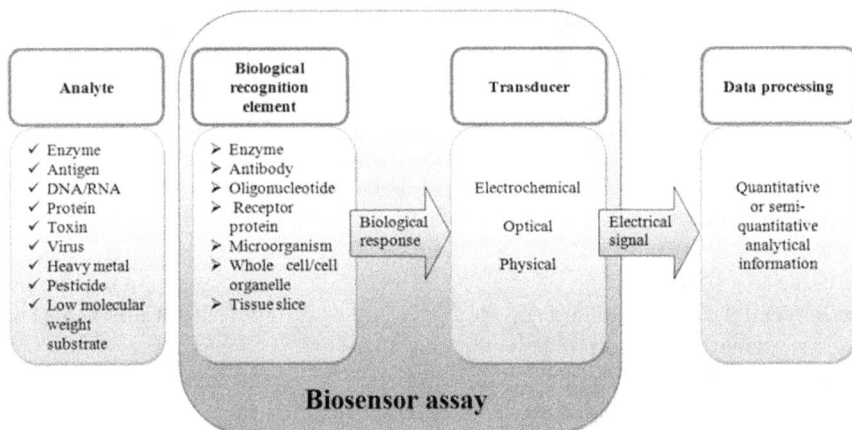

FIGURE 8.2 Schematic principle of operation of a biosensor.

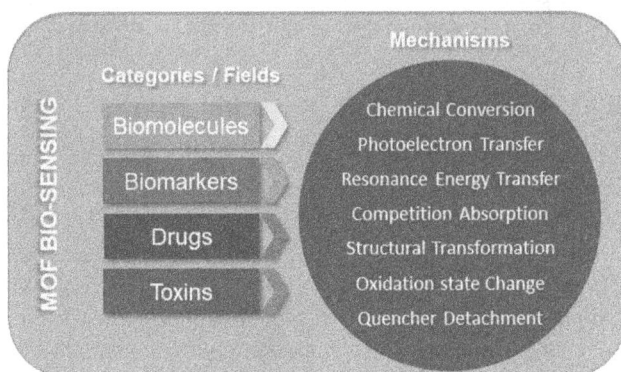

FIGURE 8.3 MOF biosensing of biomolecules, biomarker, drugs, and toxins with several detection mechanisms.

mesoporous MOF designed in silico using $Zn_4O_6^+$ nodes, ditopic and tritopic linkers. It has a specific surface area of 7,800 m^2/g with 90.3% hollow space. MOFs with a larger size profile are better functioning, such as adsorption and accumulation of sensing probes/analytes at a high level.

There are a few prevalent methods to design and synthesize the MOF biosensors.

a. One-pot synthesis, such as Cyt c protein embedded ZIF-8 MOF, is obtained by a co-precipitation method. The Cyt c ZIF-8 MOF is a far better sensor for peroxidase activity when compared to free Cyt c by a tenfold increased capacity. This MOF is quite handy in detecting peroxides even in trace amounts (Cyt c is cytochrome complex), Figure 8.4 [30].

b. Post-synthetic modification (PSM); unlike pre-synthetic approaches where the MOFs are tailor-made, PSM of MOFs is an alternative strategy to develop biosensors. Amongst several methods, the covalent and coordinate PSM are mostly in use. In the covalent PSM strategy, the organic linkers present in the MOF are modified, while new organic molecules containing metal ligating groups are introduced in the coordinative PSM strategy. For example, PSM of UiO-66-NH$_2$ after aldimine condensation with 2,3,4-trihydroxybenzaldehyde results in UIO-66-PSM. The fluorescence of UIO-66-PSM quenches readily by free bilirubin (biomarker for jaundice) via the FRET process. The structural analysis reveals that the hydroxyl groups and π-electrons of aromatic rings form weak interactions with free bilirubin, Figure 8.5 [31]. Not to forget, the functional groups of MOFs have the potential to alter/form various non-covalent interactions.

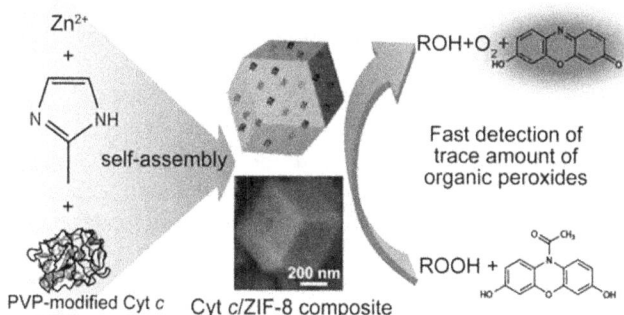

FIGURE 8.4 Detection mechanism of peroxidase by Cyt c/ZIF-8 composite. Reproduced from Refs. [30] with permission from American Chemical Society, copyright 2014.

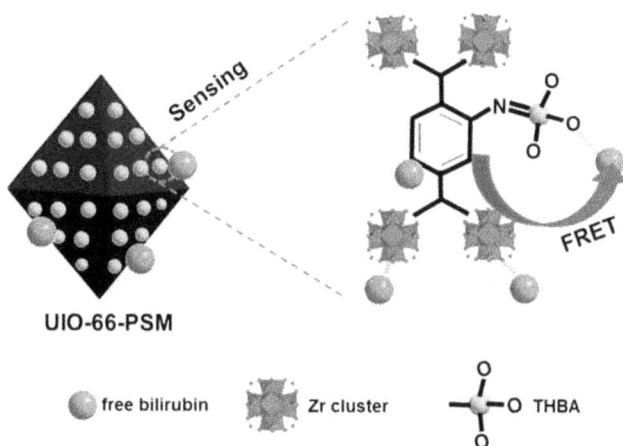

FIGURE 8.5 Fluorescence quenching in the detection of bilirubin via FRET process. Reproduced from Refs. [31] with permission from American Chemical Society, copyright 2017.

a. Post-synthetic ligand exchange (PSE); the MOF-based biosensors are developed by applying the advantages of the reversibility of coordination chemistry in the PSE. Many MOFs with different applications can also be improved through PSE. A chiral zeoliticimidazolate framework (ZIF) developed by binaphthyl-derived chiral ligands (photo-emitter derivatives of (R)- or (S)-(+)-1,1'-binaphthyl-2,2'-diamine) and ZIF-8 nanoparticles. This chiral MOF shows CPL (circularly polarized luminescence) properties reported for the enantioselective fluorescence sensing of α-hydroxycarboxylic acids with enhanced sensing potentials than the simple chiral emitters [32].

b. Encapsulation method; these biosensors result from the encapsulation of luminescent species (e.g., enzyme) with MOF without compromising the surface area hugely. The encapsulated MOF composite fabrication may be through de novo synthesis or PSM. For example, the electrochemical biosensor HRP@PCN-333(Fe) composite is developed by encapsulating (MOF) [PCN-333(Fe)] with horseradish peroxidases (HRP) (PCN = porous coordination network). PCN-333 demonstrates one of the largest structure profiles with a 5.5 nm cage size and $3.84\,cm^3\,g^{-1}$ void volumes and is considered one of the most stable and excellent MOFs for enzyme encapsulation in the aqueous solution pH range 3–9. Since the enzymes are delicate structures, they need strong supporting material for functioning under harsh catalytic conditions. In the cyclic voltammetry measurement, the immobilized HRP enzyme shows higher catalytic activities toward H_2O_2 reduction over the free enzyme. The enzyme shows almost no leaching in the process of catalysis and recycling [33,34].

Another example is sensing the 3-nitropropionic acidbelonging to mycotoxin. The 3-nitropropionic acid present in moldy sugarcane causes neurologic illness on ingestion. A dye@MOF composite developed by immobilizing fluorescein isothiocyanate (FITC) dye on MOF can detect 3-nitropropionic acid through a fluorescence quenching response (pH-responsive) to acidic species [35]. Similarly, through fluorescence quenching, a fluorescence-labeled aptamer functionalized MOF, aptamer@MOF composite, can efficiently monitor the complementary aptamer (aptamer = ssDNA or ssRNA where 'ss' stands for single-stranded) [36]. A series of Zr-based composites, aptamer@509-MOF, are developed and studied for sensing antibiotics (kanamycin), thrombin (procoagulant and anticoagulant), and some specific proteins such as CEA (carcinoembryonic antigen) [37]. Figure 8.6 shows generalized de novo and post-synthetic encapsulation of enzymes [38].

8.3 MOF BIOSENSING OF THE BIOMOLECULES, BIOMARKERS, DRUGS, AND TOXINS

As introduced above, many MOFs are available for biosensing applications. The working mechanisms involve various phenomena like binding with metal or other organic compounds by covalent or non-covalent interaction with different materials, where these MOFs act like bioreceptors. On the other hand, modified MOFs can

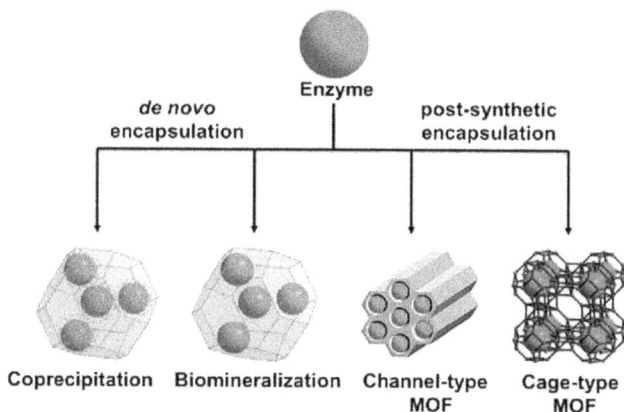

FIGURE 8.6   Encapsulation of enzyme. Reproduced from Refs. [38] with permission from Royal Society of Chemistry, copyright 2017.

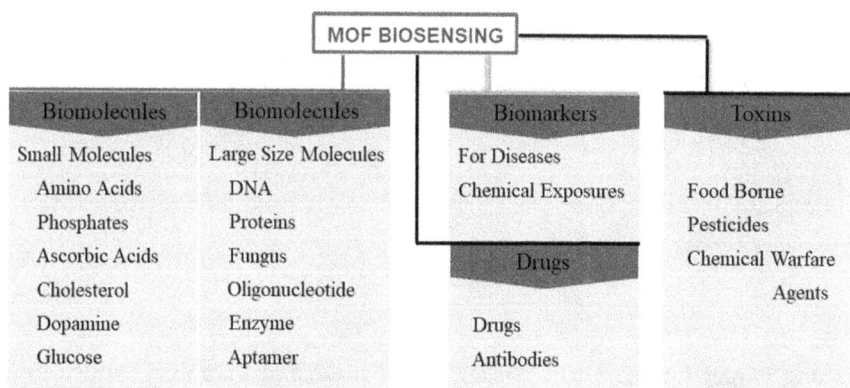

FIGURE 8.7   MOF biosensing targets.

recognize cells, cation, anions, and tissues with accuracy; are highly stable in water; and have high biodegradability. These properties of MOFs are significant in the field of biosensing. For better understanding, MOF biosensing targets are categorized into biomolecules, biomarkers, drugs, and toxins and depicted with examples in Figure 8.7.

The MOF biosensing of each category, some of their MOF platforms, and possible mechanisms are available in Tables 8.1 to 8.4.

8.4   MOF-BIOSENSING MECHANISMS

The availability of structural data (as many MOFs are crystalline) and improved theoretical calculations provide adequate information to understand the sensing mechanisms for MOF-based biosensors deeply. Furthermore, the characteristics of MOFs, such as involvement of types of organic ligands with a functional group or with functional centers, size profile, stability, etc., decide the sensing mechanisms and

TABLE 8.1

Sensing Target Category: Biomolecules

Sensing Target		MOF Platform	Detection Limit	Mechanism
Amino acids	Glutathione	Cu^{2+} CDs@ZIF-8	0.9 nM	QD
	Homocysteine	Cd(II) PPCA MOF	40 nM	CC
Phosphate	PPi, ALP	ssDNA@Ce^{3+}/Ce^{4+}MOF	55 nM	QD, CC
Ascorbic acid	Ascorbic acid	Cd-abtz MOF	75 nM	CA
Cholesterol	Cholesterol	AF-MSN-QD@ZIF-8-ChO$_x$	0.89 nM	CC
Dopamine	Dopamine	Abtz-CdI$_2$MOF	57 nM	SQ
Glucose	H_2O_2 Glucose	MIL-53 Fe(III) + GO$_x$	7.54 nM	CC
H_2O_2	H_2O_2	ChO$_x$@MOF-5 + AgNC-MoS$_2$	0.58 nM	CC
DNA	HIV dsDNA	ssDNA@Cu(II)-MOF	196 pM	QD
Protein	Target DNA	ssDNA@Co(II)-TTPP	120 pM	QD
Fungus	*S. aureus*	Bacteriophage@IRMOF-3	100 cfu/mL	Bio-Recognizers

Abbreviations: **PPi-ALP** = Pyrophosphate-Alkaline Phosphatase (in biological phenomenon, ALP converts PPi into phosphorus, which then combines with calcium to form mineralized bone tissue); **Cu^{2+} CDs@ZIF-8** = Copper doped, carbon dots MOF composite; **Cd(II) PPCA** = aldehyde-functionalized MOF containing 1H-pyrrolo-[2,3-b] pyridine-2-carbaldehyde ligand; In Cd-abtz MOF, abtz=1-(4-aminobenzyl)-1,2,4-triazole; AF-MSN-QD@ZIF-8-ChOx is 5-aminofluorescein (AF) and mesoporous silica nanoparticles (MSN) encapsulated over MOF (ChOx = cholesterol oxidase); **QD** = Quencher Detachment; **CC** = Chemical Conversion; **CA** = Competition Absorption; **SQ** = Static Quenching.

TABLE 8.2

Sensing Target Category: Biomarkers for Disease and Chemical Exposures

Sensing Target	MOF Platform	Detection Limit	Mechanism
Sarcosine	Cu^{2+}/Tb^{3+}@Ga(OH)(BTEC)	0.23 mM	QD
Prostate (P53 Gene)	ssDNA loaded	5 pM	QD
ATP, Creatine Kinase	Eu-QPTCA + BSA	1.0 U/L	CC, PET
p-Aminophenol	Tb^{3+}@Al(OH)(bpyda)	5 µg/L	CA
N-Methylformamide	Eu^{3+}@Ga(OH)(BTEC)	0.36 µM	RET

Abbreviations: **QD** = Quencher Detachment; **CC** = Chemical Conversion; **CA** = Competition Absorption; **PET** = Photoelectron transfer; **RET** = Resonance energy transfer.

sensing ability. Based on the availability of information and data on MOF biosensing, it would be easier to describe the sensing mechanism, including the following types of processes separately. PET (photoelectron transfer), CA (competition absorption), RET (resonance energy transfer), QD (quencher detachment), ST (structural transformation), CC (chemical conversion), oxidation state change (OC).

TABLE 8.3
Sensing Target Category: Drugs

Sensing Target	MOF Platform	Detection Limit	Mechanism
6-MP	NH_2-MIL-53(Al)	0.15 mM	CA
Quercetin	Tb-CBA	0.23 ppm	CA
Quercetin	Eu-MOF	34 mM	CA
NZF, NFT	BUT-12, 13	58, 90 ppb	PET, RET
Nitrofuran, quinolone	RhB@Tb-MOF	NZF 99 ppb, NFT 107 ppb	PET, CA
Antibiotics, AMP	CSMCRI-2	SDZ 65 nM, NZF 0.7 mM	PET, RET
Antibiotics	SMOF-10	ODZ 2 ppm	CA
Fluoroquinolone antibiotics	BUT-172	0.18 mM	CA

Abbreviations: **CA** = Competition Absorption; **PET** = Photoelectron transfer; **RET** = Resonance energy transfer.

TABLE 8.4
Sensing Target Category: Toxins

Sensing Target	MOF Platform	Detection Limit	Mechanism
Aflatoxines B1	LMOF-241	46 ppb	PET
Aflatoxines B1	Zr-CAU-24	19.97 ppb (64 Nm)	PET
HI	MR@Eu-MOF	0.1 mM	pH-induced ET
3-NPA	FTTC@CD-MOF	0.135 M	pH-sensitive dye
3-NPA	Eu-DNC	12.6 mM	ST
3-NPA	MPDB-PNC-222	15 mM	pH-sensitive ligand
Pesticides DCN	$[Zn_2(bpdc)_2(BPyTPE)]$	0.13 ppm	PET
Nicotine	MB@UiO-66-NH_2	0.98 mM	PET

Abbreviations: CA = Competition Absorption; **PET** = Photoelectron transfer; **RET** = Resonance energy transfer.

8.4.1 PHOTOELECTRON TRANSFER (PET)

The photoelectron/photoinduced electron transfer (PET) process between donor-acceptors is very useful in photoluminescence sensors and switches. The inter-molecular PET is also responsible for fluorescence quenching. MOFs provide a suitable platform with precise pore size and surface parameters to fit the analytes through multiple secondary bonding interactions, including hydrogen bondings. These interactions between host-guest behave as a channel for the PET. When the MOF composite (MOF with analyte) molecules are open for photoexcitation, the energy level of the LUMO of the donor (host) elevates and becomes higher than the LUMO of the acceptor (guest). At this stage, PET happens via photoelectron

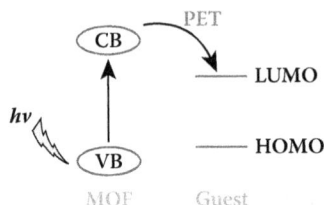

FIGURE 8.8 PET quenching mechanism.

TABLE 8.5
The Detection Limit of RhB-CDs@Cu-MOF Composite for Some Antibiotics in An Aqueous Solution

S. No.	Antibiotics	Detection Limit
1	Ciprofloxacin	25 ppb
2	Norfloxacin	44 ppb
3	Nitrofurazone	65 ppb
4	Nitrofurantoin	74 ppb
5	Tetracycline	777 ppb

transfer from the donor in the excited state to the acceptor in the ground state. The process results in emission quenching of the donor. The following Figure 8.8 illustrates the process to understand the PET fluorescence quenching. In Figure 8.8, upon photoexcitation, the conduction band (LUMO) of MOF, i.e., donor, achieves a higher energy state than the LUMO of guest, i.e., acceptor or analyte, allowing PET to occur. The quenching efficiency is inversely proportional to the analyte LUMO energy level, i.e., the lower energy level of LUMO of the probe favors better quenching.

Rhodamine-B and carbon-dot based Cu-MOF composite (RhB-CDs@Cu-MOF) upon excitation at 355 nm shows a dual-emission fluorescent platformat 430 nm and 580 nm in the sensing experiments of the RhB-CDs@Cu-MOF with several antibiotics. The sensing experiments of RhB-CDs@Cu-MOF composite with antibiotics such as aminoglycosides (gentamycin and kanamycin), quinolones (ciprofloxacin and norfloxacin), nitrofurans (nitrofurazone and nitrofurantoin), and tetracyclin shows rapid identification of quinolones, tetracyclines, and nitrofurans in water; see Table 8.5 [39,40].

8.4.2 RESONANCE ENERGY TRANSFER

After excitation, when the excited state donor returns to the ground state in the RET process by transferring its energy to the acceptor, it enters into an excited state by emission enhancement, and quenches the emission of the donor in these two non-radiative processes. The donors and acceptors that lie nearby in the MOF favor the RET process. In the next level of the switching mechanism, the porous nature of MOFs is conducive, allowing the diffusion of analytes. The analytes then interact

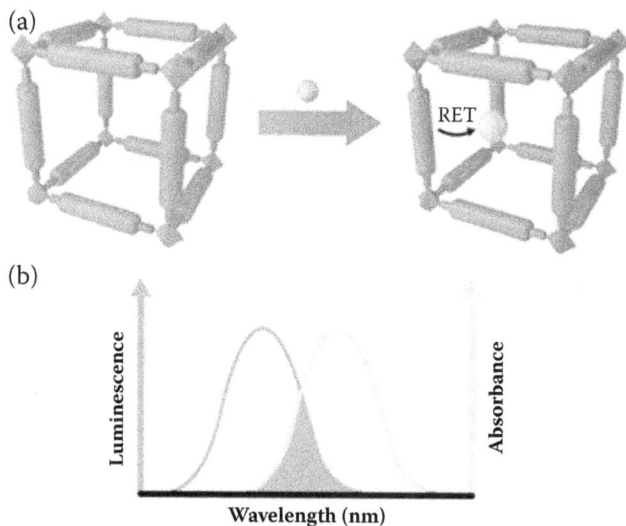

FIGURE 8.9 (a) Schematic diagram of the RET mechanism. Yellow ball: guest molecule. (b) The emission spectrum of the MOF (pink line) and the UV-vis absorption spectrum of the guest (yellow line). Blue area: spectral overlap. Reproduced from Refs. [41] with permission from Royal Society of Chemistry, copyright 2021.

with the acceptors and prohibit the RET process, resulting in the activation of MOF's fluorescence. Variation in the feeding ratios in acceptors and analytes modulates the switching mechanism; see Figure 8.9.

The RET efficiency depends upon the distance between the host and guest and their orientations, which decide the extent of emission and absorption *spectral overlap*. Another affected process in the row is a *dipole-dipole interaction*.

a. The *spectral overlap* is very important in the luminescent quenching of the MOFs. Experimental methods are available to determine spectral overlaps(between guestUV-VISIBLE absorption and MOF emission spectra). This quenching process is commonly introduced as FRET (Förster resonance energy transfer). For example, the luminescent probe, UiO-66(COOH)$_2$, Zr-MOF: Eu^{3+}, a Zr(IV)-based metal-organic framework post functionalized by Eu(III) senses bilirubin via FRET process. The bilirubin is a biomarker of jaundice hepatitis in human serum. In bilirubin sensing, the energy transfer facilitates because of spectral overlap between the absorption range of bilirubin and the emission range of Eu@UiO-66(COOH)$_2$ [85].

b. The following example of hetero-MOF explains the effect of host-guest *dipole-dipole* interaction on the efficiency of resonance energy transfer in the biosensing of anthrax. Three hetero Eu/Tb-MOFs (with three ligands, L$_1$, 2,6-naphthalenedicarboxylic acid, H$_2$NDC;L$_2$, 4,4'-biphenyldicarboxylic acid, H$_2$BPDC; and L$_3$, 1,4-benzenedicarboxylic acid, H$_2$BDC) show ratiometric sensing of 2,6-dipicolinic acid (DPA). The 2,6-dipicolinic acid (DPA) is an anthrax biomarker. The modulation of energy transfer from DPA to Ln^{3+}(Tb/Eu ions) occurs due to the competition among the ligands and DPA. Finally, the energy gap between the ligands (DPA) and Ln^{3+} relate to DPA detection performances. In the anthrax detection process, the DPA sensing performance depends upon a few factors, i.e., ligand energy mismatch and balancing, especially the competitive energy among the three ligands and DPA in human serum. The more mismatched energy differences, the worse performance. The detection limit of three Eu/Tb-MOFs(L) is 0.248 for L_1, 0.847 for L_2, and 2.277 for L_3, resulting from the matched/mismatched energy differences [42].

8.4.3 COMPETITION ABSORPTION

There are chances of overlapping the UV-visible absorption spectrum of the analyte with the fluorescence excitation spectrum of the MOF. In such a case, the absorption competition occurs for theexcitation lights of the MOF and the analyte. Such a competitive absorption results in luminescence quenching of the MOF by reducing its total available energy for the process.

For example, the biosensing of an anti-cancer drug 6-mercaptopurine (6-MP) by NH_2-MIL-53 MOF (complexed with 3-mercaptopropionic acid-capped CdTe quantum dots) follow the competition absorption quenching. The absorption range of 6-MP is 300 nm to 350 nm, which overlaps the MOF complex's excitation frequency at 332 nm. The detection limit of 6-MP is excellent, 0.15 mm, and capable of sensing even from human urine. This mechanism realizes in the sensing of Fe^{3+} ions by NTU-9-NS-MOF [NTU-9-NS is MOF $Ti_2(HDOBDC)_2(H_2DOBDC)$ where H_2DOBDC and NS abbreviate into 2,5-dihydroxyterephthalic acid and nanosheets, respectively [43]. Recognition of acetone by anionic microporous zinc-based MOFs also follows a similar quenching mechanism; see Figure 8.10. This zinc-based MOF construct is a combination of mononuclear and tetranuclear zinc clusters and two different organic

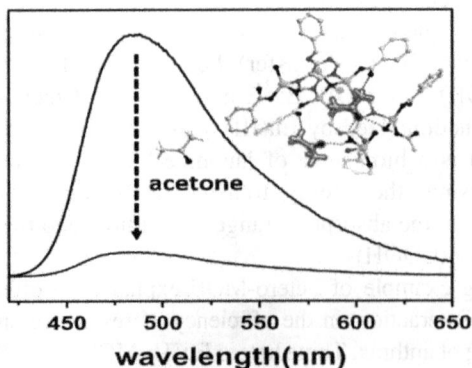

FIGURE 8.10 Detection of acetone by zinc-based MOF. Reproduced from Refs. [44] with permission from American Chemical Society, copyright 2019.

ligands, TBAPy and 4-BA or 6-BA [TBAPy = 1,3,6,8-tetrakis(p-benzoic acid)pyrene; 4-BA = 4-(methoxycarbonyl)phenyl) boronic acid] [44].

8.4.4 Structural Transformation

The MOF's versatile structure provides several ways to accommodate or interact with the analytes. Sometimes the inclusion of analytes results in remarkable structural change due to new bonds between analytes and reactive sites in host MOFs. When such analyte-induced structural transformation can bring significant variation in the luminescence of the MOF, it can apply in the detection of the analyte. The binding and mechanism of inclusion of guest molecules are easy to understand with the help of crystal structures. There is a term 'single-crystal-to-single-crystal transformation' or 'SC-SC transformation' to study the crystal structure of MOF-analytes and corresponding MOF. After including the analyte in MOF, the important structural changes such as coordination environments and locations of analyte inside the MOF array can be gathered precisely from the detailed structural analysis, including the H-bonding, secondary bonding interactions, and π interactions in the framework [45]. For example, the Eu-MOF (2,6-naphthalenedicarboxylic acid, H_2NDC) shows the luminescence changes in an acidic medium (pH range 3.00–4.00). The SC-SC transformation studies reveal such pH-induced property is due to a change in coordination states around Eu ions, resulting in emission quenching of Eu^{3+}. The short-range pH sensitivity of the Eu-MOF comes in handy for the rapid detection of acidic amino acids such as aspartic and glutamic acids, amongst others [46].

The SC-SC transformation can bring cation exchange quenching effect too. For example, Al^{3+} in fluorescent MIL-53 (Al) after exchanging with the specific metal ion, Fe^{3+} transforms the MOF into a weak-fluorescent MIL-53(Fe). This process shows high sensitivity and selectivity towards Fe^{3+} ions because there is no interference with other metal ions [47]. Zn-based HPU-1 MOF [48] reveals a similar quenching effect on transforming the framework to $Fe^{3+}@HPU-1$.

8.4.5 Chemical Conversion

There are possibilities of chemical conversions when the guest molecules interact with the MOF. The diverse pore topologies and the availability of functional sites in ligands may promote molecular-level interactions between the frameworks and analytes. Secondarily, selecting a functional organic ligand is the key to tuning MOF's properties. If the functional group can react with a target and undergo chemical conversion that simultaneously contributes to the luminescence change, the selective sensing of particular analytes is possible. The examples are available where such chemical conversion leads to ligand modification, thus significantly changing the luminescence properties of the MOF materials and becoming useful in sensing the analyte. The N-terminal cysteines undergo a condensation reaction with aldehydes, forming thiazolidine, a way to immobilize peptides and proteins on aldehyde-functionalized MOFs. An aldehyde-functionalized MOF, Cd-PPCA, consists of a Cd(II) metal center and two different ligands, 4,4',4''-tricarboxyltriphenylamine(H_3tca), and 1H-pyrrolo-[2,3-b]pyridine-2-carbaldehyde (ppca) that selectively distinguish Hcy (homo-cysteine) from natural amino acids in its condensation reaction by amplification of the luminescence signal [49].

8.4.6 OXIDATION STATE CHANGE

When chemical conversions from the reaction of MOF and analyte attribute to change in an oxidation state of metal ions, the luminescence properties are likely to change. The change in luminescence is probably due to the energy transfer or charge transfer mismatch between the ligands and metal ions. For example, cerium MOF, ZJU-136-Ce (contains Ce metal center and ligand, H_4TPTC = 1,1':4',1''-terphenyl-2',4,4'',5'-tetracarboxylic acid) is useful in fluorescence sensing of ascorbic acid. The valence fluctuation of Ce^{3+}/Ce^{4+} during the specific oxidation reaction between Ce ions forms dehydroascorbic acid (DHA). The DHA with three electron-withdrawing keto groups can hold back the LMCT process, enhancing the emission of the ligand and quenching the Ce^{3+}. Adding $KMnO_4$ oxidizes Ce^{3+} to Ce^{4+} and DHA into 2,3-diketogulonic acid (DKG), and recovers the sensing potential of MOF [50].

8.4.7 QUENCHER DETACHMENT

The strategy of the quencher detachment mechanism involves an exchange between the pre-installed quencher of the MOF matrix and the analyte. In this process, upon addition of the probe, the quencher detaches, and the luminescence of the MOF matrix realizes. For example, the encapsulation of Cu^{2+} ions on Tb@Zn-MOF quenches the luminescence of Tb ions. However, during the sensing of aspartic acid, the interplay of strong interaction between aspartic acid and Cu^{2+} detaches the latter from the MOF matrix, retrieving luminescence of Tb ions. Similarly, a fluorescein-labeled ssDNA (single-stranded DNA) quenches the fluorescence of MOF after immobilization. Further, the addition of matched ssDNA or sRNA forms a new triple complex or dsDNA (double-stranded DNA), which detaches readily, resulting in fluorescence recovery [51].

8.5 MOFS USED FOR OPTICAL BIOSENSORS

The MOF's biosensing is described under several detection mechanisms for different categories of molecules in detail in the above sections. MOFs provide unparalleled opportunities to develop numerous novel MOFs for many biosensing applications. Optical biosensing is one of the desired solutions due to its high sensitivity and rapid and accurate detection. The main aim of discussing the MOF biosensing under this section is to provide more information on luminescent MOFs, generally referred to as LMOFs. For ease of understanding, one can divide LMOFs into three categories depending upon the luminescence source inside the LMOFs: metal nodes, ligands, or guest. The ligand-based LMOFs can have extremely high luminescence efficiency. Of course, several factors can play a role behind the efficiency, such as bandgap modulation and ligand rigidification. Similarly, the luminescence efficiency of host-guest-based LMOFs depends upon the guest's location inside MOF, encapsulation (while

TABLE 8.6
Different MOFs-Based Optical Biosensors

Type of MOFs	Organic Ligand	Analytes	Ref.
Tb-MOF	Isophthalicacid	Protamine, Hep	[51]
TB-MOF/AuNPS/aptamers	Isophthalicacid	PSA	[52]
MIL-53-NH$_2$(Al) with CdTeQDS	NH$_2$-BDC	Al, 6-Mercapto-purine	[53]
EuPS@AnC/ZIF-8	–	Singlet oxygen	[54]
HMOF with FAM-labeled DNA	6-carboxyfluorescein	DNA	[54]
1-NH$_2$@THB		bilirubin	[55]
Fe-MIL-88B-NH$_2$	2-amino terephthalic acid	Bisphenol A	[56]
Tb(BTC)MOF	Benzene 1,3,5-tetracarboxylate	Fluoride	[57]
NH$_2$-MIL-53	NH$_2$-BDC	S. aureus	[58]
H$_2$dtoaCu MOF	N,N'- bis(2-hydroxyethyl) dithiooxamide	dsDNA, H$_5$N$_1$ antibody	[59,60]

in-situ preferred), and extent of ion exchange. Table 8.6 depicts some of the examples of the types.

8.6 CONCLUSION

Recently emerged, luminescent MOF materials in biosensing applications offer significant performance in detecting biomolecules, biomarkers, drugs, toxins, etc. Here, we have comprehensively discussed various properties of MOFs, and by modulating those properties, MOFs can be used as an excellent device for developing biosensors. This chapter reviews the development of various MOF-based biosensors, their functional properties, and their applications using representative examples. It is expected that ongoing advancement in research and studies in this area of MOF sensing will rapidly develop more effective biosensors.

ACKNOWLEDGMENT

AT acknowledges Banaras Hindu University, Varanasi, for providing the facility for research work. KKN acknowledges UGC-NET junior research fellowship. BM and MDP acknowledge BHU for providing a seed grant and incentive grant (IoE) under MoE Govt. India, Institute of Eminence (IoE) scheme, respectively.

REFERENCES

[1] Miller, S. E., Teplensky, M. H., Moghadam, P. Z., Fairen-Jimenez, D. Metal-organic frameworks as biosensors for luminescence-based detection and imaging. *Interface Focus*, 2016, 6 (4), 20160027. 10.1098/rsfs.2016.0027

[2] Zhao, F., Sun, T., Geng, F., Chen, P., Gao, Y. Metal-organic frameworks-based electrochemical sensors and biosensors. *Int. J. Electrochem. Sci.*, 2019, *14*, 5287–5304. 10.20964/2019.06.63

[3] Carrasco, S. Metal-organic frameworks for the development of biosensors: A current overview. *Biosensors*, 2018, *8* (4), 92. 10.3390/bios8040092

[4] Okada, K., Sawai, S., Ikigaki, K., Tokudome, Y., Falcaro, P., Takahashi, M. Electrochemical sensing and catalysis using $Cu_3(BTC)_2$ coating electrodes from Cu (OH)$_2$ films. *CrystEngComm.*, 2017, *19* (29), 4194–4200. 10.1039/c7ce00416h

[5] Zhou, J., Tian, G., Zeng, L., Song, X., Bian, X. W. Nanoscaled metal-organic frameworks for biosensing, imaging, and cancer therapy. *Adv. Healthc. Mater.*, 2018, *7* (10), 1–21. 10.1002/adhm.201800022

[6] Osman, D. I., El-Sheikh, S. M., Sheta, S. M., Ali, O. I., Salem, A. M., Shousha, W. G., EL-Khamisy, S. F., Shawky, S. M. Nucleic acids biosensors based on metal-organic framework (MOF): Paving the way to clinical laboratory diagnosis. *Biosens. Bioelectron*, 2019, *141*, 111451. 10.1016/j.bios.2019.111451

[7] Veigas, B., Giestas, L., Almeida, C., Baptista, P. V., Lisboa, N. De, Caparica, C. De, Esquillor, M., Lisboa, U. N. De, Caparica, C. De, Nova, U., Lisboa, D., Caparica, C. De. Noble metal nanoparticles for biosensing applications. *Sensors*, 2012, *12* (2), 1657–1687. 10.3390/s120201657

[8] Korzeniowska, B., Nooney, R., Wencel, D., McDonagh, C. Silica nanoparticles for cell imaging and intracellular sensing. *Nanotechnology*, 2013, *24* (44), 442002. 10.1 088/0957-4484/24/44/442002

[9] Yanez-Sedeno, P., Campuzano, S., Pingarron, J. Fullerenes in electrochemical catalytic and affinity biosensing: A review. *C*, 2017, *3* (4), 21. 10.3390/c3030021

[10] Lee, J., Kim, J., Kim, S., Min, D. H. Biosensors based on graphene oxide and its biomedical application. *Adv. Drug Deliv. Rev.*, 2016, *105*, 275–287. 10.1016/ j.addr.2016.06.001.

[11] Martynenko, I. V., Litvin, A. P., Purcell-Milton, F., Baranov, A. V., Fedorov, A. V., Gun'Ko, Y. K. Application of semiconductor quantum dots in bioimaging and biosensing. *J. Mater. Chem. B*, 2017, *5* (33), 6701–6727. 10.1039/c7tb01425b

[12] Yang, N., Chen, X., Ren, T., Zhang, P., Yang, D. Sensors and actuators B: Chemical carbon nanotube based biosensors. *Sensors Actuators B. Chem.*, 2015, *207*, 690–715. 10.1016/j.snb.2014.10.040

[13] Khan, N. A., Hasan, Z., Jhung, S. H. Beyond pristine metal-organic frameworks: Preparation and application of nanostructured, nanosized, and analogous MOFs. *Coord. Chem. Rev.*, 2018, *376*, 20–45. 10.1016/j.ccr.2018.07.016

[14] Pandey, M. D. Luminescence metal-organic frameworks as biosensor. *Materials Letter*, 2022, *308*, 131230. 10.1016/j.matlet.2021.131230

[15] Bilal, M., Adeel, M., Rasheed, T., Iqbal, H. M. N. Multifunctional metal-organic frameworks-based biocatalytic platforms: Recent developments and future prospects. *J. Mater. Res. Technol.*, 2019, *8* (2), 2359–2371. 10.1016/j.jmrt.2018.12.001.

[16] Wang, H. S. Metal–organic frameworks for biosensing and bioimaging applications. *Coord. Chem. Rev.*, 2017, *349*, 139–155. 10.1016/j.ccr.2017.08.015.

[17] Liu, C. S., Li, J., Pang, H. Metal-organic framework-based materials as an emerging platform for advanced electrochemical sensing. *Coord. Chem. Rev.*, 2020, *410*, 213222. 10.1016/j.ccr.2020.213222.

[18] Doonan, C., Riccò, R., Liang, K., Bradshaw, D., Falcaro, P. Metal-organic frameworks at the bio-interface: Synthetic strategies and applications. *Acc.Chem. Res.*, 2017, *50* (6), 1423–1432. 10.1021/acs.accounts.7b00090

[19] Riccò, R., Liang, W., Li, S., Gassensmith, J. J., Caruso, F., Doonan, C., Falcaro, P. Metal-organic frameworks for cell and virus biology: A perspective. *ACS Nano*, 2018, *12* (1), 13–23. 10.1021/acsnano.7b08056

[20] Li, P., Chen, Q., Wang, T. C., Vermeulen, N. A., Mehdi, B. L., Dohnalkova, A., Browning, N. D., Shen, D., Anderson, R., Gómez-Gualdrón, D. A., Cetin, F. M., Jagiello, J., Asiri, A. M., Stoddart, J. F., Farha, O. K. Hierarchically engineered mesoporous metal-organic frameworks toward cell-free immobilized enzyme systems. *Chem*, 2018, *4* (5), 1022–1034. 10.1016/j.chempr.2018.03.001

[21] Teplensky, M. H., Fantham, M., Poudel, C., Hockings, C., Lu, M., Guna, A., Aragones-Anglada, M., Moghadam, P. Z., Li, P., Farha, O. K., Fernández, S. B. D. Q., Richards, F. M., Jodrell, D. I., Kaminski Schierle, G., Kaminski, C. F., Fairen-Jimenez, D. A highly porous metal-organic framework system to deliver payloads for gene knockdown. *Chem*, 2019, *5* (11), 2926–2941. 10.1016/j.chempr.2019.08.015

[22] Chen, Y., Li, P., Modica, J. A., Drout, R. J., Farha, O. K. Acid-resistant mesoporous metal-organic framework toward oral insulin delivery: Protein encapsulation, protection, and release. *J. Am. Chem. Soc.*, 2018, *140* (17), 5678–5681. 10.1021/jacs.8b02089

[23] Wang, Le., Sun, P., Yang, Y., Qiao, H., Tian, H., Wu, D., Yang, S., Yuan, Q., Wang, J. Preparation of ZIF@ADH/NAD-MSN/LDH core-shell nanocomposites for the enhancement of coenzyme catalyzed double enzyme cascade. *Nanomaterials*, 2021, *11* (9), 2171. 10.3390/nano11092171.

[24] Sun, P., Li, Y., Li, J., Zhang, Y. Entrapment of horseradish peroxidase into nanometer scale metal-organic frameworks: A new nano carrier for signal amplification in enzyme-linked immunosorbent assay. *Microchimica Acta*, 2021, *188*, 409. 10.1007/s00604-021-05065-9

[25] Feng, D., Liu, T. F., Su, J., Bosch, M., Wei, Z., Wan, W., Yuan, D., Chen, Y. P., Wang, X., Wang, K., Lian, X., Gu, Gu, Z. Y., Park, J., Zou, X. Stable metal-organic frameworks containing single-molecule traps for enzyme encapsulation. *Nat. Commun.*, 2015, *6*, 5979. 10.1038/ncomms6979

[26] *MOF2020^{WEB}* conference organized together by AngewandteChemie, Chemistry-A European Journal, Chemistry – An Asian Journal, ChemNanoMat, ChemPlusChem and the European Journal of Inorganic Chemistry; DECHEMA | International Conference on Metal-Organic Frameworks and Open Framework Compounds.

[27] Thevenot, D. R., Toth, K., Durst, R. A., Wilson, G. S. Electrochemical biosensors: Recommended definitions and classification. *Biosensor and Bioelectronic*, 2001, *16* (1–2), 121–131. 10.1016/S0956-5663(01)00115-4.

[28] Turdean, G. L. Design and development of biosensors for the detection of heavy metal toxicity. *Int. J. Electrochem. Sci.*, 2011, 2011, 343125. 10.4061/2011/343125

[29] Shavanova, K., Bakakina, Y., Burkova, I., Shtepliuk, I., Viter, R., Ubelis, A., Beni, V., Starodub, N., Yakimova, R. Application of 2D nan-graphene materials and 2D oxide nanostructures for biosensing technology. *Sensors*, 2016, *16* (2), 223. 10.3390/s16020223.

[30] Lyu, F., Zhang, Y., Zare, R. N., Zare, R. N., Ge, J., Liu, Z. One-pot synthesis of protein-embedded metal-organic frameworks with enhanced biological activities. *Nano Lett.*, 2014, *14* (10), 5761–5765. 10.1021/nl5026419

[31] Du, Y., Li, X., Lv, X., Jia, Q. Highly sensitive and selective sensing of free bilirubin using Metal-organic frameworks-based energy transfer process. *ACS Appl. Mater. Interfaces*, 2017, *9* (36), 30925–30932. 10.1021/acsami.7b09091.

[32] Zhao, T., Han, J., Jin, X., Liu, Y., Liu, M., Duan, P. Enhanced circularly polarized luminescence from reorganized chiral emitters on the skeleton of a zeolitic imidazolate framework. *AngewandteChemie*, 2019, *131* (15), 5032–5036. 10.1002/ange.201900052

[33] Chen, W., Yang, W., Lu, Y., Zhu, W., Chen, X. Encapsulation of enzyme into mesoporous cages of metal-organic frameworks for the development of highly stable electrochemical biosensor. *Analytical Methods*, 2017, *9*, 3213–3220. 10.1039/C7AY00710H

[34] Feng, D., Lui, T. F., Su, J., Bosch, M., Wei, Z., Wan, W., Yuan, D., Chen, Y. P., Wang, K., Lian, X., Gu, Z. Y., Park, J., Zou, X., Zhou, H. C. Stable metal-organic frameworks containing single molecule traps for enzyme encapsulation. *Nat. Commun.*, 2015, *6*, 5979. 10.1038/ncomms6979

[35] Tian, D., Liu, X. J., Feng, R., Xu, J.l., Xu,J., Chen, R. Y., Huang, L., Bu, X. H. Microporous luminescent metal-organic framework for a sensitive and selective fluorescence sensing of toxic Mycotoxin in moldy sugarcane. *ACS Appl. Mater. Interfaces*, 2018, *10* (6), 5618–5625. 10.1021/acsami.7b15764.

[36] Song, W. J. Intercellular DNA and microRNA sensing based on metal-organic framework nanosheets with enzyme-free signal amplification. *Talanta*, 2017, *170*, 74–80. 10.1016/j.talanta.2017.02.040

[37] Zhang, Z. H., Duan, F. H., Tian, J. Y., He, J. Y., Yang, L. Y., Zhao, H., Zhang, S., Liu, C. S., He, L. H., Chen, M., Chen, D. M., Du, M. Aptamer-embedded Zirconium-based Metal-organic framework composites prepared by de nova bio-inspired approach with enhance biosensing for detecting trace analytes. *ACS Sens.*, 2017, *2* (7), 982–989. 10.1021/acssensors.7b00236.

[38] Majewski, M. B., Howarth, A. J., Li, Peng, Wasielewski, M. R., Hupp, J. T., Farha, O. K. Enzyme encapsulation in metal-organic frameworks for application in catalysis. *CrystengComm.*, 2017, *19*, 4082–4091. 10.1039/C7CE00022G

[39] Zhu, K., Fan, R., Zheng, X., Wang, P., Chen, W., Sun, T., Gai, S., Zhou, X., Yang, Y. Dual-emitting dye-CDs@MOFs for selective and sensitive identification of antibiotics and MnO_4^- in water. *J. Mater. Chem. C*, 2019, *7*, 15057–15065. 10.1039/C9TC04700J

[40] Xia, C., Xu, Y., Cao, M. M., Liu, Y. P., Xia, J. F., Jiang, D. Y., Zhou, G. H., Xie, R. J., Zhang, D. F., Li, H. L. A selective and sensitive fluorescent probe for bilirubin in human serum based on Europium(III) post-functionalized Zr(IV)-Based MOFs. *Talanta*, 2020, *15* (212), 120795. 10.1016/j.talanta.2020.120795

[41] Zhao, Y., Zeng, H., Zhu, X. W., Lu, W., Li, D. Metal-organic frameworks as photoluminescent biosensing platforms: Mechanism and applications. *Chem. Soc. Rev.*, 2021, *50*, 4484–4513. 10.1039/D0CS00955E

[42] Shen, M. L., Liu, B., Xu, L., Jiao, H. Ratiometric fluorescence detection of anthrax biomarker 2,6-dipicolinic acid using hetero MOF sensors through ligand regulation. *J. Mater. Chem. C*, 2020, *8*, 4392–4400. 10.1039/D0TC00364F.

[43] Xu, H., Gao, J., Qian.; Wang, J., He, H., Cui,Y., Yang, Y., Wang, Z., Qian, G. Metal-organic framework nanosheets for fast response and highly sensitive luminescent sensing of Fe^{3+}, *J. Mater. Chem. A*, 2016, *4*, 10900–10905. 10.1039/C6TA03065C

[44] Huang, Y., Qui, P. L., Bai, J. P., Luo, D., Lu, W., Li, D. Exclusive recognition of acetone in a luminescent BioMOF through multiple Hydrogen-bonding interactions. *Inorg. Chem.*, 2019, *58* (12), 7667–7671. 10.1021/acs.inorgchem.9b00873

[45] Wang, J., You, T., Wang, T., Liu, Q., Ma, J., Jin, G. Adsorption behavior of a Cd^{II}-triazole MOF for butan-2-one in a single-crystal-to-single-crystal(SCSC) fashion: The role of hydrogen bonding and C-H … πinteractions. *Acta Crystallogr., C Struct. Chem.*, 2019, *75* (6), 806–811. 10.1107/S2053229619006788

[46] Zhao, Y., Wan, M. Y., Bai, J. P., Zeng, H., Lu, W., Li, D. pH-Modulated luminescence switching in a Eu-MOF: Rapid detection of acidic amino acids. *J. Mater. Chem. A*, 2019, *7*, 11127–11133. 10.1039/C9TA00384C

[47] Yang, C. X., Ren, H. B., Yan, X. P. Fluorescent metal-organic frameworks MIL-53(Al) for highly selective and sensitive detection of Fe^{3+} in aqueous solution. *Anal. Chem.*, 2013, *85* (15), 7441–7446. 10.1021/ac401387z

[48] Li, H., He, Y., Li, Q., Li, S., Yi, Z., Xu, Z., Wang, Y. Highly sensitive and selective fluorescent probe for Fe^{3+} and hazardous phenol compounds based on a water-stable Zn-based metal-organic framework in aqueous media. *RSC Adv.*, 2017, *7*, 50035–50039. 10.1039/C7RA08427G.

[49] Wang, J., Liu, Y., Jiang, M., Li, Y., Xia, L., Wu, P. Aldehyde-functionalized metal-organic frameworks for selective sensing of homocysteine over Cys, GSH and other natural Amino acids. *Chem. Commun.*, 2018, *54*, 1004–1007. 10.1039/C7CC08414E

[50] Yue, D., Zhao, D., Zhang, L., Jiang, K., Zhang, X., Cui, Y., Yang, Y., Chen, B., Qian, G. A luminescent cerium metal-organic framework for the turn-on sensing of ascorbic acid. *Chem. Commun.*, 2017, *53*, 11221–11224. 10.1039/C7CC05805E

[51] Qu, F., Ding, Y., Lv, X., Xia, L., You, J., Han, W. Emissions of Terbium metal-organic frameworks modulated by dispersive/agglomerated gold nanoparticles for the construction of prostate-specific antigen biosensor. *Anal. Bioanal. Chem.*, 2019, *411* (17), 3979–3988. 10.1007/s00216-019-01883-2

[52] Jin, M., Mou, Z. L., Zhang, R. L., Liang, S. S., Zhang, Z. Q. An efficient ratiometric fluorescence sensor based on metal-organic frameworks and quantum dots for highly selective detection of 6-mercaptopurine. *Biosens. Bioelectron.*, 2017, *91*, 162–168. 10.1016/j.bios.2016.12.022

[53] Gao, J., Wang, C., Tan, H. Dual-emissive polystyrene@zeolitic imidazolate framework-8 composite for ratiometric detection of singlet oxygen. *J. Mater. Chem. B*, 2017, *5* (46), 9175–9182. 10.1039/c7tb02684f

[54] Córdova Wong, B. J., Xu, D. M., Bao, S. S., Zheng, L. M., Lei, J. Hofmann metal-organic framework monolayer nanosheets as an axial coordination platform for biosensing. *ACS Appl. Mater. Interfaces*, 2019, *11* (13), 12986–12992. 10.1021/acsami.9b00693

[55] Nandi, S., Biswas, S. A recyclable post-synthetically modified Al(III) based metal-organic framework for fast and selective fluorogenic recognition of bilirubin in human biofluids. *Dalt. Trans.*, 2019, *48* (25), 9266–9275. 10.1039/c9dt01180c

[56] Chen, M. L., Chen, J. H., Ding, L., Xu, Z., Wen, L., Wang, L. B., Cheng, Y. H. Study of the detection of bisphenol a based on a nano-sized metal-organic framework crystal and an aptamer. *Anal. Methods*, 2017, *9* (6), 906–909. 10.1039/c6ay03151j

[57] Chen, B., Wang, L., Zapata, F., Qian, G., Lobkovsky, E. B. A luminescent microporous metal-organic framework for the recognition. *J. Am. Chem. Soc.*, 2008, *130*, 6718–6719. 10.1021/ja802035e

[58] Bhardwaj, N., Bhardwaj, S. K., Mehta, J., Kim, K. H., Deep, A. MOF-bacteriophage biosensor for highly sensitive and specific detection of Staphylococcus aureus. *ACS Appl. Mater. Interfaces*, 2017, *9* (39), 33589–33598. 10.1021/acsami.7b07818

[59] Chen, L., Zheng, H., Zhu, X., Lin, Z., Guo, L., Qiu, B., Chen, G., Chen, Z. N. Metal-organic frameworks-based biosensor for sequence-specific recognition of double-stranded DNA. *Analyst*, 2013, *138* (12), 3490–3493. 10.1039/c3an00426k

[60] Wei, X., Zheng, L., Luo, F., Lin, Z., Guo, L., Qiu, B., Chen, G. Fluorescence biosensor for the H5N1 antibody based on a metal-organic framework platform. *J. Mater. Chem. B*, 2013, *1* (13), 1812–1817. 10.1039/c3tb00501a

9 Potentiality of Magnetic Metal-Organic Frameworks

Sana Ahmed and Fahmina Zafar

CONTENTS

9.1 INTRODUCTION

For a number of decades, metal-organic frameworks (MOFs) have demonstrated their usage in the field of separations, storage of gases, selective adsorption process, catalysis, and also in biomedical fields [1]. MOFs belong to porous class materials that comprise of inorganic and organic hybrid materials that have a well-organized arrangement of positively charged metal ions by organic linker molecules and render exceptional chemical and physical properties [2]. They form a cage-like structure due to their extraordinarily large internal surface area [3]. Several researchers have formed a number of common MOFs that

DOI: 10.1201/9781003252061-9

189

include Materials of Institute Lavoisier (MIL), Hong Kong University of Science and Technology (HKUST), and zeolite imidazolate framework (ZIF) have specific characteristics and attractive features for various applications [4]. However, the major drawback of MOFs are the difficulty in the separation of certain molecules due to instability in water [5]. Also, MOFs are difficult to recycle from a mixture solution [6]. Furthermore, the stability of MOFs is extremely poor and the material skeleton has the possibility to collapse at extreme acidic or water vapor conditions [7]. The mechanical strength is also fragile because it exists in the form of powder. To address these issues, functional material modification is crucial in MOFs as they solve poor chemical stability and also enhance high porosity in the large surface area [8]. A lot of functional nanoparticles such as quantum dots [9], metal nanoparticles [10], magnetic nanoparticles [11], porous silica nanospheres [12], and nano-rods [13] are extensively utilized with a combination of MOFs.

Functional nanoparticles have shown an unusual growth in commercialization of nanotechnologies. They are also utilized for the fabrication procedure in research labs frequently as they have demonstrated their capabilities in the area of thernostics, biosensing, disease detection and acting as novel materials for the treatment of various infectious diseases. In order to feature functionality in nanoparticles, the particles should have size ranging between 50–100 nm and exhibit a large surface-to-volume ratio [14]. Polymeric-, metallic- and plasmonic-based functional nanoparticles have tunable physical and chemical features which are extensively used in cell labeling and tissue engineering [15]. In addition, semiconductor nano-crystals have been used in long fluorescence lifetime and broad excitation spectroscopy [16]. However, huge manufacturing and industrialization of functional nanoparticles has concerns over biocompatibility and biosafety issues. These concerns may lead to major downfalls in clinical applications [17].

Nevertheless, combining magnetic nanoparticles (MNPs) with MOFs to form magnetic framework composites (MFCs) enhances enormous properties of the MOFs as it helps to increase the quality of thermal stability. Moreover, the separation procedures are quite easy when magnetic properties are employed, which does not need the requirement of filtration and centrifugation [18]. This method saves a lot of time on additional procedures.

In biomedical applications, issues surrounding drug release can also be rectified by applying MNP modified MOFs, which helps to release the carried substances at the specified position in the presence of an external field [19]. Hence, the combination of MOFs and MNPs has the possibility that enhances the performances in different applications. There are numerous ways we can incorporate magnetic properties in MOFs that leads to showing elevated adsorption capacity, feasible separation with an external magnetic field as well as ease of functionality by providing a large surface area. These characteristics will compensate for the shortcoming of MOFs that has been shown in various applications [20]. In this chapter, we will specifically focus on using magnetic nanoparticles incorporating MOFs, briefly discuss synthetic procedures and discuss different applications.

9.2 SYNTHESIS

There are different methods that are used for the preparation of MFCs, depending on design. The synthesis of MFCs can be done by the following methods, which we will highlight below.

9.2.1 SELF-SACRIFICIAL TEMPLATE METHOD

In this process, the morphology and properties of MFCs are controlled effectively. A metal oxide template as the metal ion source can be used. The second step to grow MOFs is to use the surface of nanoparticles (NPs) without any modification on the surface. This self-sacrificial method was efficiently proposed by Huang and his co-workers for the preparation of porous MFCs with core-shell structures [21]. One report that has mentioned the incorporation of $Fe_3O_4@SiO_2$ with $Cu(OH)_2$ serve as a self-template and transformed $Cu(OH)_2$ into HKUST-1 by using ethanol and water as the solvent at the ambient state (Figure 9.1).

It was found to be simple, selective, and simple, operated for finding mercuric ions in the water [21]. Further, there are some bimetallic alloys consisting of noble metals as well as transition metals that have low cost and high performance. This phenomenon is used as a self-sacrificial template to synthesize MFCs with definite structures. Researchers have shown the use of transition metals as well as noble metals in the preparation of hierarchical nanostructures by using-Ni-alloyed nano-flowers acting as self-sacrificial templates. The formed MFCs with this approach are better tuned and have a good encapsulation state with good thickness when the Ni content in the template is changed. When employing these composites in catalytic activity, the performance was highly enhanced in the selective hydrogenation of alkynes to cis products [22].

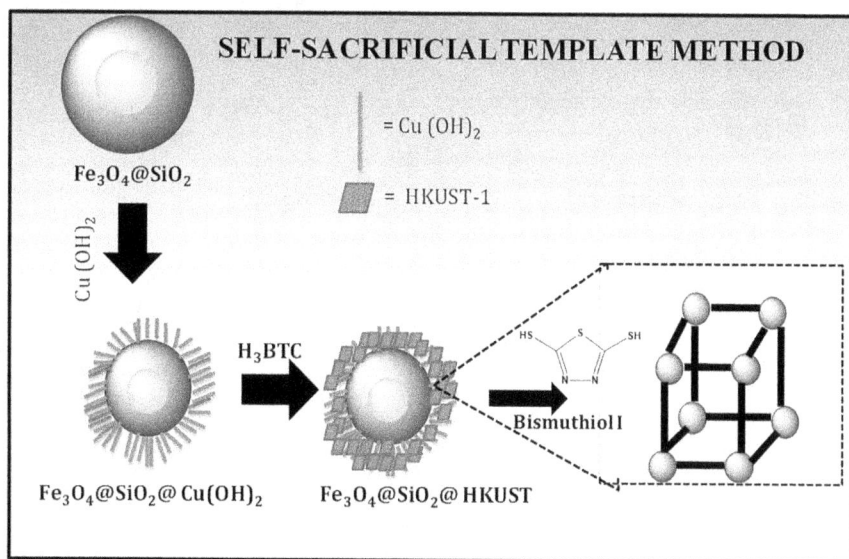

SELF-SACRIFICIAL TEMPLATE METHOD

= Cu $(OH)_2$

= HKUST-1

$Fe_3O_4@SiO_2$

$Cu (OH)_2$

H_3BTC

Bismuthiol I

$Fe_3O_4@SiO_2@ Cu(OH)_2$ $Fe_3O_4@SiO_2@ HKUST$

FIGURE 9.1 Preparation of Bi-I-functionalized magnetic HKUST composites.

9.2.2 EMULSION TEMPLATE METHOD

This template-based approach is used to obtain the porous-based material via emulsion microdroplets using micro-fluids as templates. One of the reports has used the Fe_3O_4@AgNPs, which were amalgamated as a template and ZIF-8@ZnONPs on the surface of Fe_3O_4@AgNPs [23]. Other researchers have also used the emulsion droplet template for the synthesis of MFCs via emulsion polymerization. The embedded Fe_3O_4 NPs are responsible for the synthesis of MFCs with good properties and impressive mechanical stability [24]. The same group has proposed the usage of magnetic porous polymer, which is made up of a polyacrylamide polymer using Fe_3O_4 and polyvinyl alcohol (PVA), and formed UiO-66 and Fe-MIL-101 (-NH_2) crystals of MOFs using an emulsion-based template [25] (Figure 9.2).

9.2.3 LAYER BY LAYER (LBL) SELF-ASSEMBLY METHOD

MFCs based on core shell are usually prepared by this simple and efficient methodology. In this process, the functional group has been modified on the surface of MNPs. It involves the functionalization of MNPs, using the growth of MOF with the aid of metals and organic linkers. For instance, there is a report that has prepared a triad composite consisting of Fe_3O_4-Au@MIL-100 (Fe) by using this approach [26]. In this paper, the Fe_3O_4 nanospheres are occupying the core that is responsible for the magnetic property in the composite, while plasmonic gold nanoparticles (Au NPs) are merged in the MOF. In addition, MIL-100 (Fe)

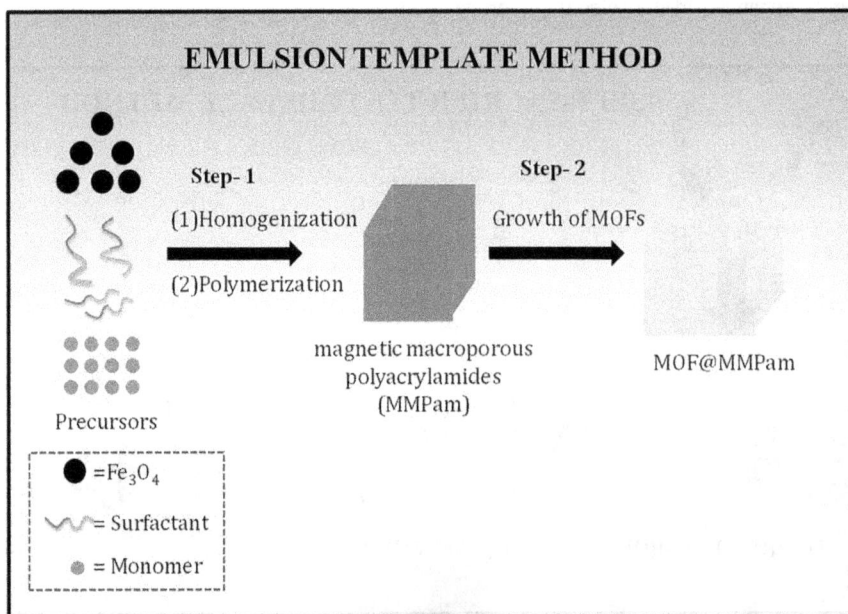

FIGURE 9.2 Synthetic routes for the synthesis of MMPam and MOF@MMPam.

comprises the composite to increase its function effectively with surface-enhanced Raman scattering (SERS) substrate. In the LBL method, HKUST-I is the most common material used. One of the reports describes the precise number of HKUST-I layers that were extended on COOH end magnetic nanoparticles [27]. By utilizing the sonication procedure, definite core-shell composites were obtained where the MOF layer is extremely homogeneous, and has precise thickness. The Brunauer-Emmett-Teller (BET) values of the MFCs were of a higher magnitude that can offer the probability of exhibiting the novel applications such as adsorptive materials as well as catalytic applications as compared to conventional magnetic adsorbents. This might be straightforward in the handling of suspension while showing higher binding capacities. Furthermore, the simple layer by layer (LBL) procedure can be employed in the synthesis of the multilevel hierarchical structure of MFCs. There has been a successful preparation of a magnetic hierarchical core-shell structure using $Fe_3O_4@$ PDA-Pd@$[Cu_3(BTC)_2]$ composites by using transmission electron microscopy (TEM) images and element mappings [28] (Figure 9.3).

There are other examples, such as amines or carboxylic acids that have been inserted on the flat layers, which promotes controlled crystal growth. In this method, MOF is grown layer by layer via a process of repetitive liquid phase epitaxy process [29]. LBL self-assembly leads to exhibiting precise control of the thickness or properties by numerous self-assembly cycles. The reaction conditions are moderate and the preparation process is quite simple and performed at ambient conditions. However, it usually requires more time for obtaining the suitable thickness of MFCs as a considerable amount of growth cycle steps are essential [30]. Secondly, only specific MOFs are synthesized using this process.

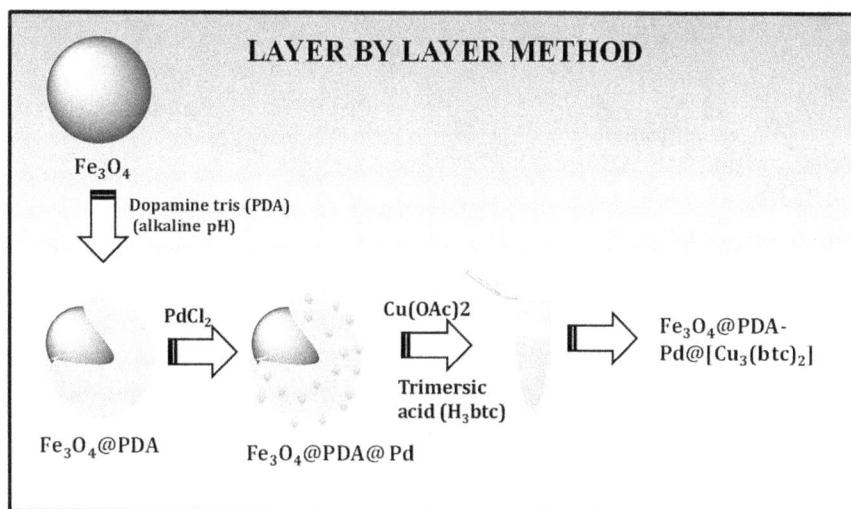

FIGURE 9.3 Synthetic routes for the synthesis of $Fe_3O_4@PDA$-Pd@$[Cu_3(btc)_2]$ composites.

9.2.4   IN-SITU GROWTH OF MOFS@MNPS AND MNPS@MOFS

Substantial efforts have been done for the preparation of magnetic framework composites (MFCs) via in-situ growth. MOF has unique properties that can provide space to MNPs to produce MFCs with novel physical and chemical characteristics. There are two probabilities to form MFCs; either MNP@MOFs or MOFs@MNPs under specific conditions. MOF@MNP methods are quite simple and completed in one step. In addition, the growth of MOFs can easily be achieved under feasible conditions and offers better performance to form MFCs. In one research, MFCs were efficiently obtained from in-situ growth of ZIF-8 on the surface of $Fe_3O_4@SiO_2$ microspheres [31]. The reaction was at ambient condition and it was found to have an efficient adsorption capacity to analyte bisphenol [31] (Figure 9.4a). Regardless, there are some drawbacks associated as it is difficult to prevent nucleation growth. Further, there is a high possibility of MNPs, as it may be embedded in the MOFs pores leading to less surface area. Also, the unreacted magnetic nanomaterials have direction to remain in the solution, resulting in a tedious separation process and demand of a large amount of organic solvents [32].

In the other MNP@MOFs based method, it demands a multistep process. This method requires the growth of MNPs in the presence of prepared MOFs and, consequently, embedding of MNPs into MOFs. The unwanted and unreacted MNPs

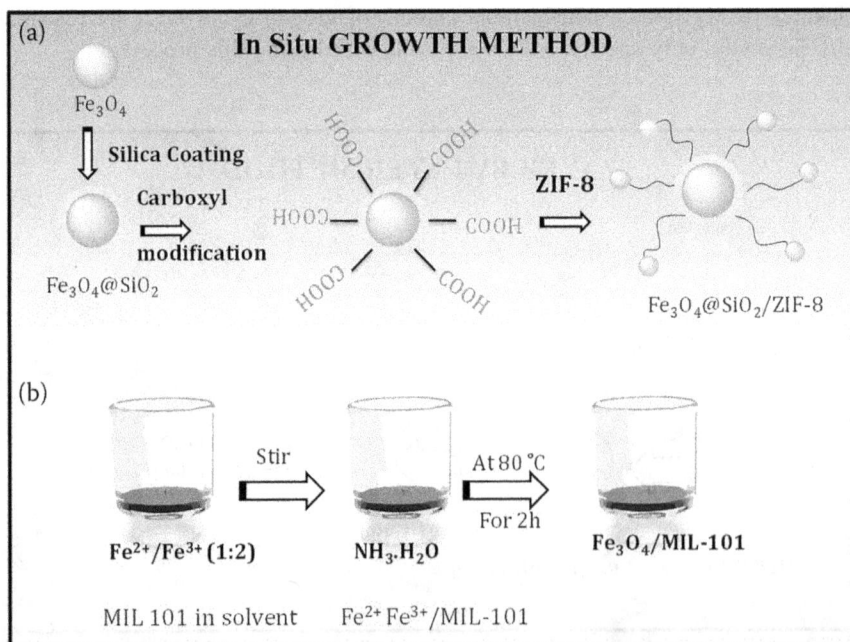

FIGURE 9.4 In-situ growth method: (a) Schematic diagram for the synthetic procedure of the $Fe_3O_4@SiO_2$/ZIF-8 via growth of MOFs on MNPs. (b) Schematic representation of the fabrication process for Fe_3O_4/MIL-101 via growth of MNPs on MOFs.

in the product are easily separated by the process of centrifugation using a magnetic field. Due to these characteristics, the MOFs' original structures are intact and they exhibit good adsorption as well as an efficient magnetic behavior. Recently, the synthesis of ferric oxide nanoparticles on MIL-101 MOFs supports the formation of nanocomposite was reported [33]. There was another report for the preparation of magnetic hybrid ferric-oxide/MIL-101 composite [34] (Figure 9.4b). The external surface of MOFs was found to be coated uniformly by prepared nanocomposites. However, this method possesses certain drawbacks as there are possibilities to show aggregation on MNPs, bad size control of MNPs and possible degradation of a MOF structure.

9.2.5 HYBRID PREPARATION METHOD

These methods involve the direct mixing of MOFs and MNPs through ultrasonic and high-temperature polymerization techniques. The MFCs are prepared through electrostatic interactions when MNP is modified on the surface of MOFs, resulting in a better magnetic effect. This procedure is one of the easiest approaches to manage the interactions between MOFs and MNPs for strong chemical bonding. Further, it is applicable in any MOF, as it does not require any specificity. A hybrid preparation method was used combining $Fe_3O_4@SiO_2$ and MOFs, where ultra-sonication procedure was applied in that step in order to disperse the solution of MNPs and MOFs. The formation of these composites was confirmed by the zeta potential values [35]. Overall, this method has offered a strong applicability for the preparation of MFCs. In other research, MFCs were synthesized via a hybrid preparation method consisting of functional MNPs, which are Fe_3O_4 incorporating high porosity and magnetization with MOF-5 ($Zn_4O(BDC)_3$ where BDC = 1,4 benzendicarboxylate and it was found to be a great potential prospect to absorb trace analytes (Figure 9.5). This hybrid magnetic composite has been shown in the extraction of various chemicals from the surroundings [36]. The MFCs produced thorough this method were found to be greater surface area, greater magnetic response characteristics, efficient stability, a good application for food analysis [37]. However, in some cases, a large amount of MOFs are required as there is an incompetency for MNPs to deposit on the surface of MOFs, resulting in detachment ultimately.

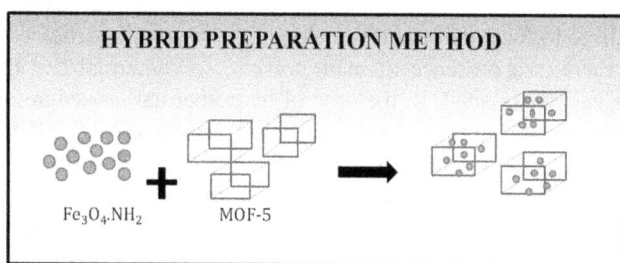

HYBRID PREPARATION METHOD

$Fe_3O_4.NH_2$ MOF-5

FIGURE 9.5 Schematic representations for the preparation of magnetic MOFs via a hybrid method.

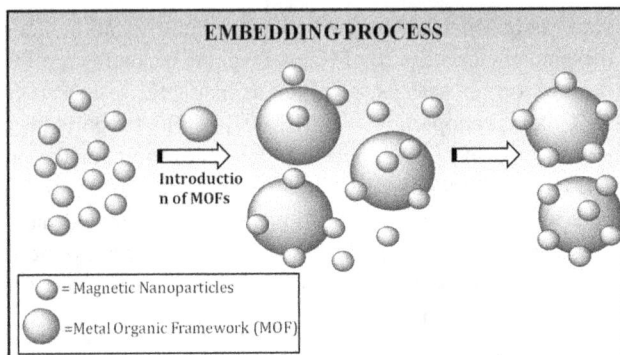

FIGURE 9.6   Schematic representation of the embedding process for synthesis of magnetic framework composites (MFCs).

9.2.6   EMBEDDING METHOD

The magnetic particle materials are embedded on the MOF's surface. The first step is to follow the nucleation step and growth mixture for the MOFs (Figure 9.6).

Further, MNPs are submerged in MOFs by using ultrasonication or hydrothermal conditions. This is widely used to propagate MOFs in a solution. Surprisingly, the morphology is quite equivalent to original MOFs. However, ligands and inorganic precursors are needed to mix the MNPs in a mixture of solvents. There is one report that Fe_3O_4/ZIF-8 composites use 2-methylimidazolate (2-MeIM) and zinc nitrate ($Zn(NO_3)_2$) in the presence of iron oxide (Fe_3O_4) nanoparticles at room temperature by using an embedding method [38]. The achieved nano-crystal embedding process was responsible for electrostatic interactions with colloidal particles that ultimately grow into Fe_3O_4/ZIF-8 composites resulting into hetero-nanostructures with magnetic properties. In order to prepare the composites, the parameters of surface charge, magnetic loading, and particle size are easily manipulated. When Zn (NO_3)$_2$ solutions was added, the other two components 2-MeIM and Zn^{2+} immediately grew as small crystals. During growth of ZIF-8, the surface charge was significantly increased, the negatively charge Fe_3O_4 was attracted to the ZIF-8 surface with the help of electrostatic interactions. In this paper, researchers have shown the amount of Zn^{2+} that was added in a controlled manner, as the average particle size was highly dependent on the amount of $Zn(NO_3)_2$ loaded. In usual cases, a sonication process is utilized to form uniform dispersion of the magnetic particles in a solution. However, if there is a presence of main solvent, solvothermal and hydrothermal processes are easily executed. In the case of hydrothermal procedure, high boiling point polar non-protic solvents such as dimethylformamide (DMF), dimethyla-cryalamide (DMA), and sometimes dimethylsulfoxide (DMSO) are used and methanol or ethanol can also be added as co-solvents. The water is the main source while other mixtures of solvents can also be used along with the water. For MFCs production, this usually occurs in a sealed vessel at a higher temperature rather than room temperature using an autoclave with higher pressure. In addition, in-situ solvothermal method synthesize MFCs materials via electrostatic interactions that can form uniformly homogeneous magnetic product [39]. It was also reported that the amount of Fe_3O_4 particles were affected their structure and magnetic properties of the composites. Moreover, when 50 mg was the loading amount, it was observed that there were no free Fe_3O_4 particles found while increasing the loading amount; the composite materials have exhibited a higher amount of Fe_3O_4 content that was confirmed by TEM. The saturation magnetization values of MFCs were reduced significantly with fewer amounts. Interestingly, the lowest saturation magnetization value showed higher magnetization characteristics. Spray drying is one of the systems that allows the formation of hollow spheres of nano-sized MOF crystals. The feed rate, flow rate, and inlet temperature are the few considered conditions for MOF preparation in less time. This technique might be helpful for the preparation of composites. For instance, magnetic nanoparticles might be embedded within HKUST-I [40,41]. This particular process has numerous advantages as it is simple, completed in a single go and the magnetization characteristics can be achieved in the MOFs. But the controlled behavior of nucleation, as well as the growth of MOFs is a difficult task to achieve. In addition, there are chances that some of the MNPs might be deposited to the MOFs pores that could result in less surface area.

9.2.7 ENCAPSULATION METHOD

In this approach, buffer interfaces are required as a carrier medium between the porous frames and MNPs to enhance the growth of MOFs to formed MFCs (Figure 9.7).

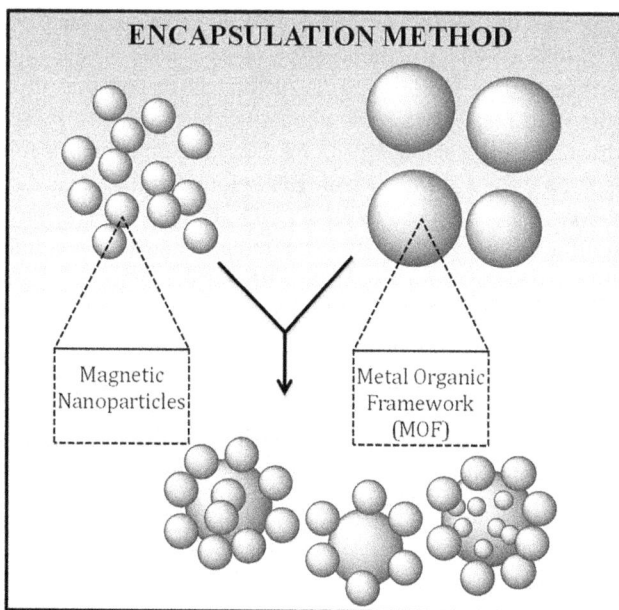

FIGURE 9.7 Schematic representation of the encapsulation process for synthesis of magnetic framework composites (MFCs).

To increase the affinity of MFCs, the MNPs are embedded into the polymer layer and afterward, growth of MOFs is initiated with nucleation. This strategy is used to prepared composite materials where MNPs are placed at the core and are encapsulated into the material. In one of the approaches, UIO-66 was demonstrated as a representative MOF for the preparation of $Fe_3O_4@SiO_2@UiO-66$ core-shell microspheres where it was modified by glutaric anhydridein the first step, to produce microspheres and then used as seeds for the growth of MOF crystals [42]. Several techniques were used to investigate whether the microspheres are embedded into the MOF crystals or not. After the preparation of composites, it was found that these materials have a thermal responsive behavior [43]. The facile synthesis of MFCs was done by Kim's group as they used a microfluidic device for the production of continuous microdroplet formation as these were used as a reactor to produce MOFs and MFCs in an extremely short time [44].

9.2.8 OTHER METHODS

The seed crystal method [45], dry gel conversion method [46], and solvent-free mechanochemical method [47] are also a few of the approaches that are being used in the preparation of MFCs. In the seed crystal approach, there are stages to grow a MOF shell on the inorganic surface with the help of the precursor as seed crystals via a hydrothermal reaction. In one of the reports, Chen and co-workers have manifested the effective synthesis where citrate functionalized Fe_3O_4 nanoparticles were dispersed in ferric chloride. In that process, $Fe_3O_4@MIL-100$ (Fe) precursors were subjected to acquisition by adsorbing Fe^{3+} and trimesic acid (H_3BTC) on the surface of nanoparticles. The remaining two were mixed and reacted using hydrothermal conditions under certain period of time where they found core-shell structure was investigated by TEM. This method is responsible to save time as compared to the LBL method and also good shell thickness can be achieved [48]. On the other hand, the dry gel conversion method is responsible for forming MOFs around magnetic NPs via a solvent vapor. But before that, the separation procedure has to proceed by combination of Fe_3O_4 and MOF precursors. The magnetic porous adsorbents are efficiently separated by the external magnetic field and also adsorb aromatic sulfur and nitrogen compounds [49]. Further, mechanochemical offers many advantages over others as it is executed in a very short time and does not require a huge amount of solvents. The research conducted by Secci and co-workers used a mechanochemical approach in which they found composites are irregularly embedded magnetic particles with a diameter ranging of 25–150 nm by using the ball milling process. Overall, due to these characteristics such as environmental sustainability and simplicity of these approaches, it might show an efficient capability in industrial applications [50].

9.3 APPLICATIONS

The MFC syntheses are simple and prepared by a variety of methods while preserving the characteristics of MOFs as well as MNPs. They contain large surface areas, variable pore size, functionality and crystalline open structures. By taking

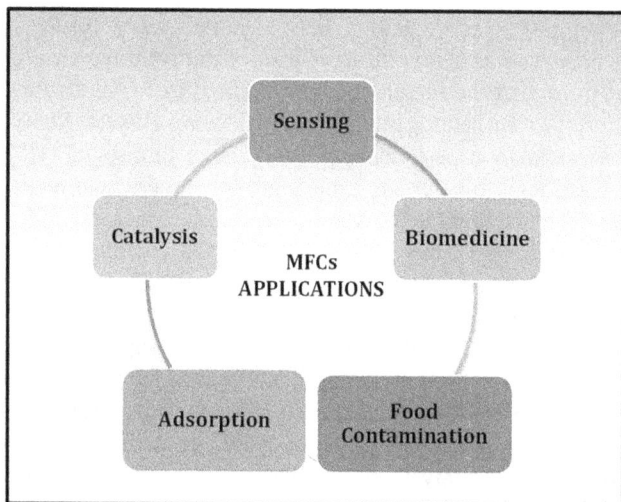

FIGURE 9.8 Possible areas for applications of metal framework composites (MFCs).

advantage of all of the attributes shown by MFCs, they can be employed in diverse applications such as catalysis, in removal of food contaminations and sensing (Figure 9.8).

9.3.1 CATALYSIS

MOFs have attracted great attention and they are being widely used in catalytic applications. Numbers of MOFs have been studied, but only few MOFs are utilized in a catalysis application such as UIO-66 (Zr_6O_8) [51], ZIF-8 (Zn^{2+} nodes) [52], MIL-101 (with Al_2O_3 [53], Fe_2O_3 [54], Cr_2O_3 [55] nodes) and MOF-5 [56] (Zn_4O or mixed oxide nodes). They offer a variety of benefits for potential application in catalysts. They have high internal surface area and active site densities for transfer large reactant and product molecules, pore structures can be tailored for shape selective catalysis, manipulative catalytic site environment. At the same time, MOFs have some drawbacks as they have less stability than inorganic solids and the possibility of minimizing the recyclability of MOF catalyst. The main disadvantage of a MOF as a catalyst is the isolation process involves a lot of extra procedures. In order to overcome this issue, MFCs could easily be separated via external magnetic fields which leads to fast separation of MOFs by chemical reaction and it is easily recycled [57]. In one report, Peng et al. have developed a nanocomposite based on Cu-based ceramic materials and Fe_3O_4 which have been utilized for heterogeneous catalyzed reaction and shows higher catalytic activity as compared to traditional heterogeneous catalysts (e.g., aluminum oxide, silicon dioxide). The formed MFCs were efficiently produced benylidenemalonitrile with 99% yield after 5 hours [58]. In another report, the MFCs were revealed for the oxidation of trimethylsilyviny-lethers to α-hydroxy-ketone [59,60]. In this paper, copper-based magnetic frame-work was utilized on carboxylic and amino-functionalized particles and have successfully exhibited an efficient yield between 75% to 99% for the synthesis of a product. This catalyst was shown to have good stability in the air after being collected several times and the effectiveness of copper-based MFCs for the reaction. There was one report in heterogeneous catalysis using Fe_3O_4@MOF-2 for breakdown of harmful diazinon pesticides with the aid of ultrasound (US) irradiation. This catalyst was found to be good, stable and showing efficient recyclability even after 15 cycles. The reaction rate of MFCs was found to increase four times that of only MNP-based systems, which shows the effective combination with magnetic and MOFs [61]. A few examples of research have shown MOFs that have been utilized for photocatalytic activity, so MFCs also exhibit this property. New MFCs were developed using Ni/Co with $BiFeO_3$ have shown superb properties for photocatalysis degradation for methylorange and 4-nitrophenol under visible light irradiation. Also, it was found that the composites were offered with elevated surface area, transfer of molecules, and adsorption properties [62]. However, there is rapid growth of MOF catalysis to manufacture on a large scale; we expect that new catalytic MOFs may require more in research and development to gather the understanding of the fundamentals of catalysis.

9.3.2 SENSING

The integration of magnetic materials into MOFs is also showing sensory-based properties. MOFs have shown a luminescence property depending on the type of metals and also depending on the type of ligands that are used [63]. In one paper, the Fe_3O_4 MNPs and Rhodamine B (RhB) modified carbon dots (CDs) were encapsulated into MOFs [64]. This composite was implied to a HClO sense, which shows stability in aqueous solutions. A phenomenon was hypothesized where the RhB-CDs were oxidized by HClO, indicating the depletion of p-electron density, ultimately RhB-CD fluorescence quenching in conjugated system [65]. Taking inspiration from that research, Lu et al. have shown the development of dual-emission fluorescent MOFs for the identification of Cu^{2+} using encapsulated fluorescein isothiocyanate (FITC) and Eu^{III} complex functionalized Fe_3O_4 into ZIF-8. The MOFs that are used in this composite provide protective layers that inhibit aggregations and also enhance the fluorescent signals for Cu^{2+} investigation. Further, they found this nanocomposite probe is quite stable for a wide range of pHs and has good prospects for bioimaging and bioanalysis [66]. The magnetic-based MOFs are also used in environmental monitoring and contaminants detection. Christus et al. have shown the detection of Hg^{2+} ions from calorimetric sensors by showing the formation of a deep blue color by the oxidation of 3, 3, 5, 5-tetramethylbenzidine. When glutathione (GSH) was added, the deep blue color disappears but showed up when mercuric ion was transformed to exhibit a blue color as a strong affinity with GSH to Hg^{2+} ions [67]. This nanocomposite was exhibiting a greater performance as a nano-based calorimetric sensor compared with standard cinnamaldehyde pyrimidine. Moreover, another type of nanocomposite Fe_3O_4@SiO_2@UiO-67 for sensing glyphosate ($C_3H_8NO_5P$) is an extra product of organophosphorous pesticides that has been showing negative effects on soil microbes and aquatic plants. The mechanism of nanocomposite based on the presence of Zr-OH group that has shown the interaction with phosphate groups that aid to recognize selectively of glycolphosphate. The excitation wavelength of fluorescent intensity was greatly changed when combined with MFCs and glyophosphate [68]. Based on all the aforementioned research, it has been shown that MFC-based sensors have extensive potential for identification. But still, the utilization of magnetic-based MOFs is still under-explored, and more elaborate studies are needed for the development of any new type of sensors.

9.3.3 NANOMEDICINES

MOFs have shown great value in biomedical applications from drug release to immunotherapy. The amalgamation of MNPs and MOFs show profound applications in biomedical characteristics as they produce a sizeable surface area for storing and releasing biomacromolecules, whereas magnetic nanoparticles are able to provide a magnetic field. Imaz and co-workers have shown the use of metal-organics transformed into nanospheres of 600 nm and show the effective encapsulation of magnetic nanoparticles and luminescent quantum dots (QDs) act as a potential for drug encapsulation [69]. Similar research has been performed by the same research group, in which they have demonstrated the use of CdSe/ZnS QDs along with magnetic nanoparticles, which lead to forming a fluorescent sphere and they have shown the effective encapsulation and release of doxorubicin that has the application as anticancer drugs [70]. Further, one research team published the result of utilizing magnetic nanoparticle modified MOFs for storage and release of the drugs. Firstly, they synthesized the nanocomposite by a precipitation method and make a nanocomplex with different properties. In this research, the HKUST was used as MOF and Fe_3O_4 as magnetic nanoparticles to form as Fe_3O_4@HKUST-2 complex and it was found that their specific surface area was reduced by 95% after incorporation of nimesulide as drugs indicate a high drug loading capacity. It was also found that 20% of drugs are released within 4 hours, and then slowly the others are released over the next 7 days. These results concluded that these kinds of nanocomposites were releasing at a moderate rate to allow drugs to be absorbed by this system. In addition, magnetic nanocomposites containing nimesulide drugs can be easily compiled by magnets less than 50s. This paper has demonstrated the capability of magnetic modified MOFs are compelled with drug loading and release, which is directly dependent on the pore size and chemical functions and practicability of using MFCs in drug delivery applications [71]. Lohe and co-workers have demonstrated the release of drugs with the incorporation of aluminum and copper-based MFCs in the presence of the magnetic field. Three MOFs were AlOH(ndc) where **ndc** = 2,6-naphthalene dicarboxylate, Dresden University of Technology (DUT-4), AlOH(bpdc) where bpdc is 4,4'-biphenyldicarboxylate, (DUT-5) and $Cu_3(btc)_2$ or HKUST-1; where **BTC** = 1,3,5-benzene tricarboxylate were prepared under solvothermal conditions using Fe_3O_4 by a co-precipitation method. The nanocomposites were found to be the same as MOFs with the same morphological features. However, the incorporation of the super-paramagnetic property of Fe_3O_4 did not show any residual magnetization and also reduces the chances of fatal blood occlusions. The formed nanocomposite was shown the presence of dynamic stimulus-induced localized heat with alternating magnetic fields. This is the first report that has shown the magnetically induced heating with an outer stimulus to activate the release of drugs [72]. Another research shows the development of magnetic bio-metal organic-framework (MBMOF) using in-situ sono-synthesis for the treatment of leishmaniasis. The use of MBMOF was shown in BALS/c mice for the treatment of leishmania and used as an ointment three times a week. During the investigation, it was found that the ointment in the mice didn't lose stability and from the outcome, it was demonstrated in the treatment of cutaneous leishmaniasis [73].

9.3.4 ADSORPTION

MFCs were found to have a great capability in the absorption process, as they have shown definite surface area and exhibit straightforward and fast separation. Due to excellent physicochemical characteristics of MFCs, they act as feasible adsorbents. The MFCs are also widely used for the adsorption of high-risk materials. We will discuss the applicability of MFCs based on adsorption technique in the following sections.

9.3.4.1 Adsorption of Dyes from Wastewater

MFCs have also been utilized in the adsorption of dyes from wastewater. The magnetic Fe_3O_4/MIL 101 (Cr) composite was formed by Wang and co-researchers by a reduction precipitation process found to be a large surface area with good magnetic properties and exhibit dispersion effect. These materials were efficient for the adsorption of acid red and orange G [74]. A new fe_3O_4-PSS@ZIF-67 magnetic core-shell composite was used for the particular adsorption of dyes and have shown the capacity for adsorption of almost more than 700 mg/g with efficient separation. The dye molecules were absorbed through the pores of magnetic composites were adsorbed due to electrostatic interactions [75]. There was another report that mentioned the use of magnetic nanoparticles ($MgFe_2O_4$) incorporated with MOF i.e $MgFe_2O_4$@MO, which enables the effective adsorption of rhodamine B (RB) and rhodamine 6G (Rh6G) confirmed from the Langmuir isotherm model. This composite provides a large surface area with super-paramagnetic properties [76]. Jiang et al. have developed Fe_3O_4@MIL-53 (Al), which was found to be efficient to remove bisphenol A, tetracycline(TC), Congo Red (CR), and MB. The presence of magnetic nanoparticles Fe_3O_4 is responsible for efficient separation in the presence of external magnetic fields. This composite might act as a potential adsorbent for extraction of dyes from water [77].

9.3.4.2 Adsorption of Gases

The MFCs were produced by using cost-effective aluminum fumarate and $MgFe_2O_4$ under the influence of alternating current magnetic field for identifying the uptake of CH_4 uptake and a good cycling performance. This magnetic induction swing adsorption process involves the dynamic triggered release for CH_4 emission sources and little time of less than 7 minutes with each cycle is required. This method was found to be great compared to other reported procedures, such as temperature and pressure swing adsorption, for the capture of CH_4 [78]. The excellent properties of MFCs such as larger surface area and magnetic properties lead to elimination of risky materials from surroundings and have the ability to selectively adsorb targeted toxic compounds from the environments by chemical modification of MFCs.

9.3.4.3   Food Contaminant Extraction

Toxins are usually generated by fungi as it is regular impurity in grains. MFCs have shown a great potential in food contamination extraction. In one of the research papers, it has been indicated that core-shell Fe_3O_4@UiO-66-NH_2@microporous composites have been used as sorbents to detach the aflatoxins. Aflatoxins are an extremely toxic type of mycotoxins and these materials were specifically designed to target aflatoxins [79]. Further, the addition of toxic additives such as plasticizer compounds and dyes are a severe warning in the safety for food. There were few types of research that were totally focused on the isolation of these contaminants from the food as one research shows the use of Fe_3O_4@ Zn-based MOF (TMU-24) materials for the removal of plasticizer compounds [80]. Another, Fe_3O_4-NH_2@ MIL-101, was used to separate certain dyes [81]. The presence of unwanted additives in foods such as plasticizers and melamine are the warnings in the food industry. Liang et al. developed magG@PDA@ZIF-8 nanosheets that were shown π–π electron system, for plasticizer detection. They were found to be highly stable, reliable, and highly sensitive [82]. This nanosheet consists of magnetic graphene (magG), polydopamine (PDA) and ZIF-8 to produce magG@PDA@ZIF-8 have unique characteristics for excellent analysis of pesticides. In other research, magnetic MOFs Fe_3O_4@PEI-MOF-5 was used as an efficient enrichment of dues in fish samples [83]. The broad application of MFCs has the ability to separate the contamination that enhances the food quality which enhances the detection throughput.

9.4   ADVANTAGES AND DISADVANTAGES OF MFCS

The amalgamation of MNPs and MOFs has shown various applications in the fields of adsorption and separation [84]. As we have mentioned in earlier sections, magnetic MOFs have porosity, large specific surface area, and also a magnetic field from nanoparticles that induce ease of separation of liquids. This has led us to conclude that MFCs hold a great promise in different applications. However, there are some critical points that are important to highlight. Biocompatibility and non-toxic synthesis are important for use in the food industry. Secondly, industrialization of composites carries a problem due to instability in water and support in hard surroundings of contaminants and pollutants. In addition, there are chances to lose its activity with reaction time/cycle, which shows a huge concern in catalysis. Long-term water and thermal stability are essential criteria to use MFCs on an industrial scale.

9.5   CONCLUSIONS

Magnetic framework composites are brand-new materials that have shown promising results with unusual properties. A variety of processes, such as the LBL methods, seed crystal encapsulation methods, and so on are used for obtaining MFC-based materials for biomedicines as well as chemical applications such as catalysts and environmental detections are attracting researchers to work in this particular area.

MFCs have demonstrated great applications in the removal of pollutants from water and can be a game-changer for sustainable development. Moreover, the use of MFCs is showing great potential for nanomedicines, too. It has already been reported that magnetic particles have manifested greatly in hyperthermia treatments and MOFs have already had the potential for drug adsorption and releases; so, by combining these two, they can play an important role in killing cancerous cells through local high temperatures with the release of drugs. Magnetic-led MOFs can also be highly applicable in catalysis applications. However, some research has indicated less activity in reaction time per cycle and the stability of MFCs needs to be improved. Sensor applicability is also revealed in MFCs and is expected to work and establish themselves as luminescent materials. To conclude, MFCs are prepared by straight-forward synthesis while more elaborate and concise studies are needed for optimizing the condition for synthesis steps. Furthermore, utilization of different MOFs is required for biomedical MOFs and water-resistant MOFs for water-based solutions. In a majority of the cases, studies were conducted only using HKUST-1 and MOF-5 involved in the development of MFCs. On that account, the exploration of combining different MOFs to magnetic nanoparticles is still restricted. We need to recognize more prospects of MFCs in other fields for more practical applications.

REFERENCES

[1] Hao, L., Libo, L., Lin, R.B., Zhou, W., Zhang, Z., Xiang, S., & Banglin, C. 2019. "Porous Metal-Organic Frameworks for Gas Storage and Separation: Status and Challenges."–*EnergyChem*, *1*(1), 1–77. 10.1016/J.ENCHEM.2019.100006

[2] Rowsell, J.L.C., & Yaghi, O.M. 2004. "Metal-Organic Frameworks: A New Class of Porous Materials." *Microporous and Mesoporous Materials*, *73*(1–2), 3–14. 10.1 016/J.MICROMESO.2004.03.034

[3] Farha, O.K., Ibrahim, E.I., Nak, C.J., Brad, G.H., Christopher, E.W., Amy, A.S., Randall, Q., Snurr, S.B.T., Nguyen, A., Özgür, Y., & Joseph, T.H. 2012. "Metal-Organic Framework Materials with Ultrahigh Surface Areas: Is the Sky the Limit?" *Journal of the American Chemical Society*, *134*(36), 15016–15021. 10.1021/JA3 055639

[4] Kong, X.J., & Jian, R.L. 2021. "An Overview of Metal-Organic Frameworks for Green Chemical Engineering." *Engineering*, *7*(8), 1115–11139. 10.1016/J.ENG. 2021.07.001

[5] Liu, B., Kumar, V., Ki-Hyun, K., Vanish, K., & Kailasa, S.K. 2020. "Critical Role of Water Stability in Metal-Organic Frameworks and Advanced Modification Strategies for the Extension of Their Applicability." *Environmental Science: Nano*, *7*, 1319–1347. 10.1039/C9EN01321K

[6] Abednatanzi, S., Parviz, G.D., Hannes, D., François-Xavier, C., Henk, V., Pascal, V.V., & Leus, K. 2019. "Mixed-Metal Metal-Organic Frameworks." *Chemical Society Reviews*, *48*, 2535–2565. 10.1039/C8CS00337H

[7] Ma, M., Li, L., Hongwei, L., Xiong, Y., & Dong, F. 2019. "Functional Metal Organic Framework/SiO2 Nanocomposites: From Versatile Synthesis to Advanced Applications." *Polymers*, *11*(11), 1823–1845. 10.3390/polym11111823

[8] Li, H., Mario, C., Kecheng, W., William, B., & Zhou, H.C. 2017. "Design and Construction of Chemically Stable Metal-Organic Frameworks." In Elaboration and Applications of Metal-Organic Frameworks, Series on Chemistry, Energy and the Environment. *World Scientific*. 10.1142/9789813226739_0001

[9] Kumagai, K., Uematsu, T., Torimoto, T., & Kuwabata, S. 2019. "Direct Surface Modification of Semiconductor Quantum Dots with Metal-Organic Frameworks." *CrystEngComm, 21,* 5568–5577. 10.1039/C9CE00769E

[10] Sun, D., & Dong-Pyo, K. 2020. "Hydrophobic MOFs@Metal Nanoparticles@COFs for Interfacially Confined Photocatalysis with High Efficiency." *ACS Applied Materials & Interfaces, 12*(18), 20589–20595. 10.1021/ACSAMI.0C04537

[11] Zhao, G., Nianqiao, Q., An, P., X., Wu., Peng, C., Ke, F., Iqbal, M., Ramachandraiah, K., & Zhu, J. 2019. "Magnetic Nanoparticles@Metal-Organic Framework Composites as Sustainable Environment Adsorbents." *Journal of Nanomaterials, 2019,* 1–11. 10.1155/2019/1454358

[12] Zhou, Z., Xiujuan, L., Wang, Y., Yi Luan, Li, X., & Du, X. 2020. "Growth of Cu-BTC MOFs on Dendrimer-like Porous Silica Nanospheres for the Catalytic Aerobic Epoxidation of Olefins." *New Journal of Chemistry, 44,* 14350–14357. 10.1039/D0NJ02672G

[13] Zhou, Z., Zhao, J., Di, Z., Liu, B., Li, Z., Wu, X., & Li, L. 2021. "Core-Shell Gold Nanorod@mesoporous-MOF Heterostructures for Combinational Phototherapy." *Nanoscale, 13,* 131–137. 10.1039/D0NR07681C

[14] Patra, J.K., Das, G., Fraceto, L.F., Campos, E.V.R., Rodriguez-Torres, M.D.P., Susana Acosta-Torres, L., Armando Diaz-Torres, L., et al. 2018. "Nano Based Drug Delivery Systems: Recent Developments and Future Prospects." *Journal of Nanobiotechnology, 16*(71), 10.1186/s12951-018-0392-8

[15] Khan, I., Saeed, K., & Khan, I. 2019. "Nanoparticles: Properties, Applications and Toxicities." *Arabian Journal of Chemistry, 12*(7), 908–931. 10.1016/J.ARABJC.2017.05.011

[16] Xavier, M., Pinaud, F., Lacoste, T.D., Dahan, M., Bruchez, M.P., Alivisatos, A.P., & Weiss, S. 2001. "Properties of Fluorescent Semiconductor Nanocrystals and Their Application to Biological Labeling." *Single Molecules, 2*(4), 261–276. 10.1002/1438-5171(200112)2:4<261::AID-SIMO261>3.0.CO;2-P

[17] Desai, Neil. 2012. "Challenges in Development of Nanoparticle-Based Therapeutics." *The AAPS Journal, 14*(2), 282–295. 10.1208/s12248-012-9339-4

[18] Aghayi-Anaraki, M., & Safarifard, V. 2020. "Fe_3O_4@MOF Magnetic Nanocomposites: Synthesis and Applications." *European Journal of Inorganic Chemistry, 2020*(20), 1916–1937. 10.1002/EJIC.202000012

[19] Cai, M., Chen, G., Qin, L., Qu, C., Dong, X., Jian, N., & Yin, X. 2020. "Metal Organic Frameworks as Drug Targeting Delivery Vehicles in the Treatment of Cancer." *Pharmaceutics, 12*(3), 232. 10.3390/PHARMACEUTICS12030232

[20] Zhao, X., Liu, S., Tang, Z., Niu, H., Cai, Y., Meng, W., Wu, F.& Giesy, J.P. 2015. "Synthesis of Magnetic Metal-Organic Framework (MOF) for Efficient Removal of Organic Dyes from Water." *Scientific Report, 5,* 11849. 10.1038/srep11849

[21] Huang, L., He, M., Chen, B., & Hu, B. 2015. "A Designable Magnetic MOF Composite and Facile Coordination-Based Post-Synthetic Strategy for the Enhanced Removal of Hg2+ from Water." *Journal of Materials Chemistry A, 3,* 11587–11595. 10.1039/C5TA01484K

[22] Chen, L., Li, H., Zhan, W., Cao, Z., Chen, J., Jiang, Q., Jiang, Y., Xie, Z., Kuang, Q.,& Zheng, L. 2016. "Controlled Encapsulation of Flower-like Rh-Ni Alloys with MOFs via Tunable Template Dealloying for Enhanced Selective Hydrogenation of Alkyne." *ACS Applied Materials & Interfaces, 8*(45), 31059–31066. 10.1021/ACSAMI.6B11567

[23] Chen,L., Zhang, M.J., Zhang, S.Y., Shi, L., Yang, Y.M., Liu, Z.,Xiao-Jie, J., Xie, R., Wang, W., & Chu, L.Y. 2020. "Simple and Continuous Fabrication of Self-Propelled Micromotors with Photocatalytic Metal-Organic Frameworks for Enhanced Synergistic Environmental Remediation." *ACS Applied Materials & Interfaces, 12*(31), 35120–35131. 10.1021/ACSAMI.0C11283

[24] Peng, J., Tan, W., Huo, J., Liu, T., Liang, Y., Wang, S., & Bradshaw, D. 2018. "Hierarchically Porous MOF/Polymer Composites via Interfacial Nanoassembly and Emulsion Polymerization." *Journal of Materials Chemistry A*, *6*, 20473–20479. 10.1039/C8TA06766J

[25] Chen, L., Ding, X., Huo, J., El Hankari, S., & Bradshaw, D. 2018. "Facile Synthesis of Magnetic Macroporous Polymer/MOF Composites as Separable Catalysts." *Journal of Materials Science*, *54*, 370–382. 10.1007/S10853-018-2835-X

[26] Lai, H., Wenjuan, S., Yuyin, Y., Danjiao, C., Liqian, W., & Fugang, X. 2019. "Uniform Arrangement of Gold Nanoparticles on Magnetic Core Particles with a Metal-Organic Framework Shell as a Substrate for Sensitive and Reproducible SERS Based Assays: Application to the Quantitation of Malachite Green and Thiram." *Microchimica Acta*, *186*, 1–9. 10.1007/S00604-019-3257-4

[27] Silvestre, M.E., Franzreb, M., Weidler, P.G., Shekhah, O., & Woll, C. 2013. "Magnetic Cores with Porous Coatings: Growth of Metal-Organic Frameworks on Particles Using Liquid Phase Epitaxy." *Advanced Functional Materials*, *23*(9), 1210–1213. 10.1002/ADFM.201202078

[28] Ma, R., Yang, P., Ma, Y., & Bian, F. 2018. "Facile Synthesis of Magnetic Hierarchical Core-Shell Structured Fe3O4@PDA-Pd@MOF Nanocomposites: Highly Integrated Multifunctional Catalysts." *ChemCatChem*, *10*(6), 1446–1454. 10.1002/CCTC.201701693

[29] Ma, Y.J., Jiang, X.X., & Lv, Y.K. 2019. "Recent Advances in Preparation and Applications of Magnetic Framework Composites." *Chemistry – An Asian Journal*, *14*(20), 3515–3530. 10.1002/ASIA.201901139

[30] Shekhah, O., Wang, H., Kowarik, S., Schreiber, F., Paulus, M., Tolan, M., Sternemann, C., et al. 2007. "Step-by-Step Route for the Synthesis of Metal–Organic Frameworks." *Journal of the American Chemical Society*, *129*(49), 15118–15119. 10.1021/JA076210U

[31] Liu, J., Qiu, H., Zhang, F., & Li, Y. 2020. "Zeolitic Imidazolate Framework-8 Coated Fe3O4@SiO2 Composites for Magnetic Solid-Phase Extraction of Bisphenols." *New Journal of Chemistry*, *44*, 5324–5332. 10.1039/D0NJ00006J

[32] Yang, J., Wang, Y., Pan, M., Xie, X., Liu, K., Hong, L., & Wang, S. 2020. "Synthesis of Magnetic Metal-Organic Frame Material and Its Application in Food Sample Preparation." *Foods*, *9*(11), 1610. 10.3390/FOODS9111610

[33] Saikia, M., Bhuyan, D., & Saikia, L. 2014. "Facile Synthesis of Fe3O4 Nanoparticles on Metal Organic Framework MIL-101(Cr): Characterization and Catalytic Activity." *New Journal of Chemistry*. *39*, 64–67. 10.1039/C4NJ01312C

[34] Jiang, Z., Li, Y., 2016. "Facile Synthesis of Magnetic Hybrid Fe3O4/MIL-101 via Heterogeneous Coprecipitation Assembly for Efficient Adsorption of Anionic Dyes." *Journal of the Taiwan Institute of Chemical Engineers*, *59*, 373–379. 10.1016/J.JTICE.2015.09.002

[35] Huo, S.H., Yan, X.P. 2012. "Facile Magnetization of Metal-Organic Framework MIL-101 for Magnetic Solid-Phase Extraction of Polycyclic Aromatic Hydrocarbons in Environmental Water Samples." *Analyst*, *137*, 3445–3451. 10.1039/C2AN35429B

[36] Hu, Y., Huang, Z., Liao, J., & Li, G. 2013. "Chemical Bonding Approach for Fabrication of Hybrid Magnetic Metal-Organic Framework-5: High Efficient Adsorbents for Magnetic Enrichment of Trace Analytes." *Analytical Chemistry*, *85*(14), 6885–3893. 10.1021/ac4011364

[37] Peña-Méndez, E.M., Mawale, R.M., Conde-González, J.E., Socas-Rodrígue, Z.B., Havel, J., & Ruiz-Pérez, C. 2020. "Metal Organic Framework Composite, Nano-Fe 3 O 4@Fe-(Benzene-1,3,5-Tricarboxylic Acid), for Solid Phase Extraction of Blood Lipid Regulators from Water." *Talanta*, *207*, 120275. 10.1016/J.TALANTA.2019.120275

[38] Pang, F., Mingyuan, H., & Jianping, G. 2015. "Controlled Synthesis of Fe3O4/ZIF-8 Nanoparticles for Magnetically Separable Nanocatalysts." *Chemistry – A European Journal, 21*(18), 6879–6887. 10.1002/CHEM.201405921

[39] Stock, N., & Biswas, S. 2011. "Synthesis of Metal-Organic Frameworks (MOFs): Routes to Various MOF Topologies, Morphologies, and Composites." *Chemical Reviews, 112*, 933–969. 10.1021/CR200304E

[40] Marquez, A.G., Horcajada, P., Grosso, D., Ferey, G., Serre, C., Sanchez, C., & Boissiere, C. "Green Scalable Aerosol Synthesis of Porous Metal-Organic Frameworks." *Chemical Communications, 49*, 3848–3850. 10.1039/C3CC39191D

[41] Sánchez, A.C., Inhar, I., Mary, C.S., & Daniel, M. 2013. "A Spray-Drying Strategy for Synthesis of Nanoscale Metal-Organic Frameworks and Their Assembly into Hollow Superstructures." *Nature Chemistry, 5*, 203–211. 10.1038/NCHEM.1569.

[42] Zhang, W., Yan, Z., Gao, J., Tong, P., Liu, W., & Zhang, L. 2015. "Metal-Organic Framework UiO-66 Modified Magnetite@silica Core-Shell Magnetic Microspheres for Magnetic Solid-Phase Extraction of Domoic Acid from Shellfish Samples." *Journal of Chromatography A, 1400*, 10–18. 10.1016/J.CHROMA.2015.04.061

[43] Jia,Y., Su, H., Wong, Y-L.E., Chen, X., & Dominic Chan, T-W. 2016. "Thermo-Responsive Polymer Tethered Metal-Organic Framework Core-Shell Magnetic Microspheres for Magnetic Solid-Phase Extraction of Alkylphenols from Environmental Water Samples." *Journal of Chromatography. A, 1456*, 42–48. 10.1016/J.CHROMA.2016.06.004

[44] Faustini, M., Kim, J., Jeong, G.Y., Kim, J.Y., Moon, H.R., Ahn, W.S., & Kim, D.P. 2013. "Microfluidic Approach toward Continuous and Ultrafast Synthesis of Metal-Organic Framework Crystals and Hetero Structures in Confined Microdroplets." *Journal of the American Chemical Society, 135*, 14619–14626. 10.1021/JA4039642

[45] Falcaro, P., Normandin, F., Takahashi, M., Scopece, P., Amenitsch, H., Costacurta, S., Doherty, C.M. et al. 2011. "Dynamic Control of MOF-5 Crystal Positioning Using a Magnetic Field." *Advanced Materials, 23*(34), 3901–3906. 10.1002/ADMA.201101233

[46] Niels, T., Gökpinar, S., Hastürk, E., Nießing, S., & Janiak, C. 2018. "Microwave-Assisted Dry-Gel Conversion – A New Sustainable Route for the Rapid Synthesis of Metal-Organic Frameworks with Solvent Re-use."*Dalton Transactions, 47*, 9850–9860. 10.1039/C8DT02029A.

[47] Singh, N.K., Gupta, S., Pecharsky, V.K., & Balema, V.P., 2017. "Solvent-Free Mechanochemical Synthesis and Magnetic Properties of Rare-Earth Based Metal-Organic Frameworks." *Journal of Alloys and Compounds, 696*, 118–122. 10.1016/J.JALLCOM.2016.11.220

[48] Yang, Q., Zhao, Q., Ren, S.S., Lu, Q., Guo, X., & Chen, Z. 2016. "Fabrication of Core-Shell Fe3O4@MIL-100(Fe) Magnetic Microspheres for the Removal of Cr (VI) in Aqueous Solution." *Journal of Solid State Chemistry, 244*, 25–30. 10.1016/J.JSSC.2016.09.010

[49] Tan, P., Xie, X.Y., Liu, X.Q., Pan, T., Gu, C., Chen, P., Zhou, J.Y., Pan, Y., & Sun, L.B. 2017. "Fabrication of Magnetically Responsive HKUST-1/Fe3O4 Composites by Dry Gel Conversion for Deep Desulfurization and Denitrogenation." *Journal of Hazardous Materials, 321*, 344–352. 10.1016/J.JHAZMAT.2016.09.026

[50] Bellusci, M., Guglielmi, P., Masi, A., Padella, F., Singh, G., Yaacoub, N., Peddis, D., & Secci, D. 2018. "Magnetic Metal-Organic Framework Composite by Fast and Facile Mechanochemical Process." *Inorganic Chemistry, 57*(4), 1806–1814. 10.1021/ACS.INORGCHEM.7B02697

[51] Dahao, J., Fang, G., Tong, Y., Wu, X., Wang, Y., Hong, D., Leng, W. et al. 2018. Multifunctional Pd@UiO-66 Catalysts for Continuous Catalytic Upgrading of Ethanol to n-Butanol." *ACS Catalysis*, *12*, 11973–11978. 10.1021/acscatal. 8b04014

[52] Oleksii, K., Chebbat, N., Commenge, J.M., Medjahdi, G., & Scneider, R. 2016. "ZIF-8 Nanoparticles as an Efficient and Reusable Catalyst for the Knoevenagel Synthesis of Cyanoacrylates and 3-Cyanocoumarins." *Tetrahedron Letters*, *57*(52), 5885–5888. 10.1016/J.TETLET.2016.11.070

[53] Pablo, S.C., Ramos-Fernandez, E.V., Gascon, J., & Kapteijin, F. 2011. "Synthesis and Characterization of an Amino Functionalized MIL-101(Al): Separation and Catalytic Properties." *Chemistry of Materials*, *23*(10), 2565–2572. 10.1021/cm103 644b

[54] Aboueloyoun, T.A., Huang, L., Ramakrishna, S., & Liu, Y. 2020. "MOF [NH2-MIL-101(Fe)] as a Powerful and Reusable Fenton-Like Catalyst." *Journal of Water Process Engineering*, *33*, 101004. 10.1016/J.JWPE.2019.101004

[55] Esmaeil, N., Panahi, F., Daneshgar, F., Bahrami, F., & Nezhad, A.N. 2018. "Metal-Organic Framework MIL-101(Cr) as an Efficient Heterogeneous Catalyst for Clean Synthesis of Benzoazoles." *ACS Omega*, *3*, 17135–17144. 10.1021/ acsomega.8b02309

[56] Song-Tao, X., Ma, C.T., Di, J.Q., & Zhang, Z.H. 2020. "MOF-5 as a Highly Efficient and Recyclable Catalyst for One Pot Synthesis of 2{,}4-Disubstituted Quinoline Derivatives." *New Journal of Chemistry*, *44*, 8414–8620. 10.1039/ D0NJ01301C

[57] Chen, L., & Xu, Q. 2019. "Metal-Organic Framework Composites for Catalysis." *Matter*, *1*(1), 57–89. 10.1016/J.MATT.2019.05.018

[58] Mao, Y., Li, J., Cao, W., Ying, Y., Hu, P., Liu, Y., Sun, L., Wang, H., Jin, C., & Peng, X. 2014. "General Incorporation of Diverse Components inside Metal-Organic Framework Thin Films at Room Temperature." *Nature Communications*, *5*, 5532. 10.1038/NCOMMS6532

[59] Takayoshi, A., Sato, T., Kanoh, H., Kaneko, K., Oguma, K., & Yanagisawa, A. 2008. "Organic–Inorganic Hybrid Polymer-Encapsulated Magnetic Nanobead Catalysts." *Chemistry – A European Journal*, *14*, 882–885. 10.1002/CHEM.2 00701371

[60] Takayoshi, A., Kawasaki, N., & Kanoh, H. 2012. "Magnetically Separable Cu-Carboxylate MOF Catalyst for the Henry Reaction." *Synlett*, *23*(10), 1549–1553. 10.1055/S-0031-1290935

[61] Saeed, S., Khatace, A., Bagheri, N., Kobya, M., Şenocak, A., Demirbas, E. Demirbas, & Karaoğlu, A.G. 2019. "Degradation of Diazinon Pesticide Using Catalyzed Persulfate with Fe3O4@MOF-2 Nanocomposite under Ultrasound Irradiation." *Journal of Industrial and Engineering Chemistry*, *77*, 280–290. 10.101 6/J.JIEC.2019.04.049

[62] Ramezanalizadeh, H., & Manteghi, F. 2017. "Immobilization of Mixed Cobalt/ Nickel Metal-Organic Framework on a Magnetic BiFeO3: A Highly Efficient Separable Photocatalyst for Degradation of Water Pollutions." *Journal of Photochemistry and Photobiology A: Chemistry*, *346*, 89–104. 10.1016/ J.JPHOTOCHEM.2017.05.041

[63] Mandel, K., Mandel, G.T., Tobias, W., Marcel, R., Werner, S., Nicolos, V., Gerhard, S., Gerhard & Klaus, M.B. 2017. "Smart Optical Composite Materials: Dispersions of Metal-Organic Framework@Superparamagnetic Microrods for Switchable Isotropic-Anisotropic Optical Properties." *ACS Nano*, *11*(1), 779–787. 10.1021/ACSNANO.6B07189

[64] Yujie, M., Xu, G., Wei, F., Cen, Y., Xu, X., Shi, M., Cheng, X., Chai, Y., Sohail, M., & Hu, Q. 2018. "One-Pot Synthesis of a Magnetic, Ratiometric Fluorescent Nanoprobe by Encapsulating Fe_3O_4 Magnetic Nanoparticles and Dual-Emissive Rhodamine B Modified Carbon Dots in Metal-Organic Framework for Enhanced HClO Sensing." *ACS Applied Materials & Interfaces*, *10*(24), 20801–20805. 10.1 021/ACSAMI.8B05643

[65] Ming, L.J., Huang, Q., Cai, P.Y., Li, C.Q., Zhang, L.H., & Zheng, Z.Y. 2014. "Design of a Highly Sensitive Fluorescent Sensor and Its Application Based on Inhibiting NaIO4 Oxidizing Rhodamine 6G." *Analytical Methods*, *6*, 5957-596. 10.1039/C4AY00853G

[66] Wang, J., Chen, H., Ru, F., Zhang, Z., Mao, X., Shan, D., Chen, J., & Lu, X. 2018. "Encapsulation of Dual-Emitting Fluorescent Magnetic Nanoprobe in Metal-Organic Frameworks for Ultrasensitive Ratiometric Detection of Cu2+." *Chemistry – A European Journal*, *24*, 3499–3505. 10.1002/CHEM.201704557

[67] Christus, A.B., Panneerselvam, P., Ravikumar, A., Morad, N., & Sivanesan, S. 2018. "Colorimetric Determination of Hg(II) Sensor Based on Magnetic Nanocomposite (Fe3O4@ZIF-67) Acting as Peroxidase Mimics." *Journal of Photochemistry and Photobiology A: Chemistry*, *364*, 715–724. 10.1016/ J.JPHOTOCHEM.2018.07.009

[68] Yang, Q., Wang, J., Chen, X., Yang, W., Pei, H., Hu, N., Li, Z., Suo, Y., Li, T., & Wang, J. 2018. "The Simultaneous Detection and Removal of Organophosphorus Pesticides by a Novel Zr-MOF Based Smart Adsorbent."*Journal of Materials Chemistry A*, *6*, 2184–2192. 10.1039/C7TA08399H

[69] Imaz, I., Jordi, H., Daniel, R.M., & Daniel, M. 2009. "Metal-Organic Spheres as Functional Systems for Guest Encapsulation." *Angewandte Chemie*, *48*(13), 2325–2329. 10.1002/ANIE.200804255

[70] Imaz, I., Rubio-Martínez, M., García-Fernández, L., García, F., Ruiz-Molina, D., Hernando. J., Puntes, V., & Maspoch, D. 2010. "Coordination Polymer Particles as Potential Drug Delivery Systems."*Chemical Communication*, *46*, 4737–4739. 10.1 039/C003084H

[71] Fei, K., Yuan, Y.P., Qiu, L.G., Shen, Y.H., Xie, A.J., Zhu, J.F., Tian, X.Y., & Zhang, L.D. 2011. "Facile Fabrication of Magnetic Metal-Organic Framework Nanocomposites for Potential Targeted Drug Delivery."*Journal of Materials Chemistry*, *21*, 72. 10.1039/C0JM01770A

[72] Lohe, M.R., Gedrich, K., Freudenberg, T., Kockrick, E., Dellmann, T., & Kaskel, S., 2011. "Heating and Separation Using Nanomagnet-Functionalized Metal-Organic Frameworks." *Chemical Communications*, *47*, 3075–3077. 10.1039/ C0CC05278G

[73] Abazari, R., Mahjoub, A.R., Molaie, S., Ghaffarifar, F., Ghasemi, E., Slawin, A.M.Z., Carpenter-Warren, L. 2018. "The Effect of Different Parameters under Ultrasound Irradiation for Synthesis of New Nanostructured Fe3O4@bio-MOF as an Efficient Anti-Leishmanial in Vitro and in Vivo Conditions." *Ultrasonics Sonochemistry*, *43*, 248–261. 10.1016/J.ULTSONCH.2018.01.022

[74] Wang, T., Zhao, P., Lu, N., Chen, H., Zhang, C., & Hou, X. 2016. "Facile Fabrication of Fe3O4/MIL-101(Cr) for Effective Removal of Acid Red 1 and Orange G from Aqueous Solution."*Chemical Engineering Journal*, *295*, 403–413. 10.1016/J.CEJ.2016.03.016

[75] Yang, Q., Ren, S.S., Zhao, Q., Lu, R., Cheng, H., Zhijun, C., & Hegen, Z. 2018. "Selective Separation of Methyl Orange from Water Using Magnetic ZIF-67 Composites."*Chemical Engineering Journal*, *333*, 49–57. 10.1016/J.CEJ.2017. 09.099

[76] Huairu, T., Peng, J.J., Lv, T.T., Sun, C., He, H., & He, H. 2018. "Preparation and Performance Study of MgFe 2 O 4 /Metal-Organic Framework Composite for Rapid Removal of Organic Dyes from Water." *Journal of Solid State Chemistry, 257,* 40–48. 10.1016/j.jssc.2017.09.017

[77] Guangpu, Z., Wo, R., Sun, Z., Hao, G., Liu, G., Zhang, Y., Guo, H., & Jiang,W. 2021. "Effective Magnetic MOFs Adsorbent for the Removal of Bisphenol A, Tetracycline, Congo Red and Methylene Blue Pollutions." *Nanomaterials, 11*(8), 1917. 10.3390/nano11081917

[78] Sadiq, M.M., Rubio-Martinez, M., Zadehahmadi, F., Suzuki, K., & Hill, M.R. 2018. "Magnetic Framework Composites for Low Concentration Methane Capture." *Industrial & Engineering Chemistry Research, 57–60* (18), 604047.10.1021/ ACS.IECR.8B00810

[79] Li, C.Y., Liu, J.M., Wang, Z.H., Lv, S.W., Zhao, N., & Wang, S. 2020. "Integration of Fe 3 O 4@UiO-66-NH 2@MON Core-Shell Structured Adsorbents for Specific Preconcentration and Sensitive Determination of Aflatoxins against Complex Sample Matrix." *Journal of Hazardous Materials, 384,* 121348. 10.1016/ J.JHAZMAT.2019.121348

[80] Yamini, Y., Safari, M., Morsali, A., & Safarifard, V. 2018. "Magnetic Frame Work Composite as an Efficient Sorbent for Magnetic Solid-Phase Extraction of Plasticizer Compounds." *Journal of Chromatography A, 1570,* 38–46. 10.1016/ J.CHROMA.2018.07.069

[81] Shi, X.R., Chen, X.L., Hao, Y.L., Li, L., Xu, H.J., & Wang, M.M. 2018. "Magnetic Metal-Organic Frameworks for Fast and Efficient Solid-Phase Extraction of Six Sudan Dyes in Tomato Sauce." *Journal of Chromatography. B, 1086,* 146–152. 10.1016/J.JCHROMB.2018.04.022

[82] Yujie, L., Wang, B., Yan, Y., & Liang, H. 2018. "Location-Controlled Synthesis of Hydrophilic Magnetic Metal-Organic Frameworks for Highly Efficient Recognition of Phthalates in Beverages." *ChemistrySelect, 3*(44), 12440–12445. 10.1002/slct.2 01802739

[83] Zhou, Z., Fu, Y., Qin, Q., Lu, X., Shi, X., Zhao, C., & Xu, G. 2018. "Synthesis of Magnetic Mesoporous Metal-Organic Framework-5 for the Effective Enrichment of Malachite Green and Crystal Violet in Fish Samples." *Journal of Chromatography A, 1560,* 19–25. 10.1016/J.CHROMA.2018.05.016

[84] Xiaoli, Z., Liu, S., Tang, Z., Niu, H., Cai, Y., Meng, W., Wu, F., & John, P. 2015. "Synthesis of Magnetic Metal-Organic Framework (MOF) for Efficient Removal of Organic Dyes from Water." *Scientific Reports, 11849,* 1–10. 10.1038/srep11849

10 Utility of Metal-Organic Frameworks in an Electrochemical Charge Storage

Anil Kumar, Usha Raju, and Jyoti

CONTENTS

10.1 INTRODUCTION

The ample use of fossil fuels by humans and consequent carbon emissions may result in an alarming situation for the future generations to come. The renewable energy sources thus need attention and efficient, reliable energy storage technologies to be devised that can store and deliver uninterrupted power for our needs, ranging from personal electronics, hybrid electrical vehicles (HEVs) to space systems. Cleaner energy storage devices such as rechargeable batteries, fuel cells, and

DOI: 10.1201/9781003252061-10

electrochemical capacitors, also termed *supercapacitors* or *ultracapacitors* come to the rescue to meet our energy needs. The supercapacitors have the merits of better energy storage ability than capacitors, longer lifetime and may be employed as stand-alone energy sources or in combination with batteries for enhanced power output. The electrodes of commercially available supercapacitors are made of porous carbons due to their high conductivity, good thermal, chemical and electrical stability and large surface area. Carbon nanotubes and carbon nanofibers are also frequently used materials for the construction of electrodes used in supercapacitors. Recently, researchers have gone to the extent of fabricating sawdust-derived activated carbon electrodes for enhancing the electrochemical performance of symmetric supercapacitors [1]. The electrolytes used in energy storage devices may be aqueous solutions or organic liquid electrolytes. The aqueous electrolytes have the advantages of low cost and high conductivity but are not suitable for use under subzero and elevated temperature conditions. On the other hand, organic electrolyte solutions suffer from the drawbacks of low conductivity, high cost, toxicity, inflammable nature and being environmentally hazardous [2]. The gel electrolytes based on an ionic liquid embodied metal-organic framework doped with lithium salts can be employed in lithium-ion secondary batteries on account of good electrochemical stability, high conductivity and ease of ion transport [3]. In asymmetric supercapacitors where two electrodes are made up of different materials, the carbon-based materials are employed for constructing negative electrodes, while metal or metal oxides are used for making anodes. The asymmetric supercapacitors are considered to have a greater energy density and longer life [4].

The supercapacitors may be broadly categorized in three types, which are electric double layer capacitors (EDLCs), pseudo-capacitors (PCs) and hybrid capacitors (HCs). An EDLC uses physical charge storage in electric double layers near electrode and electrolyte interfaces, whereas a pseudo-capacitor involves an energy storage mechanism not only through an electric double layer but also by a fast oxidation-reduction or redox reaction and intercalation in the electrode. The electrochemical capacitors exhibit features of high-power densities but low energy densities. The capacitors have the merit of efficient storage of electrical energy but suffer from limited storage capacity and high cost [5]. The supercapacitors hold an edge as they can be charged and discharged at quite a higher rate than batteries, which makes use of the phenomenon of slow diffusion of ions in electrode materials. Batteries and fuel cells have large energy densities but low power densities due to their slow reaction kinetics. A battery in simple terms consists of galvanic cells generally connected in series. A battery may be a primary or disposable battery in which chemical reactions are irreversible or secondary battery that is a rechargeable battery where chemical reactions can be reversed by the application of electrical potential. Batteries, primary or secondary, essentially contain materials required to produce electrical energy. A fuel cell is a clean energy-producing device in which electrochemical reactions take place and electric current is produced. Fuel cells continue to generate electrical energy so long as the supply of a fuel source and oxidizing agent is maintained. A variety of fuel cells are known, but the most popular hydrogen fuel cell has limitations of using highly flammable hydrogen gas as fuel. The fuel cells are still in the development stage and yet to establish their position in the market to be commercially viable because of their high cost. Improving energy density of EDLC without raising cost is still a challenging task for the scientific community before they can be fully commercialized. Metal-organic framework materials on account of their unique properties, such as high porosity and surface area, modifiable pore size, crystalline nature, vast morphology and conducive synthesis, can be tailored for efficient electrochemical charge storage mechanism to fulfill energy needs to cater to modern lifestyle. This chapter describes the applications of metal-organic frameworks for electrochemical charge storage as future generation energy materials.

10.2 ELECTROCHEMICAL CHARGE STORAGE: CAPACITORS, BATTERIES AND FUEL CELLS

10.2.1 ELECTROSTATIC CAPACITOR

A conventional capacitor consists of two parallel plates made of conducting material, separated by a dielectric, which is an insulating material (Figure 10.1).

This dielectric can be a ceramic material, mica, paper, polymer film, air or even vacuum. The capacitor usually used in high-voltage applications contains a high vacuum as the dielectric rather than air or other insulating materials. These capacitors have the capacity to store electrical energy on account of a potential difference generated between the plates because of a charge separation created by mobility of electrons from one metallic plate to the other. The charge-holding capacity or capacitance of a capacitor depends upon the size and nature of the material of plates, distance between the plates and dielectric properties of the insulator [6]. These ordinal capacitors may be further classified as ceramic capacitors, paper capacitors or oil capacitors, etc. depending on the nature of dielectric material employed [7].

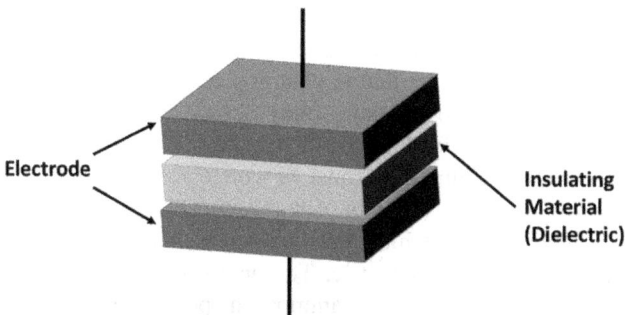

FIGURE 10.1 Depicts electrostatic capacitor showing three plates stacked, with two electrode plates at the ends and the middle plate representing a dielectric.

10.2.2 ELECTROLYTIC CAPACITORS

An electrolytic capacitor, as shown in Figure 10.2, is a capacitor consisting of a metal anode capable of forming an oxide layer through the process of

Oxide Film (Dielectric)

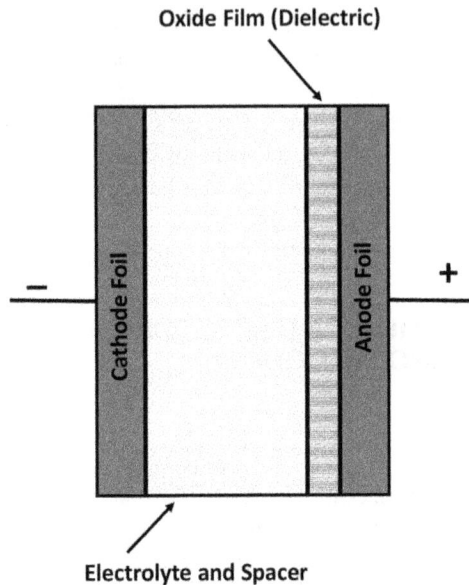

Electrolyte and Spacer

FIGURE 10.2 Schematic representation of an Electrolytic capacitor comprising of two thin layers of foil at two ends separated by the electrolyte. An oxide film coating, acting as a dielectric, is used to anodize the anode foil.

anodization. The electrolyte used is a solid, liquid or gel containing a high concentration of ions and serves as the cathode. The thin insulating oxide layer serves as a dielectric material of the capacitor. Some metals, for example aluminum, tantalum and niobium, also called valve metals, have the ability to form a very thin film of their own oxide on the surface when voltage is applied. Thus, these special metal capacitors with aluminum, tantalum or niobium as an anode are named aluminum, tantalum and niobium capacitors [7,8]. Though they have high capacitance in smaller size due to enlarged anode area, much care has to be taken in their handling as overheating may occur due to reverse voltages and safety valves may be provided to avoid a condition of overheating. As these capacitors are polarized, thus they must be connected correctly. A typical aluminum electrolytic capacitor consists of highly purified aluminum foil etched to increase the effective area, which gets oxidized and the resulting insulating aluminum oxide serves as a dielectric. The enclosure made up of aluminum is filled with a mild acidic solution of ammonium perborate jelly. The outer aluminum shell and electrolyte serve as a cathode [8,9]. Because of their large capacitance, they are quite useful but have a limited life span.

10.2.3 ELECTROCHEMICAL CAPACITORS

Electrochemical capacitors form a link between dielectric capacitors and batteries as their charge storage capacities are many fold higher than traditional dielectric capacitors but they suffer from lower energy densities as compared to other storage

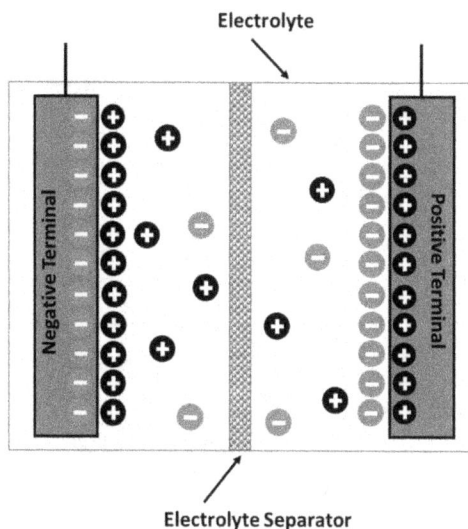

FIGURE 10.3 Electrochemical capacitor comprising primarily of two electrodes, an electrolyte and an insulating separator between the two electrodes.

and energy conversion devices [10]. The electrochemical capacitors are also called supercapacitors or ultracapacitors. The supercapacitors are characterized by high power density, long cyclic stability, quick charge and discharge ability and safety [11,12]. An electrochemical capacitor ordinarily is made up of two electrodes of porous material segregated by a separator to avoid short circuit of the device (Figure 10.3).

The separator allows free movement of ions passing through it during working of the device. The assembly is immersed in an aqueous or organic electrolyte contained in a suitable button or spiral wound cylinder container to prevent leakage or spillover and consequent damage to the environment [13–15].

10.2.4 Electric Double Layer Capacitors (EDLCs)

An electrical double layer capacitor is a charge storage device that involves physical separation of positive and negative charges in the form of an electric double layer at the electrode/electrolyte interface (Figure 10.4).

The charge storage process that occurs through physical ion adsorption on the surface is non-Faradaic as no charge transfer takes place through the electrode material to ions of the electrolyte. EDLCs have high capacitance due to the large surface area of electrodes and charge separation of only a few angstroms. Usually, EDLCs involve electric double layer charge storage techniques, and thus have poor energy densities but are able to supply higher power as compared to batteries [10]. As the workings of EDLCs involve physical charge storage that is reversible, these devices have a remarkably longer cycle life [15].

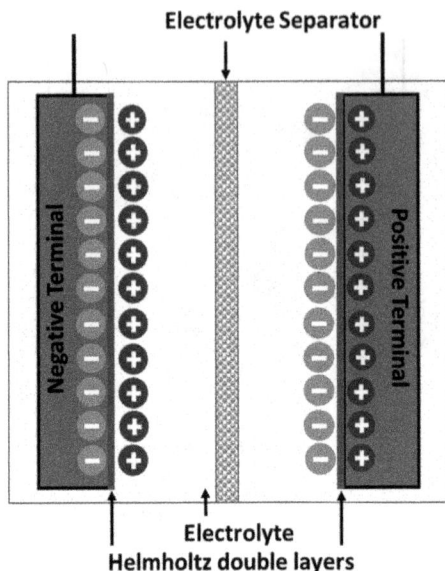

FIGURE 10.4 Depicts basic EDLC type supercapacitor with electric double layers formed on both sides of the electrodes due to non-Faradaic interactions.

10.2.5 ELECTROCHEMICAL PSEUDO-CAPACITORS (PCS)

The electrochemical pseudo-capacitors store energy both by non-Faradaic electric double layer technique similar to EDLC and a major portion of their pseudo-capacitance arises due to Faradaic mechanisms involving fast surface redox reactions that occur on electrode, electro sorption and intercalation of ions [16]. Pseudo-capacitors involve a Faradaic charge transfer across electrodes and electrolytes (Figure 10.5).

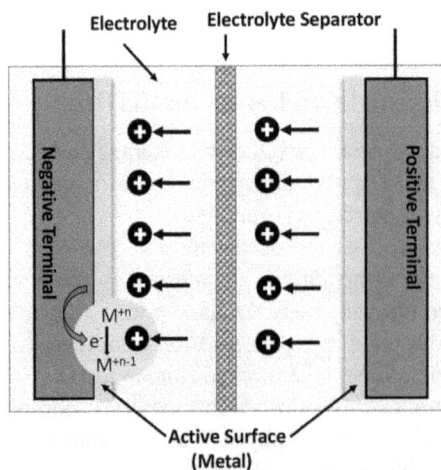

FIGURE 10.5 Representation of a pseudo-capacitor exhibiting redox reaction at the active surface of the electrode while positive charges moving towards negative terminal [17].

The charging and discharge process of pseudocapacitors resembles that of batteries. Helmholtz was the first to identify in 1853 that when an electrical conductor is in contact with an electrolyte, a common boundary is created between two phases and two layers of opposite polarity exist at the interface. Thus, electrical charges can not only be restricted on the conductor surface but extend to electrode-electrolyte double layer interface [18]. Energy density and power density are two major criteria to be considered for determining the suitability of an energy storage device. Since the mechanism of charge storage in pseudo-capacitors not only involves the surface of electrodes but bulk material as well, thus they exhibit enormously higher energy densities as compared to electric double layer capacitors. However, Faradaic supercapacitors exhibit comparatively lesser power density as compared to EDLC due to relatively slower faradaic reactions and they lack stability during cycling because of redox reactions occurring at the electrode [19]. Electrochemical pseudo-capacitors containing metal oxide as electrode material exhibit a great amount of electrochemical pseudo-capacitance. Oxide-type electrochemical double-layer capacitors can exhibit cyclic life up to 1 million under suitable conditions; for instance, ruthenium dioxide pseudo-capacitance has a life cycle of several hundred-thousand cycles but it has limitations of high cost and scarce availability. Another class of materials exhibiting a highly reversible nature is the conducting polymers such as polyaniline, which are cheaper than ruthenium dioxide but are less stable [20].

10.2.6 HYBRID CAPACITORS

An asymmetric hybrid capacitor is an energy storage device that has a much higher capacitance and energy density than a conventional symmetric supercapacitor. A hybrid asymmetric capacitor generally consists of two electrodes that behave differently (Figure 10.6).

These energy storage devices have merits of having one electrode with high value of Faradaic pseudo-capacitance to provide high specific energy and a second electrode with non-Faradaic high double layer capacitance for high specific power. Hybrid capacitors formed by the combination of multivalent metal ion batteries like aluminum ion batteries or zinc ion batteries with supercapacitors have the merits of delivering high energy density, high power density and enormously high cycle life [21]. A hybrid capacitor formed by an electrochemical capacitor and battery combination with suitable electrodes can be of help for improving performance of electric vehicles as a mode of transport and minimize the emission of carbon dioxide and oxides of nitrogen in the environment [22]. A simple hybrid lithium-ion capacitor, a combination of lithium-ion battery and supercapacitor consists of activated carbon as cathode and carbon material doped with lithium ions acting as anode. A wide range of nanostructured materials, such as a variety of metal oxides with a high specific surface area, nanotubes, nanowires and nanobeads have evolved with time for use as electrode materials in these hybrid capacitors to improve their performance [23,24]. The hybrid electrochemical energy storage systems [HEESS] consisting of a battery and supercapacitor are known to diminish the battery degradation rate by approximately 40% at the cost of just 1/8th of the system cost as

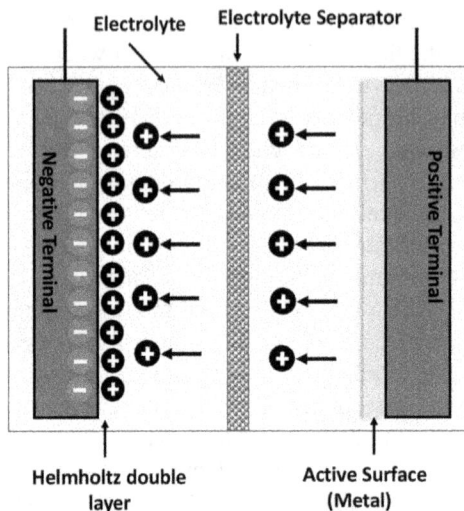

FIGURE 10.6 Depicts a hybrid capacitor showing positive terminal with active surface and Helmholtz double layer at negative terminal.

compared to battery alone energy storage system by redefining size and energy management system [25].

10.2.7 BATTERIES

A battery is a charge storage device generally made up of electrochemical cells that convert chemical energy into electrical energy (DC current) by virtue of electrochemical reactions taking place in the device. A battery is made of two electrodes called an anode and cathode made up of conducting material and an electrolyte that allows the flow of ions through it (Figure 10.7).

The type of batteries that can be used only once, as they cannot be recharged on account of irreversible chemical reactions taking place in them, are known as primary batteries. A dry cell is frequently used in households to power wall clocks, television remote, toys, etc. and is a typical example of primary battery, in which the outer zinc container behaves as an anode and a graphite rod placed in the center acts as cathode that is surrounded by a paste of manganese dioxide separated from an electrolyte by a separator. A paste of ammonium chloride and zinc chloride filled in the space between anode and cathode functions as electrolyte. It produces a voltage of about 1.5 volt. A second category, where electrochemical reactions taking place in a battery are reversible in nature so that it can be recharged and reused, is known as a secondary battery. The first practical version of lead-acid battery, invented by Gaston Plante in 1860, is an example of the most popular rechargeable or secondary battery [26]. It consists of a negative electrode made of a metallic lead and a positive electrode made of lead dioxide dipped in an electrolyte that is a dilute sulfuric acid solution. Lithium-ion batteries most commonly used in portable electronics namely smartphones, laptops,

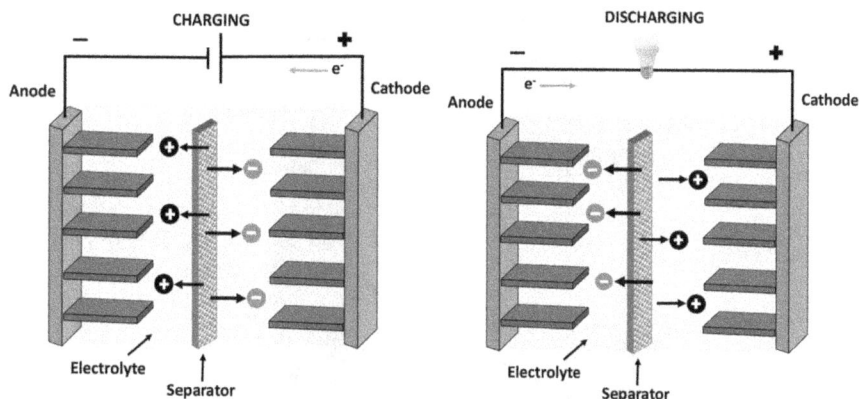

FIGURE 10.7 Schematic representation of a battery. (a) Charging and (b) discharging mechanism. Electrons flow from positive to negative electrode during charging and vice versa during discharging when a load is connected.

etc. and electrical vehicles, though credited with high energy density, have a risk of getting overheated leading to onboard battery fires. Further problems of aging and cost factors limit their widespread application in the automobile industry and an adequate cooling thermal management system needs to be adopted for their effective commercialization [27].

10.2.8 FUEL CELLS

A fuel cell produces electric current through electrochemical reactions that take place in a fuel and an oxidizing agent. It can continuously supply electrical energy so long as fuel and oxygen are fed into it from some suitable source. A fuel cell is an assembly of two electrodes, called an anode and cathode, immersed in a medium called an electrolyte for flow of charged ions (Figure 10.8).

Fuel cells being efficient and clean energy converter devices can come to the rescue of mankind to reduce environmental pollution and greenhouse gas emission resulting from the combustion of fossil fuels. Depending upon the electrolyte used, the fuel cells are categorized in different types, which further describe the type of fuel used, catalyst required, temperature range and electrochemical reaction which takes place for powering the system. Most popular proton exchange membrane fuel cells involve a simple chemical reaction between hydrogen used as a fuel and an oxidant (oxygen from air) to produce electric current for automobile powering systems. Natural gas, biogas and even carbon monoxide can also be used as a fuel in different fuel cells. Issues related to improvement in performance, lifetime and cost reduction need to be addressed before their widespread application [28]. Challenges still remain in the development of solid-oxide fuel cells (SOFCs) that can be operated at a lower temperature range, as high-temperature operations are not feasible for mobile applications in transportation [29].

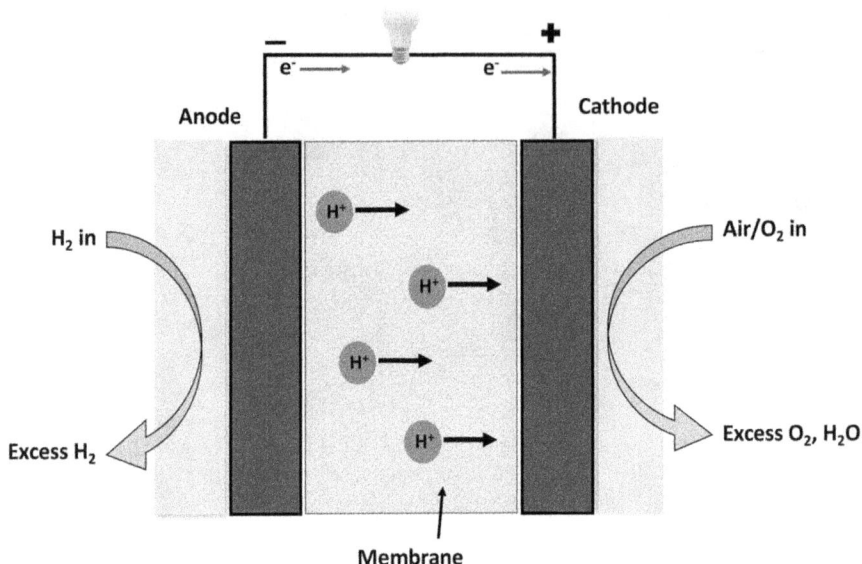

FIGURE 10.8 Representation of a simple proton exchange membrane fuel cell and its components.

10.3 ELECTRODES FOR ENERGY STORAGE DEVICES

The electrochemical capacitors by virtue of their higher power density hold an edge over their competitor candidate batteries. To increase the energy density and reduce cost of supercapacitors is still a challenging task before they may replace or augment batteries in capture and storage of electrical energy. The problem can be addressed by manipulating supercapacitors by the use of composite metal-organic frameworks impregnated with highly conducting materials like carbon nanotubes, reduced graphene oxide or activated carbon fibers, as electrode material in these energy storage devices for improved energy storage capacity, higher power output, swift charge and discharge process and longer cyclic life.

Performance of an electrochemical capacitor or supercapacitor basically depends on electrostatic interactions at the electrode/electrolyte interface, which in turn is mainly decided by the nature of electrode material and composition of electrolyte besides operating conditions of the system. Thus, choice of electrode material and electrolyte becomes crucial for efficient storage of electrical energy by a device. A perfect combination of appropriate electrode material and suitable electrolyte for stupendous performance of an energy delivering device with improved energy storage capacity and intense power density is in the wish list of the scientific community to boost the energy economy. Combination of an electrode made of highly porous material with vast surface area and an electrolyte with ions size matching with pore size can provide high energy density [30]. The electrode material must have pore size perfectly tuned to sub-angstrom accuracy with electrolyte ions, as smaller or larger pores just by a difference of an angstrom scale with ions result in much decline in capacitance. Largeot et al. have proposed designing supercapacitors for maximum energy density for solvent-free liquid electrolytes and solvated organic salts. They have demonstrated that maximum double layer capacitance is displayed when the pore size is equal to the ion size, while capacitance values decline significantly when carbon electrodes having pores with size larger or smaller size even on an angstrom scale than ions are employed [31]. Graphene, because of its unique physical and chemical properties, has been a favorite electrode material employed in supercapacitors. Cao et al. have revealed that the presence of oxygen functional groups can improve specific capacitance of graphene electrodes in supercapacitors. Graphene nanosheets with functional groups bearing oxygen were designed via thermal reduction of graphene oxide powder at varied temperatures from 200°C to 800°C under controlled conditions and studied their electrochemical performance in supercapacitors (Figure 10.9) [32].

A good variety of electrode materials have been developed and employed with an aim to increase efficiency of charge storage devices. The EDLC in which charge is stored in an electrical double layer without Faradaic reactions occurring in it, involves accumulation of charge on the electrode surface and movement of ions to the electrode pores forming a double layer. Therefore, EDLCs display significantly higher energy density than electrolytic capacitors on account of increased surface area available on the electrode and decrease in distance between the electrodes. Because of non-Faradaic reactions involved in charge storage, the electrode material does not undergo any chemical change in its composition, which results in a much longer cycle lifetime of super capacitors. Batteries on the other hand involve Faradaic redox reactions, which limits their lifetime. In pseudo-capacitors only a part of charge storage occurs through EDLC, while Faradaic oxidation-reduction reactions play a great role in energy storage. Performance of a pseudo-capacitor and electrical double layer capacitor is thus directly related to the surface area and porosity of electrodes. To get maximum output from these electrical energy storage devices, the most crucial factor that needs attention is higher surface-to-volume ratio by introducing new materials [33,34].

10.4 METAL-ORGANIC FRAMEWORK MATERIALS FOR ELECTROCHEMICAL CHARGE STORAGE

MOFs, the hybrid organic-inorganic crystalline materials, due to their unique set of properties, tailor-made morphology, high crystalline nature, splendid porosity, magnificent specific surface area, tunable pore size, conducive fabrication and post-synthetic modifications, can be exploited for use as electrode material in energy storage devices. The MOFs and their composites can serve either as sacrificial templates and precursors for the design of nanostructures and nanocomposites for supercapacitor applications or they may be employed directly as electrode material in energy storage and conversion applications by an appropriate choice of metal nodes and organic linkers [35–37].

Metal-organic framework materials composited with graphene, a highly conductive material, can play a great role in electrochemical charge storage and conversion in a battery or a supercapacitor [38].

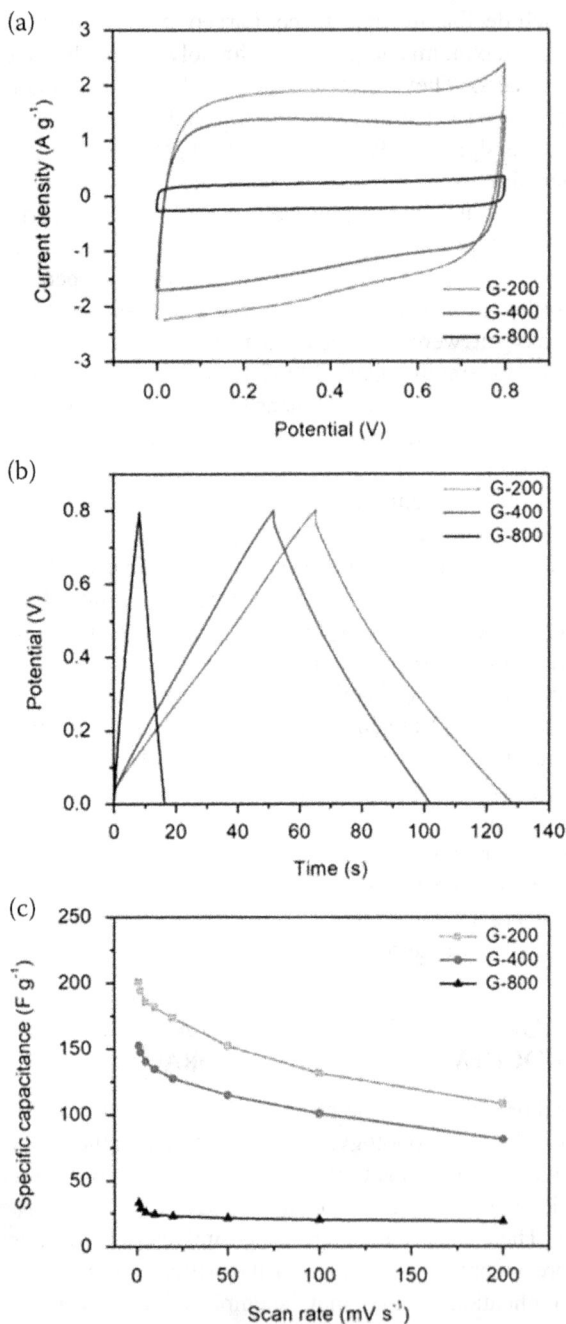

FIGURE 10.9 Depicts the electrochemical capacitive behavior of graphene with different oxygen content. (a) Cyclic voltammetry at a scan rate of 20 mV s^{-1}. (b) Galvanostatic charge-discharge curves at a current density of 1 A g^{-1}. (c) Specific capacitance at different scan rates (reproduced from [32]).

A simple and quick method which involves simple mixing of metal-organic frameworks and graphene oxide has been devised by the researchers for sizable production of 3D graphene oxide/metal-organic framework (GO-MOF) composite hydrogels with controllable composition, which serve as precursor for synthesis of MOF-based composite aerogels like rGO/Fe_2O_3. Such materials exhibit elevated specific capacitance and longer cyclic life. Based on rGO/Fe_2O_3 composite aerogel, a flexible all-solid-state supercapacitor device with an excellent mechanical flexibility, large volumetric capacitance and a good capacity retention power has been fabricated by the group [39]. Wang et al. have fabricated a unique threaded Co-MOF adopting a combination of a flexible N-donor and a rigid multi-carboxylate linker. The performance of the fabricated material in bulk crystalline form as well as in nano-rod powder form has been assessed for use as electrode material in a supercapacitor and concluded that gravimetric capacitance of device with electrode designed from nano-rod powder form is significantly greater than the one designed with bulk-crystal electrode. The much superior electrochemical performance of nano-rods has been attributed to a 3D network structure and smaller size of nano-rods [40]. Quite recently, Ashourdan et al. have reported an excellent electrochemical performance of electrode made by using nanocompositeG-HQG@NiCoMn-MOF, synthesized from nanocomposite of glucose (G), high quality graphene (HQG), and NiCoMn-MOF for supercapacitor applications. The asymmetric type exhibits an extra ordinary electrochemical performance, excellent rate capability and brilliant cycle life [41]. Tang et al. have claimed that distorted crystal structure of metal-organic frameworks, having highly ordered crystal lattice otherwise, can enhance their electrochemical performance significantly. The researchers have demonstrated the role of structurally distorted metal-organic framework material as an anode and electrocatalyst for in-situ electrochemical energy conversion and storage through fast surface Faradaic reactions [42].

10.4.1 BIMETALLIC NI/CO-MOF FOR ENERGY STORAGE APPLICATIONS

Recently researchers have shown keen interest in the development of bimetallic Ni/Co duo amalgamated with different noble metal-organic frameworks for superior electrochemical performance when used as electrode material for supercapacitors. A bimetallic metal-organic framework incorporating nickel and cobalt when employed as electrode materials in supercapacitors, displays a better specific capacitance and good stability even after a large number of charging-discharging cycles. Moreover, an asymmetric supercapacitor consisting of nickel-cobalt bimetallic metal-organic framework and reduced graphene oxide (rGO) electrodes display a high-energy density and an excellent electrochemical stability after a large number of charge discharge cycles [43]. Bimetallic Ni/Co-MOFs nanosheets that display excellent electrochemical performance have also been synthesized by Chi et al. [44]. Liang et al. have synthesized a series of bimetallic metal-organic frameworks with varied ratios of Ni/Co through a hydrothermal technique and have claimed that bimetallic Ni/Co-MOF exhibit remarkable electrochemical performance as compared to corresponding monometallic Ni-MOF and Co-MOF [45]. Ni/Co bimetallic-MOF (Ni/Co-MOF-74) has also been synthesized by Guo et al.

through a one-step solvothermal method, which they have claimed to have high specific capacity and better rate capability [46]. Chen et al. have synthesized bimetallic nickel-cobalt-MOFs nanosheets and studied their electrochemical behavior as an electrode material. The researchers have demonstrated that Ni-Co-S electrode fabricated by sulfurization of bimetallic Ni/Co-MOF-5 exhibits enhanced specific capacity and better cyclic stability, and thus may play a great role in supercapacitor as electrode material. Further, Ni–Co–S displays an excellent cyclic stability and around 94% of initial capacitance was maintained even after 3,000 charge/discharge cycles [47]. Nanostructured Ni/Co-MOF with semi-hollow spheres synthesized through a wet route by Ojha et al. has a high effective surface area and open pores to facilitate rapid movement of large number of electrolyte ions during a charge–discharge process. The supercapacitor constructed using this material as a cathode in combination with a highly porous carbon anode and thin interlayer of porous carbon exhibits almost four times higher electrochemical performance compared to a supercapacitor with similar configuration without interlayers [48]. A fast room-temperature solution-phase–based method has been devised by Xu et al. for scalable synthesis of nickel/cobalt MOFs that exhibit excellent charge storage capacity and serve as a promising electrode material for high-performance SCs [49]. To improve the conductivity of Ni-MOFs, mixed-metal organic frameworks (M-MOFs) have been designed by partial substitution of Ni^{2+} ions with Co^{2+} or Zn^{2+} ions in the framework. These M-MOFs exhibit excellent electrochemical performance on account of free pores that facilitate charge transport and enhanced electrochemical double-layer capacitance (EDLC) due to a large specific surface area of the electrode [50].

Sanati et al. have reviewed the comparison of employing bimetallic metal-organic frameworks materials rather than monometallic MOFs for improved electrical conductivity, high charge storage capacity and tunable electrochemical activity in supercapacitors. The bimetallic MOFs are known to exhibit better performance, higher electrical conductivity and good stability [51].

10.4.2 Waste to Energy Via Metal-Organic Frameworks

A good piece of work has been done by the researchers; by converting waste polyethylene terephthalate (PET) bottles into highly beneficial nano porous MOFs [52]. A facile method has been adopted for producing mesoporous carbon, ZnO@ MC and Co_3O_4@MC composites with very high specific surface area, by incorporating metal-oxide nanoparticles, through carbonization of PET bottles derived MOFs for supercapacitor application. The Co_3O_4@MC composite exhibits extraordinary supercapacitor performance, excellent stability and rate capability [53]. MOFs synthesized from waste metal source for metal center and recycled polymer for organic linker, using green solvents extracted from waste can be an added advantage for energy storage applications [54]. Deleu et al. carried out one pot synthesis of MOFs with a large surface area; MIL-53(Al) and MIL-47(V) using salts of metal ions and PET bottles [55]. Recently, NH_2-MIL-53(Al) has been employed as a precursor for preparation of hydrangea-like nitrogen-doped porous carbons (HNPCs). With a large surface area, these PCs show magnificent electrochemical performance enabling them to be used as a potential electrode material for capacitive deionization (CDI) technology [56]. Al-based MOF, MIL-53 with purity and porosity comparable to the one synthesized in literature, has been prepared by the researchers by an easy method using PET bottles and Li-ion-battery waste [57]. Hydrogen, a future generation fuel, along with fuel cell technologies, can resolve energy crisis and environmental degradation issues once onboard storage limitations of hydrogen as fuel are overcome. Ren et al. have claimed that Cr-MOF materials synthesized using a BDC acid linker derived from waste poly-ethylene terephthalate bottles display better hydrogen storage properties than the one obtained from commercial BDC [58].

10.4.3 Design Strategies of Metal-Organic Frameworks for Energy Storage Applications

The selection of electrode material is of paramount importance for improving the electrochemical performance of supercapacitors. Designing electrochemical capacitors with new materials to enhance their efficiency in terms of higher capacitance and energy density, fast charge/discharge mechanism and longer cyclic life can solve energy crises and dependence on fossil fuels. The performance of an energy storage device can be substantially enhanced by the integration of nano-porous materials ranging from zero-dimensional to three-dimensional with porosity varying from microporous to mesoporous for electrodes into their design. Electrochemical capacitors containing nanosized inorganic metal oxide material electrode exhibit greater specific capacitance, higher power and energy density and have low cost. Moreover, composite transition metal oxides formed by combining two or more oxides on account of synergistic effect may enhance electrochemical performance of pseudocapacitors. Nanostructures, namely nanotubes, nanorods, nanoribbons, nanoneedles, nanoarrays, etc., can be utilized in the fabrication of double layer capacitor and hybrid capacitors [59]. To put it in a nutshell, the electrode material for maximum efficiency of an energy storage device should have merits of excellent electrical conductivity, good temperature stability, large surface area, tunable pore size, corrosion resistance and low cost.

Baumann et al. have prescribed the design strategies for metal-organic frameworks to be used for charge storage utility (Figure 10.10) [17].

For example, the electrochemical stability of MOFs can be augmented by incorporating redox-inactive nodes and shorter and rigid linkers. Further, by introducing flexible linkers, multi-metals in the framework and controlling crystal size can improve mechanical properties of MOF employed for electrochemical energy storage [60]. Kung et al. have devised a method to reproduce electronically conductive mesoporous MOF material, NU-1000 besides retaining its crystallinity and porosity, by introduction of molecular tin on the MOF nodes [61]. Sun et al. have demonstrated that iron-based MOFs display much higher electrical conductivity amongst other metal ions: Mg^{2+}, Mn^{2+}, Co^{2+}, Ni^{2+}, Cu^{2+}, Zn^{2+} and Cd^{2+} ions, due to valence electrons with high energy and mixed valency of Fe^{2+}/Fe^{3+} [62]. Thus, introducing Fe^{2+} in a MOF framework and ions with variable valency may

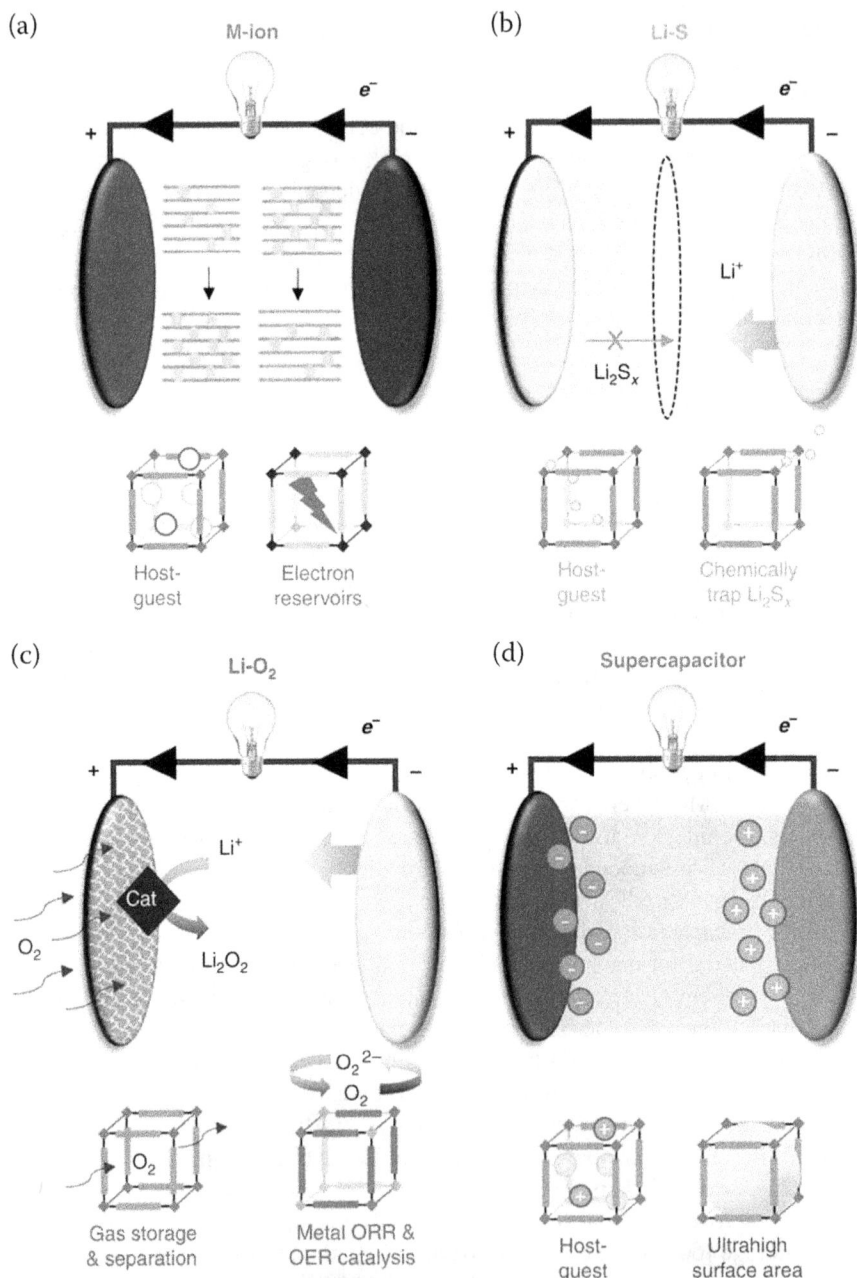

FIGURE 10.10 Depicts the unique MOF properties for targeting specific challenges in energy storage devices. (a) Metal-ion batteries rely on host-guest interactions to store ions while installation of electron reservoirs can improve charge conduction and increase deliverable capacity. (b) Lithium-sulfur batteries use host–guest interactions to store lithium and sulfide ions. Functional groups can be also used to trap polysulfides and diminish capacity loss. (c) MOFs can serve as selective gas sieves and as gas storage media in lithium-oxygen batteries. The metal nodes can also serve as oxygen reduction (ORR) or evolution catalysts. (d) The high porosity and host–guest nature.

prove boon to enhance electrical conductivity of porous materials. Table 10.1 summarizes the characteristic properties of various materials utilized in super-capacitor applications in terms of specific capacitance and the number of cycles involved.

10.4.4 MOFs FOR ENERGY STORAGE IN BATTERIES

A battery is a charge storage device with high energy density but lower power density than a capacitor. Lithium-ion batteries (LIBs), an advanced battery technology, on account of their high energy density and light weight, find applications in mobile phones, digital cameras, computers, hybrid electric vehicles (HEV), etc. to name a few.

Having highest energy density, LIBs provide a better alternative over nickel-cadmium (Ni-Cd) and nickel-metal-hydride (Ni-MH) batteries in portable electronic items. However, use of LIBs involves safety issues due to overheating at high voltage, have much less energy density than gasoline and high cost. Sodium-ion batteries (SIBs) can serve as low-cost energy technology because of the relative abundance of sodium sources in sea and Earth's crust [64–67]. Maiti et al. have designed a bar-shaped metal-organic framework Mn-1,3,5-benzenetricarboxylate, MOF (Mn-BTC MOF), which when used as anode material, exhibits high specific capacity and maintains stability of the MOF framework [68]. Besides electrical conductivity of MOFs, specific capacity and rate performance, thermal, chemical, and structural stability of MOFs are crucial for their role in LIBs for practical applications. Thus, MOFs for their utility as electrode material in LIB, should have high electrical conductivity and structural stability [69]. Ni-based metal organic frameworks (Ni-MOFs) are synthesized by the solvothermal technique, followed by successive carbonization and oxidation yield hierarchical NiO/Ni/graphene composites that exhibit promising performance as the anode for LIBs and SIBs [70]. Advancements in zinc-ion batteries may change the future of the energy storage market as ZIBs are potential alternatives to LIBs on account of their merits of excellent safety and low cost compared to LIBs. Mao et al. have achieved MOF-based synthesis of α-Mn_2O_3 for use as a cathode in ZIB, which displays a high specific capacity [63].

10.5 CONCLUSION

The electrochemical energy storage and conversion technologies have evolved with time, involving several energy storage devices, which when optimized to their full potential, can resolve energy crises and environmental concerns worldwide. A good deal of research has been done on electrochemical capacitors and batteries as a storehouse of energy and their conversion by the use of metal-organic frameworks on account of their unique properties such as crystalline nature, highly porous structure, extensive surface area, variable pore dimensions, easy fabrication and post-synthetic modifications.

The commercially used electrochemical capacitors or supercapacitors mostly adopt an electric double layer charge storage mechanism, which results in lower

TABLE 10.1

Performance of Some Recently Reported Metal-Organic Frameworks

S. No.	MOF Electrode Material	Method of Synthesis	Precursor Used	Capacitance Retention (%)	Specific Capacity/Capacitance	No. of Cycles	Ref.
1	3D Co-MOF nanorod powder	Hydrothermal	Co-MOF	58.3	1394 F g^{-1}	3,000	[40]
2	G-HQG@NiCoMn-MOFnanocomposite	Graphite peeling	NiCoMn-MOF	89.2	1263.6 C g^{-1}	5,000	[41]
3	lamellae-like Ni/Co-MOF	Hydrothermal	Ni/Co-MOF	75.5	568 C g^{-1}	3,000	[43]
4	NiCo-MOF (Ni:Co = 2:1) nanosheet	Hydrothermal	NiCo-MOF	81	901.60 F g^{-1}	3,000	[44]
5	NiCoMOF//AC (Ni:Co = 1:1)	Hydrothermal	NiCo-MOF	83	1333 F g^{-1}	2,000	[45]
6	(Ni/CoMOF-74) (Ni: Co = 5: 10)	Solvothermal	Ni-doped MOF	82.5	359.1 C g^{-1}	1,000	[46]
7	PC/PC@FP//PC@ FP/NiCo-MOF	Hydrothermal	NiCo-MOF	88	350 F g^{-1}	10,000	[48]
8	Ni$_2$Co-MOF	Solution-phase	Ni/Co-MOF	79	116.3 F g^{-1}	5,000	[49]
9	ZnO@MC and Co$_3$O$_4$@MC	Carbonization	ZnO and Co3O4-MOF	94.8	~97 and ~180 F g^{-1}	5,000 (for Co$_3$O$_4$@MC)	[53]
10	HNPC-900	Carbonization	NH$_2$-MIL-53 (A1)	86.49	136.65 F g^{-1}	15	[56]
11	α-Mn$_2$O$_3$	MOF-derived	Mn$_2$O$_3$-MOF	53.3	225 mAh g^{-1}	1,700	[63]

energy storage per unit weight compared to a battery, despite having faster charge ability, power density and long cycle life. The supercapacitors therefore cannot be used for continuous power supply. The energy density of ultracapacitors can be optimized by increasing operating voltage or by incorporating supercapacitive electrodes. However, increasing operating voltages generally results in lower power densities for these devices. The charge storage capacity of EDLCs relies on specific surface area of their electrode and average pore size of the active material. A material to be used as an electrode in electrochemical energy storage device must have attributes of good conductivity, high specific surface area, suitable pore size, chemical/thermal/mechanical stability and inertness [10]. Metal-organic frameworks, on account of their compositional and structural diversity, can be employed as templates or precursors to generate electrode materials. MOFs with unique morphology, mixed metal nodes, and conductive characteristics can be used as such or can be composited with other functional materials for synergistic effect to improve electrochemical performance [71].

The electrochemical batteries suffer from serious issues of short cycle life and lower efficiency in spite of having high energy densities. Metal-organic frameworks and their composites have shown much potential in energy storage batteries because of their unique properties, but cyclic life stability of MOF materials-based electrodes are still not up to the mark when compared with metal oxides. Though the development of lithium-based batteries and sodium-ion batteries (SIBs) have drawn much attention, many technical issues still need to be resolved [72,73].

The Li-metal anode, one of the most promising electrode materials with the highest theoretical capacity of 3,860 mA h g^{-1}, facees challenges of low Coulombic efficiencies and dendrite growth upon cycling, causing the threat of a fatal short-circuit and raises serious safety issues [74]. A zinc-ion–based dendrite-free battery having good electrochemical stability has been designed by Wang et al. by post-synthetic modification of MOF [75]. Solid-state electrolytes may be devised for rechargeable batteries to prevent dendrite formation and improve their efficiency and safety. Qian et al. have proposed a porous coating of MOF-199 to restrain the development of lithium dendrites which diminish columbic efficacy and raise serious safety issues [76].

Metal-organic frameworks may prove boon for clean and efficient energy conversion and storage in fuel cell technology. MOF-based materials are highly promising candidates for catalysis in proton-exchange membrane fuel cells (PEMFC) and may serve as precursors for the preparation of electrocatalyst. Copper/iron-MOFs display high catalytic performance and stability [77].

By adopting hybrid storage systems involving a combination of two or more technologies, such as a high energy density battery or fuel cell coupled with a high-power supercapacitor, design flexibility and improved performance can be achieved along with longer system life. Although rechargeable lithium-ion batteries used in smartphones, toys, medical equipment, electric vehicles, digital cameras and computers, are credited with the merits of having high energy storage capacity per unit mass and volume, durability, light weight and environment-friendly nature, still suffer from issues of high cost, reliability and safety concerns [78]. The electrolytes most commonly used in commercial lithium-ion batteries are liquid electrolytes containing lithium salts and organic solvents that are volatile and flammable in nature. Thus, researchers have proposed the use of gel-polymer electrolytes, alternative solvents and additives to address the safety issues [79]. Molten ionic electrolytes or ion gels have been proposed to replace toxic and hazardous organic solvents for application in rechargeable lithium batteries [80]. However, certain ionic liquids, for example those containing chloroaluminate ($AlCl_4^-$) anions, are highly water sensitive and corrosive in nature [81].

The electrochemical capacitors hold an edge over their competitor candidate batteries, by virtue of their higher power density. To increase the energy density and reduce cost of supercapacitors is still a challenging task. The problem can be addressed by manipulating supercapacitors by the use of composite metal-organic frameworks impregnated with highly conducting materials like carbon nanotubes, reduced graphene oxide or activated carbon fibers, as electrode material in these energy storage devices for improved energy storage capacity, fast charging/discharging phenomenon and long cyclic life. However, problems like poor electrical conductivity and fragile stability of metal-organic frameworks need to be addressed for their applications in electrochemical energy storage devices. MOF structures generally collapse in electrolyte systems, which diminish the availability of active sites and inhibit mobility of ions. Designing MOFs with structural stability under stringent thermal, chemical, mechanical and electrochemical conditions is highly desired. The judicious choice of linkers, appropriate solvents, manipulating design features, synthetic strategy and optimization of synergistic effects by introducing multi-metallic nodes in the framework of MOFs or their composites can lead these materials to win over their competitors for use in energy storage devices to improve their electrochemical performance [82]. The porosity of MOF electrodes must be adjusted to obtain excellent volumetric energy density, coulombic efficiency and rate capability. By optimizing porosity to match the electrolyte ion size and high surface area with more active sites of electrode materials can result in high energy density of the device. Further downsizing metal-organic frameworks to a nano-scale facilitates better electrochemical performance as an electrode in battery applications [83].

The electrical conductivity of MOFs can be enhanced by designing novel MOF composites with the introduction of functional components such as carbon nanostructures, metal oxides, metal nanoparticles, conducting polymers etc. in their framework. Quite recently, Ehrnst et al. have reported an ultrafast synthesis of freestanding quasi 2D MOF/graphene oxide composite by templating MOF directly on the graphene oxide sheets. The crystal structure of MOF is maintained even during the reduction of GO to conductive rGO and MOF composite exhibits marvelous supercapacitor performance [84].

The application of MOFs in energy storage and conversion devices is still an emerging field; thus, the challenges related to operating conditions, rational design, high cost, structural stability and environmental concerns need to be addressed before these devices could be fully commercialized for energy storage and conversion applications.

REFERENCES

[1] Zhou Y, Li J, Hu S, Qian G, Shi J, Zhao S, et al. Sawdust-derived activated carbon with hierarchical pores for high-performance symmetric supercapacitors. *Nanomaterials* 2022;12:810. 10.3390/nano12050810

[2] Castro-Gutiérrez J, Celzard A, Fierro V. Energy storage in supercapacitors: Focus on tannin-derived carbon electrodes. *Front Mater* 2020;7:1–25. 10.3389/fmats.2020. 00217

[3] Singh A, Vedarajan R, Matsumi N. Modified metal organic frameworks (MOFs)/ionic liquid matrices for efficient charge storage. *J Electrochem Soc* 2017;164:H5169–H5174. 10.1149/2.0191708jes

[4] Nagarajarao SH, Nandagudi A, Viswanatha R, Basavaraja BM, Santosh MS, Praveen BM, et al. Recent developments in supercapacitor electrodes: A mini review. *ChemEngineering* 2022;6. 10.3390/chemengineering6010005

[5] Sterner M, Stadler I. Handbook of Energy Storage – Demand, Technologies, Integration. 2019. Springer Berlin Heidelberg. 10.1007/978-3-662-55504-0

[6] Sudhakar YN, Selvakumar M, Bhat DK. Biopolymer Electrolyte for Supercapacitor. 2018. Elsevier. 10.1016/b978-0-12-813447-4.00003-0

[7] Ono S. *Passive Films for Electrolytic Capacitors*, Elsevier; 2018. 10.1016/B978-0-12-409547-2.13412-2

[8] Rawlins JC. *Capacitance. Basic AC Circuits*, Elsevier; 2000, pp. 171–215. 10.1016/ B978-075067173-6/50007-9

[9] Sinclair I. *Capacitors. Passive Components Circuit Design*, Elsevier; 2001, pp. 89–124. 10.1016/B978-075064933-9/50004-2

[10] Abbas Q, Mirzaeian M, Hunt MRC, Hall P, Raza R. Current state and future prospects for electrochemical energy storage and conversion systems. *Energies* 2020;13:5847. 10.3390/en13215847

[11] Wang Y, Song Y, Xia Y. Electrochemical capacitors: Mechanism, materials, systems, characterization and applications. *Chem Soc Rev* 2016;45:5925–5950. 10.1039/ c5cs00580a

[12] Simon P, Gogotsi Y. Materials for electrochemical capacitors. *Nat Mater* 2008;7:845–854. 10.1038/nmat2297

[13] Burke A. Ultracapacitors: Why, how, and where is the technology. *J Power Sources* 2000;91:37–50. 10.1016/S0378-7753(00)00485-7

[14] Miller JR, Simon P. Materials science: Electrochemical capacitors for energy management. *Science (80-)* 2008;321:651–652. 10.1126/science.1158736

[15] Meng C, Liu C, Chen L, Hu C, Fan S. Highly flexible and all-solid-state paperlike polymer supercapacitors. *Nano Lett* 2010;10:4025–4031. 10.1021/nl1019672

[16] Conway BE, Birss V, Wojtowicz J. The role and utilization of pseudocapacitance for energy storage by supercapacitors. *J Power Sources* 1997;66:1–14. 10.1016/S03 78-7753(96)02474-3

[17] Baumann AE, Burns DA, Liu B, Thoi VS. Metal-organic framework functionalization and design strategies for advanced electrochemical energy storage devices. *Commun Chem* 2019;2:86. 10.1038/s42004-019-0184-6

[18] Sarno M. Nanotechnology in energy storage: The supercapacitors. *Stud Surf Sci Catal* 2019;179:431–458. 10.1016/B978-0-444-64337-7.00022-7

[19] Viswanathan B. Supercapacitors. Energy Sources 2017; ISBN 9780444563538:315–328. 10.1016/b978-0-444-56353-8.00013-7

[20] Trasatti S, Buzzanca G. Ruthenium dioxide: A new interesting electrode material. Solid state structure and electrochemical behaviour. *J Electroanal Chem Interfacial Electrochem* 1971;29:A1–A5. 10.1016/S0022-0728(71)80111-0

[21] Dong L, Yang W, Yang W, Li Y, Wu W, Wang G. Multivalent metal ion hybrid capacitors: A review with a focus on zinc-ion hybrid capacitors. *J Mater Chem A* 2019;7:13810–13832. 10.1039/c9ta02678a

[22] Conway BE, Pell WG. Double-layer and pseudocapacitance types of electrochemical capacitors and their applications to the development of hybrid devices. *J Solid State Electrochem* 2003;7:637–644. 10.1007/s10008-003-0395-7

[23] Ding J, Hu W, Paek E, Mitlin D. Review of hybrid ion capacitors: From aqueous to lithium to sodium. *Chem Rev* 2018;118:6457–6498. 10.1021/acs.chemrev.8b00116

[24] Sun Y, Tang J, Qin F, Yuan J, Zhang K, Li J, et al. Hybrid lithium-ion capacitors with asymmetric graphene electrodes. *J Mater Chem A* 2017;5:13601–13609. 10.1039/C7TA01113J

[25] Zhang L, Hu X, Wang Z, Ruan J, Ma C, Song Z, et al. Hybrid electrochemical energy storage systems: An overview for smart grid and electrified vehicle applications. *Renew Sustain Energy Rev* 2021;139:110581. 10.1016/j.rser.2020.110581

[26] Kurzweil P. Gaston Planté and his invention of the lead–acid battery—The genesis of the first practical rechargeable battery. *J Power Sources* 2010;195:4424–4434. 10.1016/j.jpowsour.2009.12.126

[27] Kim J, Oh J, Lee H. Review on battery thermal management system for electric vehicles. *Appl Therm Eng* 2019;149:192–212. 10.1016/j.applthermaleng.2018.12.020

[28] Scherer GG. Fuel Cell types and their electrochemistry. In: Meyers RA, editor. *Fuel Cells*, New York: Springer New York; 2013, pp. 97–119. 10.1007/978-1-4614-5785-5_5

[29] Types of Fuel Cells | Department of Energy n.d. https://www.energy.gov/eere/fuelcells/types-fuel-cells (accessed May 9, 2022)

[30] Simon P, Gogotsi Y. Capacitive energy storage in nanostructured carbon-lectrolyte systems. *Acc Chem Res* 2013;46:1094–1103. 10.1021/ar200306b

[31] Largeot C, Portet C, Chmiola J, Taberna PL, Gogotsi Y, Simon P. Relation between the ion size and pore size for an electric double-layer capacitor. *J Am Chem Soc* 2008;130:2730–2731. 10.1021/ja7106178

[32] Cao H, Peng X, Zhao M, Liu P, Xu B, Guo J. Oxygen functional groups improve the energy storage performances of graphene electrochemical supercapacitors. *RSC Adv* 2018;8:2858–2865. 10.1039/c7ra12425b

[33] Muzaffar A, Ahamed MB, Deshmukh K, Thirumalai J. A review on recent advances in hybrid supercapacitors: Design, fabrication and applications. *Renew Sustain Energy Rev* 2019;101:123–145. 10.1016/j.rser.2018.10.026

[34] Zhang Y, Mei H xin, Cao Y, Yan X hua, Yan J, Gao H li, et al. Recent advances and challenges of electrode materials for flexible supercapacitors. *Coord Chem Rev* 2021;438:213910. 10.1016/j.ccr.2021.213910

[35] Salunkhe RR, Kaneti Y V, Yamauchi Y. Metal-organic framework-derived nanoporous metal oxides toward supercapacitor applications: Progress and prospects. *ACS Nano* 2017;11:5293–5308. 10.1021/acsnano.7b02796

[36] Liang Z, Qu C, Guo W, Zou R, Xu Q. Pristine Metal-Organic Frameworks and their Composites for Energy Storage and Conversion. *Adv Mater* 2018;30:1702891. 10.1002/adma.201702891

[37] Yap MH, Fow KL, Chen GZ. Synthesis and applications of MOF-derived porous nanostructures. *Green Energy Environ* 2017;2:218–245. 10.1016/j.gee.2017.05.003

[38] Zhang M, Shan Y, Kong Q, Pang H. Applications of metal-organic framework-graphene composite materials in electrochemical energy storage. *FlatChem* 2022;32:100332. 10.1016/j.flatc.2021.100332

[39] Xu X, Shi W, Li P, Ye S, Ye C, Ye H, et al. Facile fabrication of three-dimensional graphene and metal-organic framework composites and their derivatives for

flexible all-solid-state supercapacitors. *Chem Mater* 2017;29:6058–6065. 10.1021/acs.chemmater.7b01947

[40] Wang K, Wang X, Zhang D, Wang H, Wang Z, Zhao M, et al. Interpenetrated nano-MOFs for ultrahigh-performance supercapacitors and excellent dye adsorption performance. *CrystEngComm* 2018;20:6940–6949. 10.1039/c8ce01067f

[41] Ashourdan M, Semnani A, Hasanpour F, Moosavifard SE. Synthesis of nickel cobalt manganese metal organic framework@high quality graphene composites as novel electrode materials for high performance supercapacitors. *J Electroanal Chem* 2021;895:115452. 10.1016/j.jelechem.2021.115452

[42] Tang Y, Zhang H, Jin Y, Shi J, Zou R. Boosting the electrochemical energy storage and conversion performance by structural distortion in metal–organic frameworks. *Chem Eng J* 2022;443:136269. 10.1016/j.cej.2022.136269

[43] Zhang X, Wang J, Ji X, Sui Y, Wei F, Qi J, et al. Nickel/cobalt bimetallic metal-organic frameworks ultrathin nanosheets with enhanced performance for super-capacitors. *J Alloys Compd* 2020;825:154069. 10.1016/j.jallcom.2020.154069

[44] Chi Y, Yang W, Xing Y, Li Y, Pang H, Xu Q. Ni/Co bimetallic organic framework nanosheet assemblies for high-performance electrochemical energy storage. *Nanoscale* 2020;12:10685–10692. 10.1039/D0NR02016H

[45] Liang Y, Yao W, Duan J, Chu M, Sun S, Li X. Nickel cobalt bimetallic metal-organic frameworks with a layer-and-channel structure for high-performance su-percapacitors. *J Energy Storage* 2021;33. 10.1016/j.est.2020.102149

[46] Guo S, Xu X, Liu J, Zhang Q, Wang H. In situ growth of Ni-Doped Co-MOF-74 on Ni foam for high-performance electrochemical energy storage. *J Electrochem Soc* 2020;167:020539. 10.1149/1945-7111/ab6bbc

[47] Chen C, Wu MK, Tao K, Zhou JJ, Li YL, Han X, et al. Formation of bimetallic metal-organic framework nanosheets and their derived porous nickel-cobalt sulfides for supercapacitors. *Dalt Trans* 2018;47:5639–5645. 10.1039/c8dt00464a

[48] Ojha M, Wu B, Deepa M. NiCo Metal-organic framework and porous carbon interlayer-based supercapacitors integrated with a solar cell for a stand-alone power supply system. *ACS Appl Mater Interfaces* 2020;12:42749–42762. 10.1021/acsami.0c10883

[49] Xu F, Chen N, Fan Z, Du G. Ni/Co-based metal organic frameworks rapidly syn-thesized in ambient environment for high energy and power hybrid supercapacitors. *Appl Surf Sci* 2020;528:146920. 10.1016/j.apsusc.2020.146920

[50] Jiao Y, Pei J, Chen D, Yan C, Hu Y, Zhang Q, et al. Mixed-metallic MOF based electrode materials for high performance hybrid supercapacitors. *J Mater Chem A* 2017;5:1094–1102. 10.1039/C6TA09805C

[51] Sanati S, Abazari R, Albero J, Morsali A, García H, Liang Z, et al. Metal-organic framework derived bimetallic materials for electrochemical energy storage. *Angew Chemie* 2021;133:11148–11167. 10.1002/ange.202010093

[52] Lo SH, Senthil Raja D, Chen CW, Kang YH, Chen JJ, Lin CH. Waste polyethylene terephthalate (PET) materials as sustainable precursors for the synthesis of nano-porous MOFs, MIL-47, MIL-53(Cr, Al, Ga) and MIL-101(Cr). *Dalt Trans* 2016;45:9565–9573. 10.1039/c6dt01282e

[53] Al-Enizi AM, Ahmed J, Ubaidullah M, Shaikh SF, Ahamad T, Naushad M, et al. Utilization of waste polyethylene terephthalate bottles to develop metal-organic frameworks for energy applications: A clean and feasible approach. *J Clean Prod* 2020;248:119251. 10.1016/j.jclepro.2019.119251

[54] El-Sayed ESM, Yuan D. Waste to MOFs: Sustainable linker, metal, and solvent sources for value-added MOF synthesis and applications. *Green Chem* 2020;22:4082–4104. 10.1039/d0gc00353k

[55] Deleu WPR, Stassen I, Jonckheere D, Ameloot R, De Vos DE. Waste PET (bottles) as a resource or substrate for MOF synthesis. *J Mater Chem A* 2016;4:9519–9525. 10.1039/C6TA02381A

[56] Zong M, Huo S, Liu Y, Zhang X, Li K. Hydrangea-like nitrogen-doped porous carbons derived from NH2-MIL-53(Al) for high-performance capacitive deionization. *Sep Purif Technol* 2021;256:117818. 10.1016/j.seppur.2020.117818

[57] Lagae-Capelle E, Cognet M, Madhavi S, Carboni M, Meyer D. Combining organic and inorganic wastes to form metal-organic frameworks. *Materials (Basel)* 2020;13:441. 10.3390/ma13020441

[58] Ren J, Dyosiba X, Musyoka NM, Langmi HW, North BC, Mathe M, et al. Green synthesis of chromium-based metal-organic framework (Cr-MOF) from waste polyethylene terephthalate (PET) bottles for hydrogen storage applications. *Int J Hydrogen Energy* 2016;41:18141–18146. 10.1016/j.ijhydene.2016.08.040

[59] Boddula R, Ahmer MF, Asiri AM. *Morphology Design Paradigms for Supercapacitors.* CRC Press; 2019. 10.1201/9780429263347

[60] Bon V, Kavoosi N, Senkovska I, Müller P, Schaber J, Wallacher D, et al. Tuning the flexibility in MOFs by SBU functionalization. *Dalt Trans* 2016;45:4407–4415. 10.1 039/c5dt03504j

[61] Kung C-W, Platero-Prats AE, Drout RJ, Kang J, Wang TC, Audu CO, et al. Inorganic "conductive glass" approach to rendering mesoporous metal-organic frameworks electronically conductive and chemically responsive. *ACS Appl Mater Interfaces* 2018;10:30532–30540. 10.1021/acsami.8b08270

[62] Sun L, Hendon CH, Park SS, Tulchinsky Y, Wan R, Wang F, et al. Is iron unique in promoting electrical conductivity in MOFs? *Chem Sci* 2017;8:4450–4457. 10.1039/ C7SC00647K

[63] Mao M, Wu X, Hu Y, Yuan Q, He Y-B, Kang F. Charge storage mechanism of MOF-derived Mn2O3 as high performance cathode of aqueous zinc-ion batteries. *J Energy Chem* 2021;52:277–283. 10.1016/j.jechem.2020.04.061

[64] Chen W, Liang J, Yang Z, Li G. A review of lithium-ion battery for electric vehicle applications and beyond. *Energy Procedia* 2019;158:4363–4368. 10.1016/ j.egypro.2019.01.783

[65] Yamaki J. Thermal Stability of Materials in Lithium-Ion Cells. *Lithium-Ion Batteries*, Elsevier; 2014, pp. 461–482. 10.1016/B978-0-444-59513-3.00020-0

[66] Nigl T, Baldauf M, Hohenberger M, Pomberger R. Lithium-ion batteries as ignition sources in waste treatment processes—A semi-quantitate risk analysis and assessment of battery-caused waste fires. *Processes* 2020;9:49. 10.3390/pr9010049

[67] Kubota K, Komaba S. Review—Practical issues and future perspective for Na-Ion batteries. *J Electrochem Soc* 2015;162:A2538–A2550. 10.1149/2.0151514jes

[68] Maiti S, Pramanik A, Manju U, Mahanty S. Reversible lithium storage in manganese 1,3,5-Benzenetricarboxylate metal-organic framework with high capacity and rate performance. *ACS Appl Mater Interfaces* 2015;7:16357–16363. 10.1021/ acsami.5b03414

[69] Xu G, Nie P, Dou H, Ding B, Li L, Zhang X. Exploring metal organic frameworks for energy storage in batteries and supercapacitors. *Mater Today* 2017;20:191–209. 10.1016/j.mattod.2016.10.003

[70] Zou F, Chen Y-M, Liu K, Yu Z, Liang W, Bhaway SM, et al. Metal organic frameworks derived hierarchical hollow NiO/Ni/Graphene composites for lithium and sodium storage. *ACS Nano* 2016;10:377–386. 10.1021/acsnano.5b05041

[71] Tian D, Wang C, Lu X. Metal-organic frameworks and their derived functional materials for supercapacitor electrode application. *Adv Energy Sustain Res* 2021;2:2100024. 10.1002/aesr.202100024

[72] Zhao R, Liang Z, Zou R, Xu Q. Metal-organic frameworks for batteries. *Joule* 2018;2:2235–2259. 10.1016/j.joule.2018.09.019

[73] Etacheri V, Marom R, Elazari R, Salitra G, Aurbach D. Challenges in the development of advanced Li-ion batteries: A review. *Energy Environ Sci* 2011;4:3243. 10.1039/c1ee01598b

[74] Xu W, Wang J, Ding F, Chen X, Nasybulin E, Zhang Y, et al. Lithium metal anodes for rechargeable batteries. *Energy Environ Sci* 2014;7:513–537. 10.1039/C3EE4 0795K

[75] Wang Z, Hu J, Han L, Wang Z, Wang H, Zhao Q, et al. A MOF-based single-ion Zn2+ solid electrolyte leading to dendrite-free rechargeable Zn batteries. *Nano Energy* 2019;56:92–99. 10.1016/j.nanoen.2018.11.038

[76] Qian J, Li Y, Zhang M, Luo R, Wang F, Ye Y, et al. Protecting lithium/sodium metal anode with metal-organic framework based compact and robust shield. *Nano Energy* 2019;60:866–874. 10.1016/j.nanoen.2019.04.030

[77] Zhao Y, Song Z, Li X, Sun Q, Cheng N, Lawes S, et al. Metal organic frameworks for energy storage and conversion. *Energy Storage Mater* 2016;2:35–62. 10.1016/j.ensm.2015.11.005

[78] Lu L, Han X, Li J, Hua J, Ouyang M. A review on the key issues for lithium-ion battery management in electric vehicles. *J Power Sources* 2013;226:272–288. 10.1016/j.jpowsour.2012.10.060

[79] Montanino M, Passerini S, Appetecchi GB. *Electrolytes for Rechargeable Lithium Batteries: Rechargeable Lithium Batteries*, Elsevier; 2015, pp. 73–116. 10.1016/B978-1-78242-090-3.00004-3

[80] Tripathi AK. Ionic liquid–based solid electrolytes (ionogels) for application in rechargeable lithium battery. *Mater Today Energy* 2021;20:100643. 10.1016/j.mtener.2021.100643

[81] Watanabe M, Thomas ML, Zhang S, Ueno K, Yasuda T, Dokko K. Application of ionic liquids to energy storage and conversion materials and devices. *Chem Rev* 2017;117:7190–7239. 10.1021/acs.chemrev.6b00504

[82] Li S, Lin J, Xiong W, Guo X, Wu D, Zhang Q, et al. Design principles and direct applications of cobalt-based metal-organic frameworks for electrochemical energy storage. *Coord Chem Rev* 2021;438:213872. 10.1016/j.ccr.2021.213872

[83] Zhong M, Kong L, Zhao K, Zhang Y, Li N, Bu X. Recent progress of nanoscale metal-organic frameworks in synthesis and battery applications. *Adv Sci* 2021;8:2001980. 10.1002/advs.202001980

[84] Ehrnst Y, Ahmed H, Komljenovic R, Massahud E, Shepelin NA, Sherrell PC, et al. Acoustotemplating: Rapid synthesis of freestanding quasi-2D MOF/graphene oxide heterostructures for supercapacitor applications. *J Mater Chem A* 2022;10:7058–7072. 10.1039/D1TA10493D

11 Potential Redox Functions for Catalytic Hybrid Materials of Dimensional Cyanide-Bridged MOFs and Laccase Protein

Takashiro Akitsu, Yoshiyuki Sato, and Daisuke Nakane

CONTENTS

11.1 INTRODUCTION

11.1.1 METAL-ORGANIC FRAMEWORKS (MOFS) AS FUNCTIONAL MATERIALS

For approximately the past two decades, metal-organic frameworks (MOFs) have been one of the most important functional organic/inorganic materials composed of metal complexes [1,2]. MOFs can be regarded as functional materials having designed

DOI: 10.1201/9781003252061-11

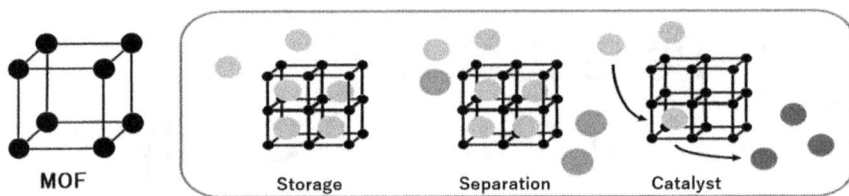

FIGURE 11.1 MOF and their typical functions: (left) storage, (middle) separation, and (right) catalysis.

structures comprising organic ligands and metal ions. Such crystal structures generally result in porous structures and large surface areas both inside and outside of the crystals. Prepared by self-assembling processes of metal sources and organic ligands, structures of MOFs can be adjusted precisely at the molecular level. This characteristic makes it possible to apply them to a wide range of functional materials using not only actual atoms, but also void spaces. For example, storage (including gas or small molecules), separation (distinguishing size or recognizing molecular shapes for inclusion), or catalysis (reaction inside of the pores by the substrates included) can be realized by using MOFs, which are difficult for conventional catalysts (Figure 11.1) [3].

Among such functions, redox or electron transfer functions may depend on steric structures made by organic linkers and metal nodes of MOFs (i.e., "distances" about electronic reactions) in a different way. For instance, heterogeneous supramolecular catalysis may be expected to enhance its activity depending on conditions of the surface of MOFs [4]. Electrochemical energy storage devices of MOFs should be designed to meet suitable conditions for size or mobility of carrier metal ions, as well as redox conditions of metal ions at the nodes of MOFs [5].

In the past, indeed, there were also the following opinions: There are lots of demerits of enzyme immobilization such as immobilization requires additional time, equipment and materials so is more expensive to set up; immobilized enzymes may be less active as they cannot mix freely with the substrate; Any contamination is costly to deal with because the whole system would need to be stopped.

11.1.2 DOCKING OF MOF AND ENZYME

Reactions involving electron transfer and the resulting improved functions of docking of MOF and enzyme may be attributed to steric factors and electrochemical factors in a simple manner. As for docking, moreover, inclusion (of small molecules) or adsorption (onto large molecules) should be distinguished from the viewpoint of distances of electron transfer. In general, so-called "docking", which may sometimes be a broad term, there are at least four methods of receptor-ligand docking such as rigid ligand-rigid receptor, rigid ligand- flexible receptor, flexible ligand-rigid receptor, flexible ligand- flexible receptor. However, in this chapter, adsorption of enzyme (changeable molecular orientation) onto large and rigid MOF should be imagined essentially regardless of size or contact fashion of MOF.

Immobilization of enzymes (protein molecules) by encapsulation in MOFs or adsorption on MOFs will improve reaction activity and chemical stability of

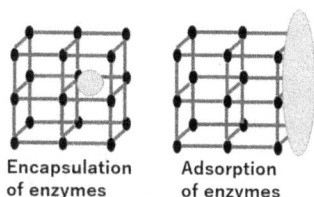

FIGURE 11.2   Typical ways to assemble enzymes with MOFs: (left) encapsulation inner space and (right) adsorption on the surface.

enzymes generally [6]. It is also expected that the ability to improve the function, while maintaining the stability, of the enzyme substance will lead to cost reduction of the functional material using the enzyme. From this perspective, so-called "artificial metalloproteins" (proteins including catalytic metal complexes) should be developed to combine not discrete metal complexes, but MOFs providing additional merits.

There are two promising or possible patterns of MOF docking to enzymes: (1) encapsulation type; and (2) adsorption type (Figure 11.2). In the encapsulation type, an enzyme is trapped inside of the MOF structure, whose void space is larger than the enzyme. In contrast, the adsorption type is a method in which the enzyme is immobilized and docked by adsorbing the enzyme on the surface of the MOF structure, whose pore size is not important for docking. Appropriate docking types (enzyme@MOF) should be assessed depending on required functions, because both types possess advantages and disadvantages. In a later section, for example, we will address electron transfer from cathode, metal complex, and protein. Moreover, not only can one-electron transfer pathways be proposed, depending on steric or electrochemical conditions for the reactions, but electrochemically-possible and sterically-advantageous electron transfer mechanisms are realized. However, a certain metal complex or MOF will select its preferable ones, depending on overall conditions. In this way, to find a deterministic mechanism or appropriate factors themselves constitutes an urgent problem for such systems at present.

Many preparation methods and features exist to assemble enzymes with MOFs, e.g., synthesis of MOF in the presence of enzyme, confinement of enzyme inside of MOF (one-pot), adsorption or binding of enzyme on MOF using non-covalent interactions or covalent bonds, etc. In fact, each docking or preparation method offers its own merits and demerits. Therefore, a method that takes optimal advantage of the characteristics of each method is strongly desired.

The pore confinement (one-pot) method is carried out inside of MOFs, since the size of pores or MOFs themselves may be restricted. Protection of protein molecules provided by (large) MOFs makes the enzyme less susceptible to influences of the external environment. It is necessary, however, for internalization to have sufficient space to accommodate the enzyme. In the method of internalization into the pore, the enzyme must be sufficiently small to fit into the pore scale of the MOF with large organic "pillars" ligands or have large void space in a "supramolecular" MOF prepared by thermodynamic self-assembly processes.

Adsorption on the surface of MOFs using both non-covalent or covalent interactions may require that the enzyme is exposed, which will weaken the protective effect of MOFs for the enzyme. As a consequence, preparations by adsorption may be susceptible to elution and degradation by enzymes, and are not suitable for reuse many times (e.g., repeating electrochemical reactions).

In both cases, such preparation should be practically carried out in aqueous buffer solutions under quite mild conditions. In addition, since there is a possibility that the enzyme may be denatured in the solution, permissible conditions are markedly constrained. Actually, sometimes only "luck" can provide successful conditions for preparation. Sufficient data on interactions between enzymes and MOFs have not yet been obtained. Regarding overcoming such issues, the study of enzyme@MOF composites is a relatively novel topic, and it will be actively pursued in the future due to its enormous potential [7–9].

11.1.3 Some Examples of Enzyme@MOF

In addition to the cases mentioned above, numerous conditions also exist under which the original functionality of enzymes cannot be easily utilized due to their fragile structure and susceptibility to environmental effects (e.g., temperature, solvent). Therefore, for using MOFs with proteins, such requirements must be addressed. For instance, by immobilizing the enzyme, it is possible to improve both structural fragility and chemical stability (e.g., robustness, thermal stability, resistance to pH effects, prevention of enzyme aggregation, etc.). Increasing the stability of an enzyme leads to reusability and improvement of enzyme activity. By docking to MOFs, in other words, it serves not only as a catalyst, making it highly efficient, but also as a protective substance for enzymes.

The introduction of MOFs as catalysts for enzymes to enhance their functionality or performance remains an active area of research. One of the major advantages of MOFs is their large surface area, which is not matched by other materials, as well as the ease of structural design. This means that they have the optimal structural space for the target enzyme, and the large surface area allows more enzymes to be adsorbed. Consequently, they are gaining popularity as materials for immobilizing enzymes. As previously mentioned regarding the improvement of enzyme activity when used as an enzyme catalyst, it is expected to be applied in various fields.

Herein, we introduce some examples using the co-precipitation method, which is a method that confines the enzyme inside of the MOF by one-pot synthesis. A complex material in which glucose oxidase was trapped inside of an amorphous MOF showed approximately 20 times higher activity than that in an MOF with a crystalline structure. Moreover, the confinement of glucose oxidase inside of the MOF protects the enzyme from the effects of temperature, pH and other environmental factors, and provides high stability. The high activity and high stability of this enzyme@MOF complex is expected to be applied to the biosensing of enzymes. This will facilitate the sensing of glucose in living cells without eroding them, and distinguish between cancer cells and normal cells. MOF is attracting attention as a substance with versatility not limited to the field of chemistry, and is being considered for various applications in the field of biology, such as cancer diagnosis and tumor detection [10].

To investigate the effect of substituents on reactivity, the UiO-66 was used as the MOF for enzyme immobilization. Glucose oxidase (GOx) and horseradish peroxidase (HRP) were immobilized in the MOFs, and changes in their activities were examined. In addition, UiO-66-NH_2 with substituents attached to the MOF was also synthesized in the same manner, and the effect of the substituent change on the enzyme system was also determined. The addition of glucose to HRP can act in conjunction with it to promote its activation. The experimental results demonstrated that HRP/GOx@UiO-66-NH_2 immobilized 6% more than HRP/GOx@UiO-66. Furthermore, only 36% of the immobilized enzymes were eluted. The enzymatic activity of the enzyme@MOF complex was also increased compared to the enzyme alone, with an activity of 143 U/mg in HRP/GOx@UiO-66, compared to 100 U/mg in the enzyme alone. Moreover, when HRP/GOx@UiO-66-NH_2 was used, the enzyme activity was 189 U/mg, which was particularly high compared to other enzymes. The reason for this increase in stability and activity is that the amino group of the MOF ligand strengthens the electrostatic interaction at the enzyme@ MOF interface. HRP/GOx@UiO-66-NH_2 was stable for more than one month at room temperature, indicating that the ligand has a significant effect on the enzyme accepting the catalyst [11].

In one study of adsorption by one-pot synthesis, NH2-MIL-53(Al) (NMOF), an MOF with amino substituents, was used as a support for enzyme fixation. Since most of the MOFs reported so far are unstable under weakly acidic conditions, a need has existed to develop MOFs with stability under low pH environments. Since NMOFs have high stability under pH 4–6, their functionality can be utilized in environments that are challenging for conventional MOFs. In addition, the structure of the MOFs with coordinated amino groups and higher surface area may allow for better adsorption of enzymes. After successful immobilization, the amount of laccase immobilized on NMOF was found to be 625 mg/g. The immobilized laccase exhibited better pH stability and thermal stability compared to laccase in its natural state. Furthermore, the immobilized laccase showed excellent reusability, retaining approximately 63% of its initial activity after 10 times of reuse. This result confirms that MOF not only protects the enzyme, but also possesses catalytic properties. The results suggest that MOFs can be used not only for laccase, but also for other enzymes in a broad range of applications [12].

An example of the use of the MOF internalization docking approach is the study of immobilizing ABTS (2,2'-azino-bis (3-ethylbenzothiazoline-6-sulfonic acid)) inside of MOFs. Laccase combined with ABTS is known as a bioelectrocatalyst for the four-electron reduction reaction (ORR). In this study, ABTS was immobilized on MOF MIL-100(Fe) as the immobilization matrix, and laccase was further immobilized on the ABTS immobilized in the pores. As a consequence of docking only ABTS, the X-ray powder diffraction (PXRD) pattern is highly similar to that of MOF alone, indicating that the MOF structure is maintained after docking. As anticipated from the change in MOF porosity, the relative intensity ratio (R) of the peaks at $2\theta = 3.4°$ and $4.2°$ changed remarkably after docking (R = 0.73–0.37). Furthermore, comparison of the N_2 adsorption isotherms showed a decrease in the amount of adsorbed N_2 due to docking (from 1332 m^2g^{-1} to 1179 m^2g^{-1}). This decrease in specific surface area confirms that a significant amount of ABTS was incorporated into the porous material. As a subsequent step, a comparison was made for the complex of ABTS@ MOF docked with laccase. Cyclic voltammetry results revealed that the redox was higher than that of the other two forms, confirming the high electrocatalytic current density. In addition, the long-term stability of the current densities showed that the redox activity was sustained over a long period of time [13].

In summary of this introductory section, the main adsorption methods are the physical adsorption method (immobilization by intermolecular force between the enzyme and the MOF), the covalent bonding method (immobilization by the bond between the functional group of the enzyme and the MOF), the encapsulation method (enzyme is confined inside of the MOF), and the co-precipitation method (the enzyme and the MOF precursor are mixed hydrothermally, and crystal formation and enzyme inclusion are performed simultaneously). Because each adsorption method possesses distinct advantages and disadvantages, researchers must select the most appropriate method by considering the following factors: (1) binding between laccase and MOF (the stronger is the bond, the better are the functionality of the MOF and the stability of the enzyme); (2) high accessibility to the catalytic site of laccase; and (3) high accessibility to the catalytic site of the immobilized laccase. According to these three factors, it may be relatively straightforward to select appropriate MOFs for a certain target protein.

11.2 DIMENSIONAL STRUCTURES OF CYANIDE-BRIDGED BIMETALLIC COMPLEXES

11.2.1 PRUSSIAN BLUE ANALOGS (PBAs) AS CYANIDE-BRIDGED COMPLEXES

In this section, we will focus on the subject structures of MOFs whose metal atoms are bridged by "cyanide" groups. Among numerous metal complexes containing cyanide groups, the coordination chemistry of $[Fe(CN)_6]^{3-/4-}$ units and related metal complexes have been known to exhibit unique electrochemical [14] and magnetic behavior [15], e.g., Prussian blue analogous (PBA) classically (Figure 11.3). Cyanide-bridged complexes, which can develop 0 to 3 dimensional structures, will be described depending on organic ligands or the number of cyanide ligands under suitable preparation procedures. The diverse electrochemical properties of PBAs are attributed to relatively strong mixed-valence properties observed in their structural arrangements and grid-type complexes, as well as to the cluster structure that is sufficiently stable to accommodate changes in bond length due to oxidation and reduction of the metal center.

Recently, hexacyanidemetalates have been employed as a key basis for designing cyanide-bridged two-metal assemblies with various dimensions (0D: non-bridged, 1D: chain-like, 2D: ladder-like, 3D: jungle gym-like) (Figure 11.3). Among them, magnetic materials based on cyanide-bridged molecules have attracted increasing attention for their application in functional devices because of their various functionalities, including optical magnets, chiral magnets, and magnetic conductors. However, the molecular design of cyanide-bridged metal complexes, including their electronic functions, remains a challenging task in the field of crystal engineering of organic-inorganic mixed frameworks and coordination polymers.

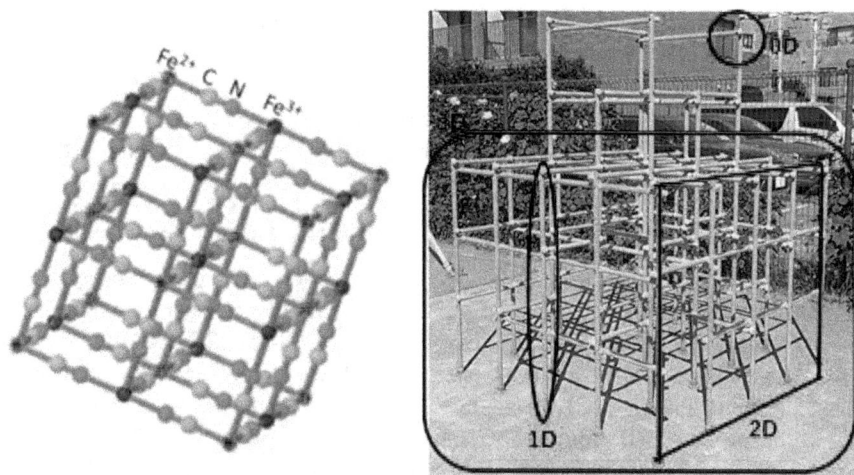

FIGURE 11.3   (left) 3D-bridged framework of Prussian blue (analogs) like a "jungle gym" structure. (right) A jungle gym showing 0–3D moieties.

Such a wide range of properties has led to extensive research on the synthesis of PBA having cyanide-bridged 3D networks. The following report addresses the dimensionality of cyanide-bridged complexes and investigates the functionality of each dimension. A typical example of a bimetallic compound commonly found in the structure of MOFs is the PBA and its analogs mentioned previously. The magnetism of "original" Prussian blue $Fe_4[Fe(CN)_6]_3 \cdot nH_2O$ ($Tc = 5.6$ K) was first reported in 1928 [16]. Subsequent literature has confirmed the magnetic change with temperature in PBAs, and it is known as magnetic metal complexes. The study of the magnetic structure of PBAs is particularly crucial for the development of molecule-based magnets that exhibit a broad range of magnetic properties, depending on temperature change. In order to reduce the complex structural network of PBAs, it is necessary to make the metal complexes coordinated to a simple structure.

The molecular quadrilateral with a cyanide group bridging structure is the simplest unit model of PBA. This structure allows access to a wide range of physical properties, including rich electrochemical properties, control of valence and spin states, and spin crossover phenomena. Prussian blue itself is a ferromagnet, but some derivatives of different metals have been found to be ferrimagnetic, even at room temperature. In fact, a broad range of physical properties, such as photo-magnetism, spin crossover and electrochromism, have been observed from PBAs in which different metals were introduced instead of iron. These functionalities are expected to be applied to hydrogen storage, molecular sieves, and nanoscale devices [17].

Studies on the charge transfer behavior of cyanide-bridged heterometallic materials and molecules have been actively conducted. As a result of the observation of charge transfer induced spin transition (CTIST) using a Prussian blue analog, $K_{0.2}Co_{1.4}[Fe(CN)_6] \cdot 6.9H_2O$, electron transfer from Co(II) ion to Fe(III) ion, and subsequent transition of Co(III) ion from high-spin state to low-spin state, were confirmed. This CTIST behavior was observed in the material, and was induced by light irradiation to the inter-valence charge transfer (IVCT) band in the material. Since this report, attention has been focused on the development of molecular species that exhibit CTIST behavior upon thermal or optical stimulation [18].

11.2.2 Dimensional Cyanide-Bridged Bimetallic Complexes

In addition to structural aspects, the magnetism of cyanide-bridged complexes divided into different dimensions is discussed for 2D or 3D complexes incorporating organic chelate ligands. For example, bimetallic complex $[Ni(en)_2]_3[M(CN)_6]_2 \cdot 2H_2O$ (M(III) = Fe, Mn, Cr, and Co) has been synthesized [16]. These compounds exhibited ferromagnetic interactions in M(III) metal ions. This can be predicted from the orthogonality of the orbitals of the M(III) ion and Ni(II). From the values of Wise's constant, it is found that the ferromagnetic interaction tends to increase in the order of Fe, Mn, and Cr, due to the difference in magnetism between Ni and cyanide-bridged metal ions, in which the nearest Ni(II) ion is responsible for ferromagnetic spin exchange. A ferrimagnetic 3D complex $[Ni(tren)]_3[Fe(CN)_6]_2 \cdot 6H_2O$ can be explained by structural distortions inducing weak antiferromagnetic interactions between Fe(III) and Ni(II) ions at the intersection of 1D chains, resulting in a 3D lattice from the antiferromagnetically coupled 1D ferromagnetic chains. Bonding properties and structural flexibility of MOF depends on MOF used.

The study of the structure and functionality of cyanide-bridged complexes classified into structural dimensions has also long been a topic of interest. By changing the metal ions, substituents, and dimensions inside of the structure of the formed MOFs, it will be possible to utilize the MOFs with various functions other than magnetism.

Concerning the structure of cyanide-bridged complexes, we have previously examined the change in the structural distortion of metal-organic structures of Cu(II) coordinated to cyanide-metallic complexes as a function of temperature [19]. Cu(II) complexes of the $3d^9$ electron configuration possess a flexible coordination structure due to the Jahn-Teller effect, which allows them to flexibly change their coordination structure. As a consequence, they have been attracting increasing attention in the fields of structural inorganic chemistry and crystal chemistry. This condition, however, does not apply to all Cu(II) complexes. Indeed, the Jahn-Teller strain was not observed in $[Cu(H_2O)_6](NO_3)_2$ with a different counter ion, indicating that the Jahn-Teller strain is affected by the coordination conditions (Figure 11.4).

A mononuclear complex, $[Cu(en)_2](ClO_4)_2$ (en = ethylenediamine) was used as a 0D complex. From this complex, significant effects on both axial coordination bond length and lattice volume were obtained from temperature-dependent measurements. For instance, the crystal structure at 297 K, 274 K, and 120 K showed a 2.26% decrease in lattice volume with temperature change from 297 K to 120 K.

Regarding the 1D structure, both linear and zigzag structures could be identified. Typical examples of a zigzag 1D structure are $[Cu(N\text{-}Eten)_2][Pt(CN)_4]$ (E-Eten = N-ethylethylenediamine) and $[Cu(chxn)_2][Ni(CN)_4] \cdot 2H_2O$ (chxn = 1,2-cyclohexanediamine). The Cu(II) moieties and the tetracoordinated cyanide

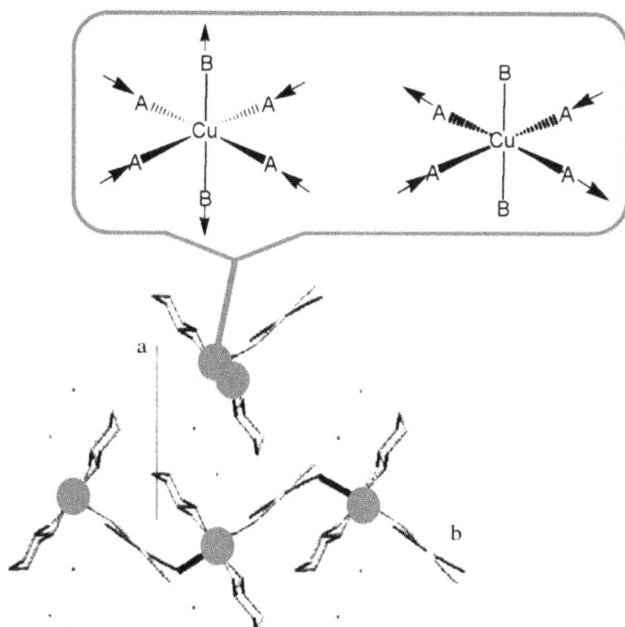

FIGURE 11.4 Cyanide-bridged Cu(II)-M bimetallic assemblies and Jahn-Teller effect around Cu(II) ions.

complexes alternately form a zigzag chain structure along the b-axis direction. As in the case of the linear structure, the effect of the temperature-dependent structural change of the lattice was relatively small; whereas, metal ion radii of the tetra-coordinate cyanide moiety were more significant than the introduction of substituents on the coordinating nitrogen atoms of the chelate ligands.

Concerning the 2D bridged structure, $[Cu(N-Eten)_2]_3[Fe(CN)_6]_2 \cdot 4H_2O$ and $[Cu(N-Eten)_2]_3[Co(CN)_6]_2 \cdot 4H_2O$ could be found, in which the N-Eten Cu(II) complex and the hexa-coordinated cyanide complex form a 2D ladder-shaped bridged structure. Relatively, the Jahn-Teller strain at the interchain site was less affected than within the long chain bridge, resulting in a larger temperature variation (Fe: interchain Cu-N = 2.843 Å (299 K), 2.819 Å (100 K)). The effect of metal substitution in the hexa-coordinated cyanide complex was found to be relatively small [20]. A 2D $\{[Cu(N-Eten)_2]_3[Fe(CN)_6]_2\}_n$ has -Fe1-CN-Cu1-NC-Fe2- interchains to form a network of cyanide-bridges. Cross-linked -Fe1-CN-Cu1-NC-Fe2- along the ac-plane and the a-axis are present in the crystal. On the other hand, 2D $\{[Cu(trans-(1R, 2R)-chxn)_2]_3\}_n$ was a cyanide-bridged bimetallic assembly of a 2D network alternating $[Fe(CN)_6]^{3-}$ and $[Cu(chxn)_2(H_2O)_2]^{2+}$ included $[Cu(chxn)_2(H_2O)_2]^{2+}$ in the networks. The mononuclear Cu(II) moiety, which is less elongated by the Jahn-Teller effect, exhibits a large structural change by changing the temperature between 263 K and 100 K. In this way, a typical example of 0–3D cyanide-bridged Cu(II)-M bimetallic complexes is depicted in Figure 11.5.

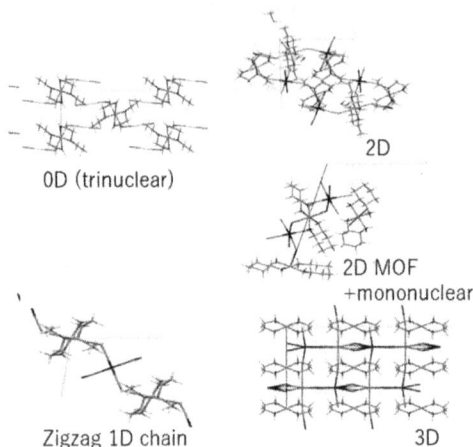

FIGURE 11.5   Cyanide-bridged Cu(II)-M bimetallic assemblies of various dimensionality.

11.2.3 CRYSTAL STRUCTURE DATABASE FOR REDOX CU-FE MOFS

Artificial intelligence (AI) is gaining increasing attention as a way to design chemical compounds using the Cambridge Crystal Structure Database (CSD) [21] with the support of computational chemistry. In order to design the structure of metal complexes, it is difficult to predict optimized structures considering the environment with high accuracy because it requires advanced calculation techniques at present. To overcome this challenge, by assuming that the structures of MOFs and proteins are identical as stored in CSD and Protein Data Bank (PDB), respectively, we virtually predict docking structures of MOFs and proteins by docking programs (e.g., GOLD by CCDC). Included (discrete) metal complexes in protein molecules can be processed with such programs. Indeed, even if optimized structures of metal complexes are different from crystal structures, DFT can estimate optimal structures in protein molecules. It is usually difficult, however, to simulate structural features of adsorption protein on the MOFs. Therefore, to search redox active cyanide-bridged Cu-Fe bimetallic assemblies of suitable structures in a data-driven manner, we keep both MOFs' structure and well-oriented contact with the shortest distance between MOF and protein molecules. In the future, our goal is to achieve highly accurate prediction of optimized structures of "adsorbed" MOFs based on CSD data with AI methods.

With our aim of identifying suitable redox materials, searched cyanide-bridged Cu-Fe MOFs of 0D, which means "discrete" herein, (Figures 11.6–11.10) [22–26], 1D (Figures 11.11–11.13) [27–29], and 2D (Figures 11.14–11.17) [30–33] with CCDC codes and brief structural features in captions are depicted as follows.

As described above, the study of the structure and functionality of cyanide-bridged complexes by dimension has been a topic of interest for some time. By changing the metal ions, substituents, and dimensions inside of the structure of the formed MOFs, it will be possible to utilize the MOFs with various functionalities. We believe that magnetism and catalysis are the most notable functionalities of

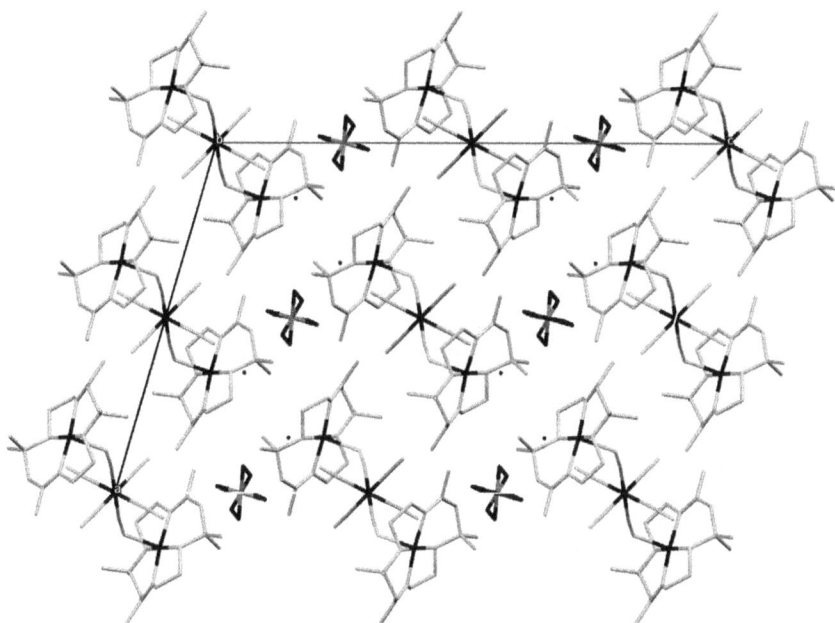

FIGURE 11.6   0D Cu-Fe-Cu+anion BACNEA. It crystallized in the $C2$ space group. The crystal structure is symmetrical, with Fe and Cu atoms connected by cyanide-bridges.

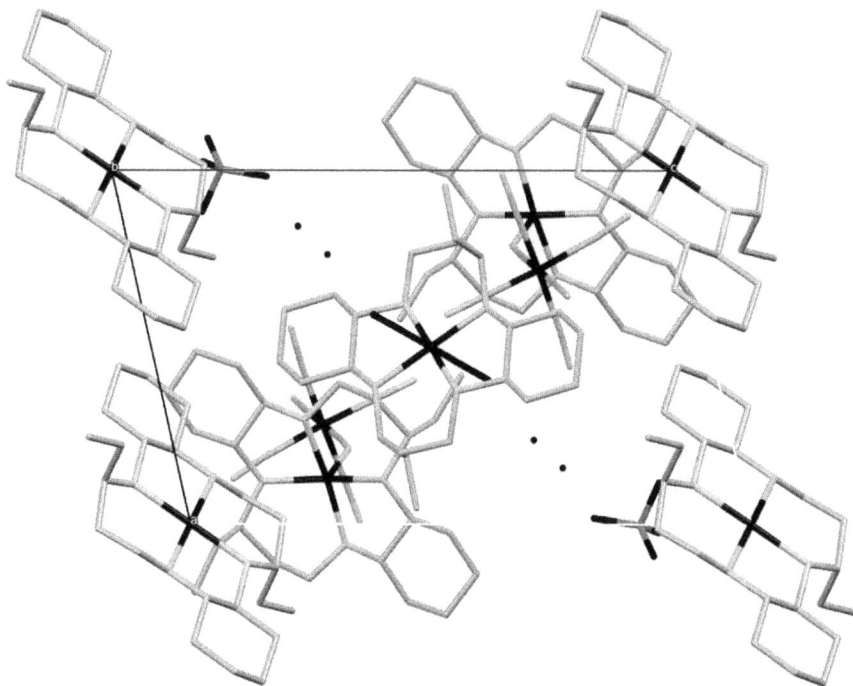

FIGURE 11.7   0D Cu-Fe+cation IXAKIC. It shows a 2D honeycomb structure. The Fe atoms form the edges of the structure, and the Cu precursors are connected to the Fe atoms as edges, resulting in a hexagonal structure.

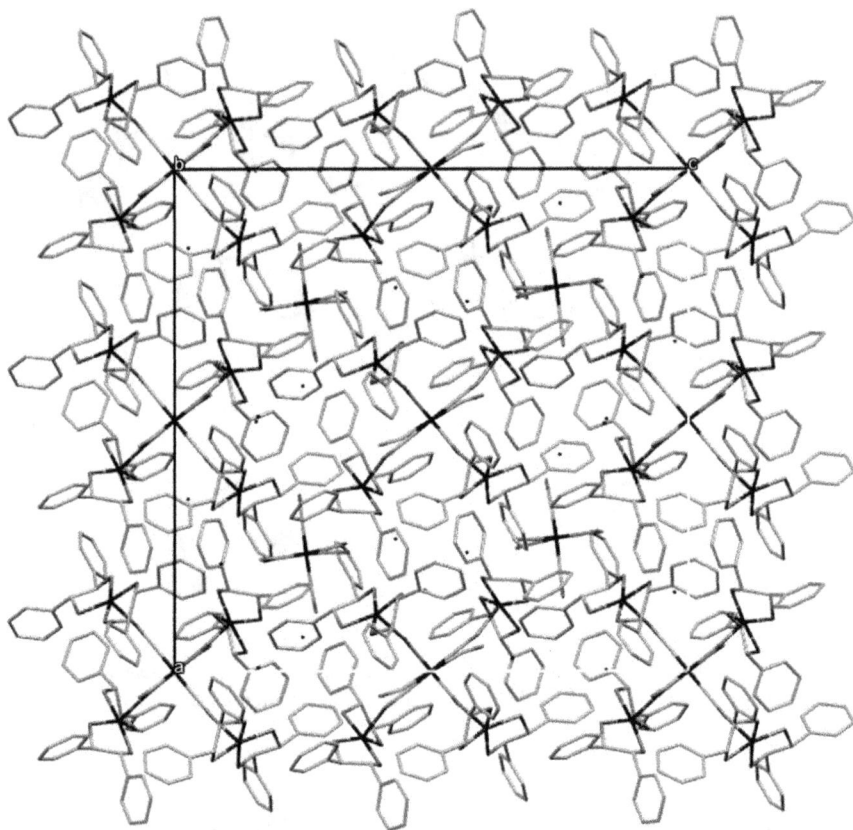

FIGURE 11.8 0D Cu-Fe+anion LUWYEJ. The Cu and Fe atoms crystallize in the monoclinic space group $C2$. The asymmetric unit consisting of Cu and Fe atoms is composed of a five-nuclear $FeCu_4$ cation, four $[Fe(CN)_6]^{4-}$ sites with complex bridging structures, and 10 water molecules. A 3D network structure of hydrogen bonds is formed, with large gaps around the mononuclear $[Fe(CN)_6]^{4-}$ anions.

MOFs. In the next section, we will focus on the use of MOFs as catalysts for the functional activity of enzymes.

11.3   COMPOSITE MATERIALS OF DIMENSIONAL CYANIDE-BRIDGED MOF AND LACCASE

11.3.1   Mononuclear Cyanide Metal Complexes and Laccase

Laccase is an enzyme that is often treated in a wide range of fields, including biological and biochemical fields. It exhibits catalytic oxidative power and can be applied *in vitro* by introducing mediators that are biocatalysts with water as the only product as an oxygen-acceptor. One area of research that incorporates laccase is its application in biofuel cells. When laccase is present in a biofuel cell, it acts as a

FIGURE 11.9 0D Cu-Fe+anion GIWFIC. Four N atoms bonded to the Fe atom of the central element are connected by cyanide-bridges, showing an octahedral structure.

catalyst for the four-electron reduction of oxygen to water at the cathode. Laccase contains three copper sites that mediate the reduction of oxygen to water. Proteins that contain several types of copper sites are called multi-copper oxidases [34–36]. The four copper atoms in the active site of laccase are divided into three types (T1-T3), and the T2 and T3 copper atoms form a trinuclear cluster. The T1 copper first receives electrons from the substrate and oxidizes the substrate, followed by a four-electron reduction at the T2 and T3 sites. While many possible applications exist, the stability and activity of MOFs are influenced by environment factors in which they exist, such as pH, humidity, and solubility. For this reason, we focused on enhancing the stability and activity of laccase by combining MOFs as mediators with laccase. Among three active sites of laccase, only electron receiving T1 site may be important for the discussion of this chapter.

In the cathode of biofuel cells, some redox-active compounds are used as a mediator (e.g., $[Fe(CN)_6]^{3-/4-}$ and $[Os(CN)_6]^{3-/4-}$ as cyanide complexes) between

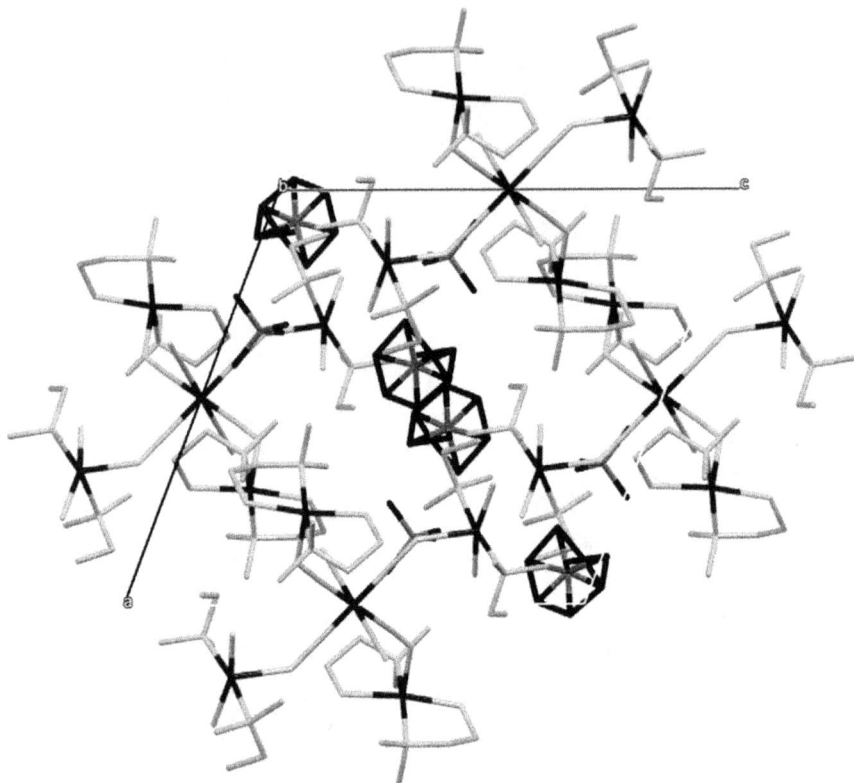

FIGURE 11.10 0D Cu-Fe+anion OBOPUR. Four {Cu(dmen)$_2$} are bridged to the cyanide group coordinated to the Fe atom. The overall structure forms an octahedral structure.

electrode and multi-copper oxidase (Figure 11.18). In order to extract appropriate steric and electrochemical factors of cyanide complexes for laccase, we have systematically investigated cyanide metal complexes and laccase. Because multicopper oxidase [34–36] based biofuel cells [37–40] use bio-based catalysts, they possess the advantage to generate electricity at room temperature and low environmental load [41–43]. A mediator that assists the donation and reception of the electron has been also used for electron transfer between the electrode and laccase [44–46] to improve electron transfer from cathode to laccase (near the T1 site). Typical examples of cyanide metal complex mediators for multicopper oxidase have been reported [47]. Bilirubin oxidase (BOD) from *Myrothecium verrucatia*, a family of multicopper oxidase, with E^o = 460 mV, could be mediated by $[W(CN)_8]^{3-/4-}$ (E^o = 0.320 V), $[Os(CN)_8]^{3-/4-}$ (E^o = 0.448 V), and $[Mo(CN)^8]^{3-/4-}$ (E^o = 0.584 V). In this way, 1D electron transfer from cathode to oxygen could be realized generally.

To improve electron transfer between the electrode and laccase (e.g., from *Trametes versicolor* with E^o = 580 mV), several types of metal complexes have been investigated as mediators [48–54], in addition to known good mediators, such as $[Cu(bby)_3]^{2+}$ and $[Fe(CN)_6]^{3-}$, and polymers, including Os ions [55,56].

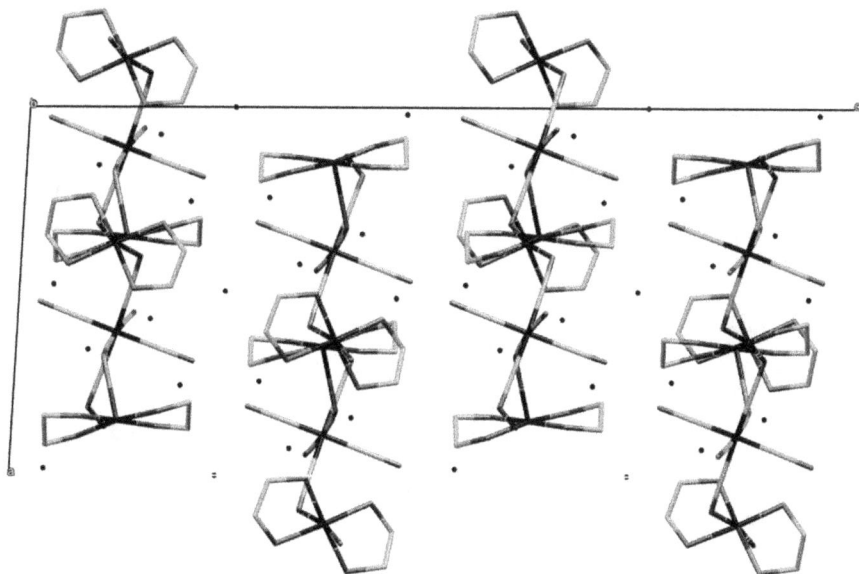

FIGURE 11.11   1D Cu-Fe BOFREU. Cu precursors are coordinated to Fe atoms with six cyanide groups to form 1D structure. The intermolecular hydrogen bonds form a 3D network.

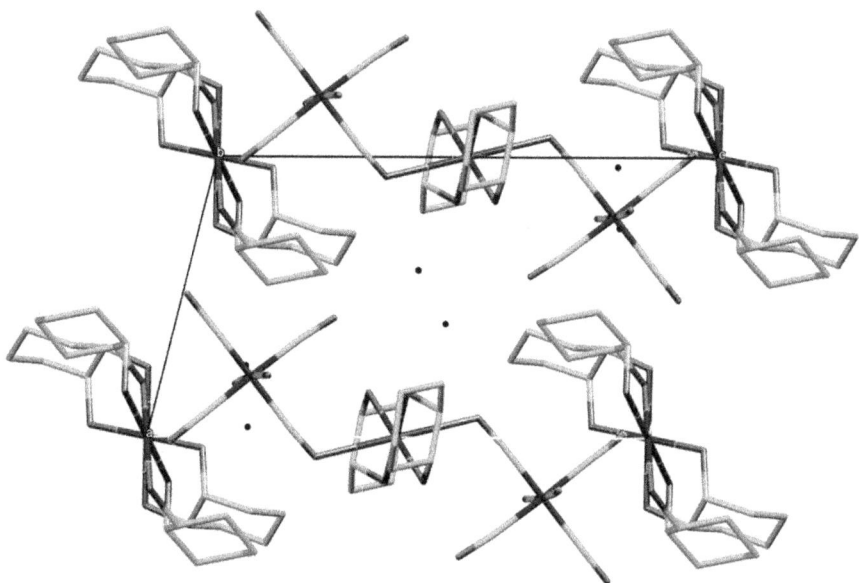

FIGURE 11.12   1D Cu-Fe AREBOO. It crystallized in the *P*1 space group. It exhibits a zigzag structure due to cyanide-bridges with Cu precursors. It forms a 2D network due to hydrogen bonding by water molecules existing within and between molecules.

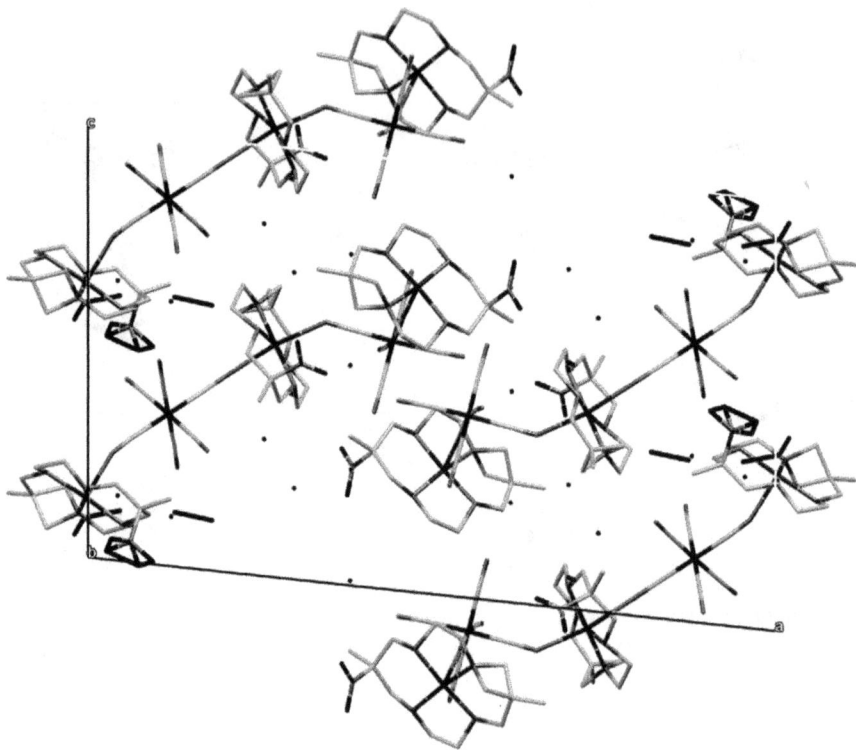

FIGURE 11.13 1D Cu-Fe GUJTAH. The coordination polymers, in which Fe and Cu atoms are cross-linked by cyanide groups, are linked together like branches to form a 1D structure.

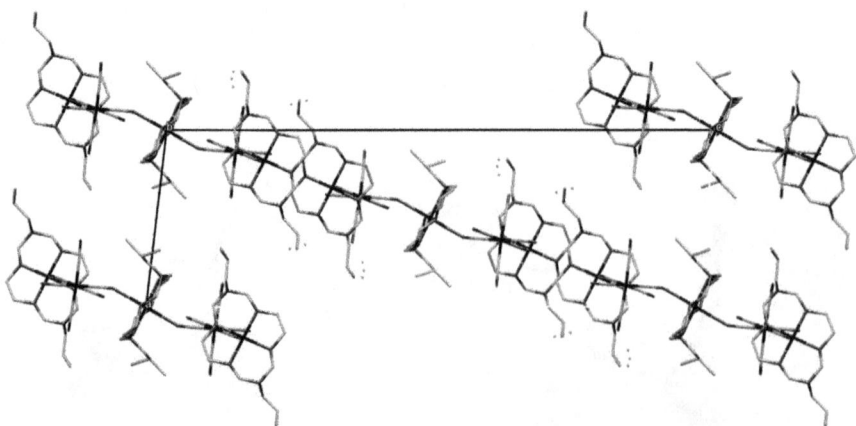

FIGURE 11.14 2D Cu-Fe EDOZUV. The structure is the monoclinic system of space group $P2_1/n$. Fe and Cu atoms are bridged on the same plane, and a 3D honeycomb structure is formed by the Cu precursor.

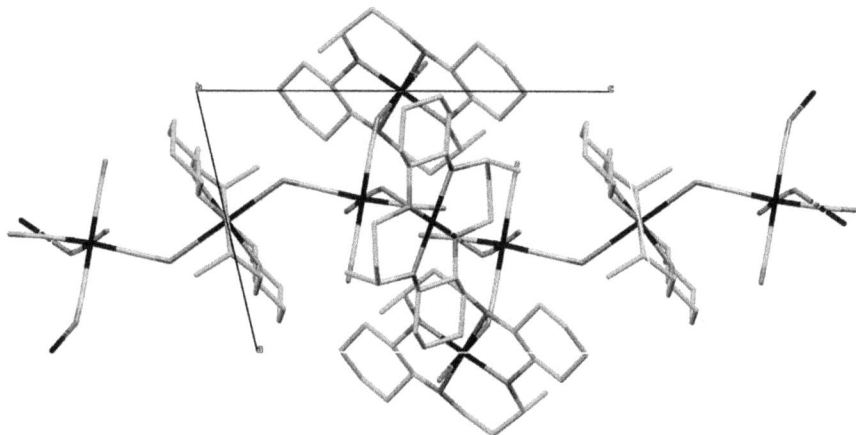

FIGURE 11.15 2D Cu-Fe IXAKEY. Octahedral structure with six cyanide groups co-ordinated to Fe. Jahn-Teller distortion occurs in the Cu precursor with methyl groups, and the axial Cu-N bond length is longer than that in the equatorial direction.

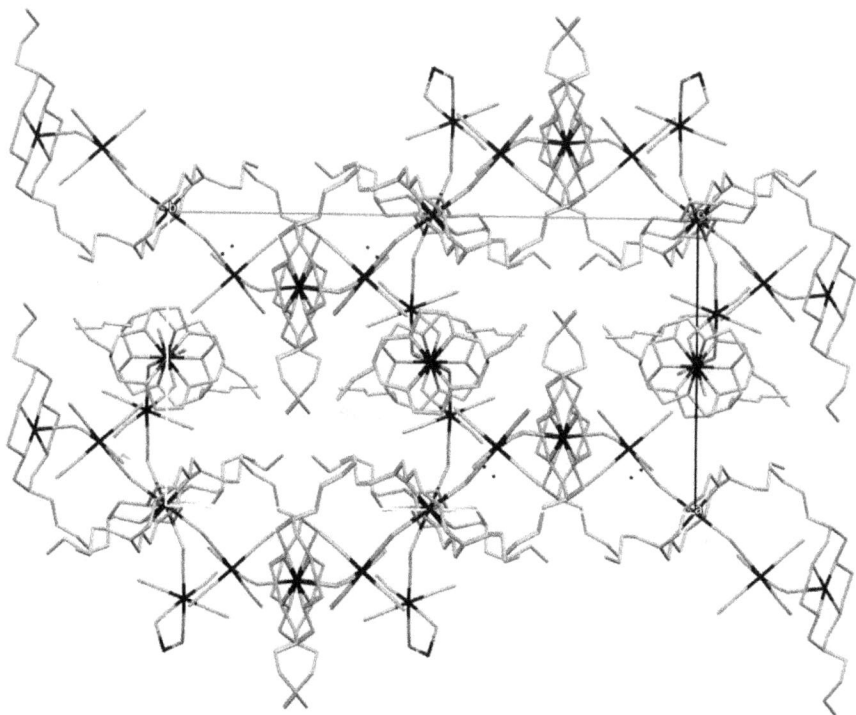

FIGURE 11.16 2D Cu-Fe WOJWOI. The crystal is composed of a cyanide-bridged complex forming a 2D structure and a five-nuclear structural material of Cu_3Fe_2 with a 0D structure. The cyanide-bridges from $[CuL_4]^{2+}$ cation to $[Fe(CN)_6]^{3-}$ anion form a 2D structure.

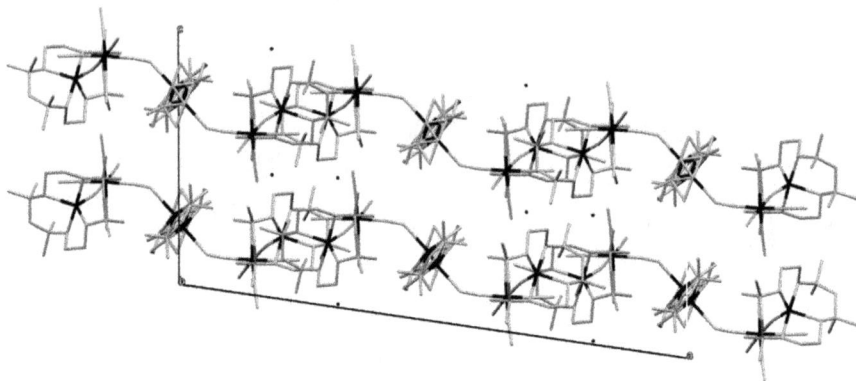

FIGURE 11.17 2D Cu-Fe NICZIJ. Five-nuclear Cu precursors and crystalline water molecules form a complex intertwined network. The Cu precursors and Fe atoms are alternately cyanide-bridged. At this time, the cyanide bridges are in cis-coordination sites.

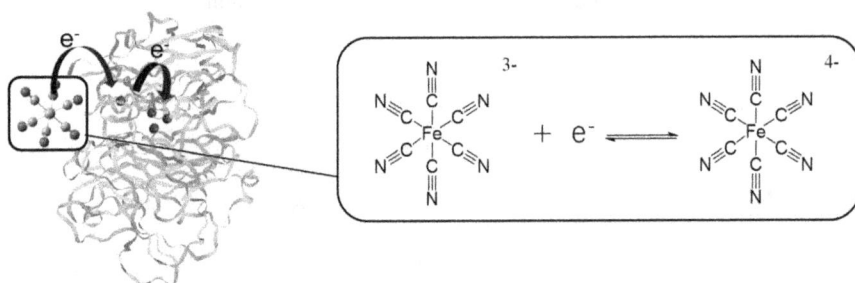

FIGURE 11.18 Electron transfer from redox $[Fe(CN)_6]^{3-/4-}$ mediator to laccase.

However, can such one-directional electron flow occur under any condition? Even though $[Mo(CN)_8]^{3-/4-}$ has the same values of redox potentials under a certain condition [57], it can also act as an opposite side redox mediator for another enzyme [58]. In this way, in addition to steric docking factors for the hydrophobic pocket of laccase and appropriate redox potential under a certain condition, the possibility of an unexpected function (beyond inhibition [59–70]) of additional metal complexes should be examined, as well, even under unfavorable conditions. In this context, herein, we have investigated some cyanide metal complexes as mediators for laccase (from *Trametes versicolor*) under non-ideal conditions.

Laccase exhibits an intense emission peak at 340 nm by excitation of UV light at 280 nm. When cyanide metal complexes were added to laccase, fluorescence intensity around 340 nm due to laccase significantly decreased, which indicated that they can act as a quencher (not luminescent spices in this situation). Unfortunately, computational simulation could not determine their docking sites (later section), including that these ions into laccase could be confirmed by this quenching behavior. The Stern-Volmer equation below was also employed to analyze the quenching behavior [71]:

$$\frac{F_0}{F} = 1 + k_{sv}[Q] \tag{11.1}$$

$$\ln \frac{F_0 - F}{F} = \ln k_a + n \ln [Q] \tag{11.2}$$

where F_0: fluorescence intensity without quenching agent; F: fluorescence intensity with quenching agent; K: Stern-Volmer constant; and $[Q]$: quenching agent concentration.

On the other hand, the addition of $[Mo(CN)_8]^{4-}$ resulted in little spectral change, which suggests: (1) long-distance docking to laccase (unfavorable steric situation), i.e., the distance between laccase and the anion from $[Mo(CN)_8]^{4-}$ was very far even if they are docking near the molecular surface of laccase; or (2) unexpected direction of energy transfer due to redox potentials (unfavorable electrochemical situation) besides electron transfer under this experimental condition.

Previously, $[Au(CN)_2]^-$ is known to exhibit quenching with electron acceptor [72], while $[Mo(CN)_8]^{4-}$ is known to quench fluorescent metal complex [73]. $[Au(CN)_2]^-$ shows unique geometric and charge-transfer properties, as the $5d^{10}$ configuration of Au(I) ion favors aurophilic interactions, which results in tuning of the optical properties. Due to the strong electron-accepting potential of viologens, luminescence of $[Au(CN)_2]^-$ is weakened by quenchers of triplet metal-to-ligand charge transfer (MLCT) [72]. Based on redox potentials of $[Mo(CN)_8]^{4-/3-}$ (-0.77 V vs. NHE) as a quencher and photo-excited $[Pt_2(pop)_4]^{4-/5-}$ (pop = pyrophosphate$^{(2-)}$) (1.34 V) as a sensitizer, it is possible that (not long-distance) electron-transfer quenching takes place, which is much smaller than the diffusion-controlled limit [73].

In this way, emission or quenching of cyanide metal complex also depends on the co-existence of electron donor/acceptor generally, which may be sensitive to the condition of laccase varying valence states easily in this study. The experimental data are shown in Figure 11.19.

In order to investigate its functionality as a mediator, cyclic voltammetry measurements of laccase and metal complexes were performed under oxygen conditions, and the results were compared with those of laccase alone. In most cases, a decrease in the current and potential of the laccase/metal complex was observed. It was further found that the hexa-coordinated cyanide-bridged metal complexes were suitable as mediators.

Subsequently, a study was conducted to consider the functionality of metal complexes as mediators from two aspects: steric and electrochemical properties. Several types of metal complexes were selected as mediators to be used between the electrode and laccase. The mediators of the metal complexes allow more current to flow, even though they require long distance electron transfer as opposed to direct electron transfer. This mechanism stems from the 3D structure of the metal complexes. Six-coordinate cyanide-bridged metal complexes other than $[Fe(CN)_6]^{3-/4-}$, which is a typical structure, were chosen as the metal complexes [74].

The selected metal complexes were $[Cr(CN)_6]^{3-}$, $[Co(CN)_6]^{3-}$, and $[Pt(CN)_6]^{2-}$. Cyclic voltammetry measurements were carried out to determine the change in redox

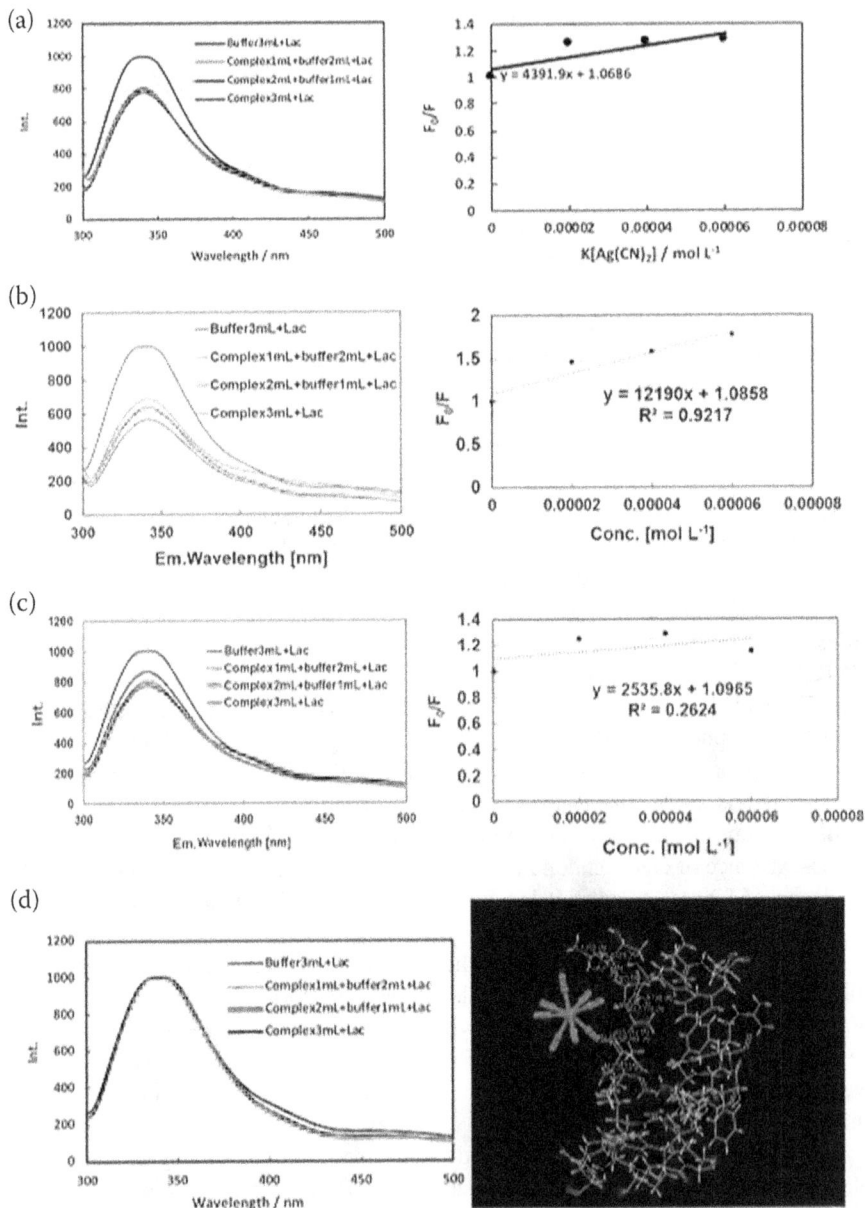

FIGURE 11.19 (left) Fluorescence intensity change and (right) Stern-Volmer plots (except for $[Mo(CN)_8]^{4-}$, instead of docking simulation) for some mononuclear cyanide complexes (a) $[Au(CN)_2]^-$, (b) $[Cr(CN)_6]^{3-}$, (c) $[Pt^{IV}(CN)_6]^{2-}$, and (d) $[Mo(CN)_8]^{4-}$.

potential upon docking of laccase with metal complexes. It was found that [Cr$(CN)_6]^{3-}$ showed an oxidation peak at -0.713 V and a reduction peak at -0.250 V, [Co$(CN)_6]^{3-}$ showed an oxidation peak at -0.713 V and a reduction peak at -0.249 V, and [Pt$(CN)_6]^{2-}$ showed an oxidation peak at -0.692 V and a reduction peak at -0.237 V. The fluorescence spectra of [Cr$(CN)_6]^{3-}$, [Co$(CN)_6]^{3-}$, and [Pt$(CN)_6]^{2-}$ exhibited significant quenching in the fluorescence spectrum around 340 nm. Next, we examine the behavior of the quenching.

11.3.2 Dimensional Cyanide-Bridged MOFs and Laccase

Although discrete cyanide metal complexes were investigated as mediators for laccase, cyanide-bridged metal complexes were too large to be included on molecular surface of laccase like other MOFs. Moreover, rigid bridging coordination polymers of cyanide-bridged metal complexes may be difficult to construct "surrounding" MOF as supramolecular units with enzyme. Therefore, there were few studies of cyanide-bridged metal complexes. In addition, a certain Cu(II) complex and cyanide-bridged [Fe$(CN)_6]^{3-}$ complex may be potentially good mediators. Thus, a hypothesis that Cu(II) and Fe(III) bimetallic complexes may be suitable redox units can be thought from the viewpoint of electrochemistry.

We have been studying cyanide-bridged bimetallic assemblies as mediators in order to improve electron transfer efficiency between the cathode electrode of biofuel cells and the oxygen reduction catalyst metal enzyme (laccase). As part of the optimization of the oxidation-reduction potential of metal ions and the complex-laccase bond mode, we attempt adsorption of laccase to the MOF complex surface and enzyme stabilization, in addition to the laccase surface-encapsulating complex. In this context, the properties of various cyanide-bridged Cu(II) complexes, such as 3D, 2D and 1D, and the cyanide-bridged complexes of the elements as mediators, were evaluated (Figure 11.20) [75].

We have focused so far on the structure of MOFs, and now examine MOFs from the viewpoint of their functional properties, which are metallo-organic structures with a 2D structure, $\{[Cu(N-Eten)_2]_2[Fe(CN)_6]\}_n$ and $\{[Cu(trans-(1R,2R)-chxn)_2]_3[Fe(CN)_6]_2\}_n$. These substances are [Fe$(CN)_6]^{3-}$, in which an iron atom is

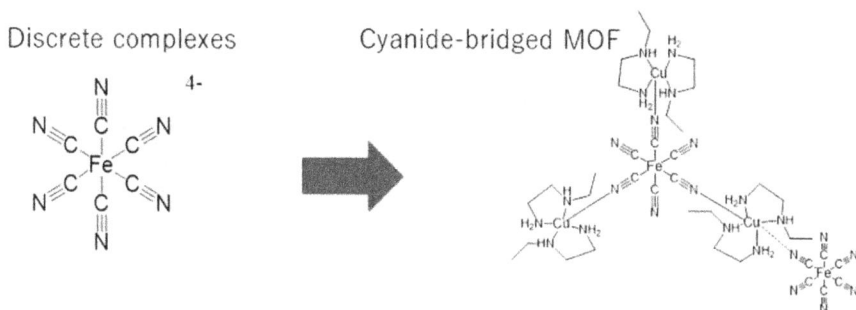

FIGURE 11.20 Mediator complexes (left) conventional discrete cyanide complexes and (right) cyanide-bridged MOF.

bonded to a cyanide group. The $[Fe(CN)_6]^{3-}$ ion has a standard electrode potential of approximately 0.36 V and is relatively inexpensive as a metal complex, making it suitable for use as a mediator. Some of the mediators used in cathodes are capable of smoothing the electron transfer between the electrode and the enzyme. By using two metal organic structures as mediators in a biofuel cell containing the enzyme laccase, respectively, the four-electron reduction of oxygen by laccase in the battery can be promoted. This is thought to be due to the fact that Fe(III) and Cu(II) ions are cross-linked by cyanide groups and behave like a molecular magnet with mixed valence, which enables gradual electron transfer.

As mentioned earlier in this chapter, we have been working on the subject of metal complexes that enhance the activity of laccase. We have studied Cu(II)-Fe(III) bimetallic complexes in order to improve the electron reduction function of laccase using cyanide-bridged metal complexes as mediators (Figure 11.21). The two metals are inequivalent ions to each other, and we asserted that electron transfer would be enhanced by using a cyanide-bridged metal complex as a mediator. As a sample for the mediator, $[Cu(N\text{-}Eten)_2]_3[Fe(CN)_6]_2$ or $[Cu(chxn)_2]_3[Fe(CN)_6]_2$ was synthesized. First, cyclic voltammetry measurements of electrodes with thin films of metal complexes and laccase on glassy carbon were carried out. As a result, a slow redox peak was observed around 0.5–0.6 V, suggesting that these complexes act as intermediaries between the cathode electrode and laccase. Furthermore, the same measurement was performed by adding carbon nanotubes to the glassy carbon electrode, and an increase in the reduction peak was confirmed in the presence of oxygen. The increase in the reducing power of laccase was definitively confirmed by the metal complex containing N-ethylethylenediamine, which was found to contribute to the improvement of the reducing power of laccase as a mediator [76].

As described above, the study of the structure and functionality of cyanide-bridged complexes by dimension has long been a topic of interest. By changing the metal ions, substituents and dimensions inside of the structure of the formed MOFs, it will be possible to utilize the MOFs with various functionalities. We believe that magnetism

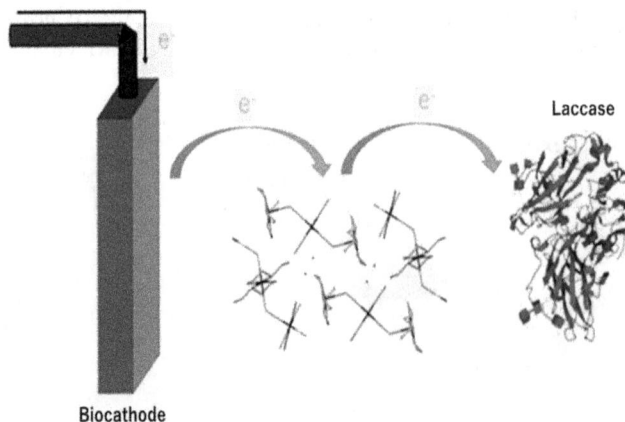

FIGURE 11.21 Model of electron transfer pathway at cathode, 2D cyanide-bridged Cu(II)-Fe MOF, and laccase.

and catalysis are the most notable functionalities of MOFs. In this textbook, we will focus on the use of MOFs as catalysts for the functional activity of enzymes.

We aimed to identify MOFs that could serve as mediators for optimizing the activity of laccase, and decided to use metal complexes that fit the requirements as samples from the dimensionally-organized structures investigated from CSD. In the selection of metal complexes to be used as mediators, we focused on cyanide-bridged metal complexes, which have been confirmed to improve the redox function of laccase, among those that have been previously treated as research themes. Since it has been reported that the redox function is enhanced in the inequivalent bimetallic structure of Fe(III)-Cu(II) among cyanide-bridged metal complexes, we focused on the structure of Cu(II)-M. As a consequence of the above considerations, the following cyanide-bridged metal complexes were selected (see also Figure 11.5):

0D: $[Cu(N\text{-}Eten)_2][Ag(CN)_2]_2$ (Figure 11.22) [77]
1D: $\{[Cu(N\text{-}Eten)_2][Pd(CN)_4]\}_n$ (Figure 11.23) [78]
2D: $\{[Cu(N\text{-}Eten)_2]_3[Fe(CN)_6]_2\}_n$ and $\{[Cu(chxn)_2]_3[Fe(CN)_6]_2\}_n$ (Figures 11.20 and 11.21, and also Figure 11.5) [76]
3D: $\{[Cu(en)_2][KFe(CN)_6]\}n$ (Figure 11.24) [79]

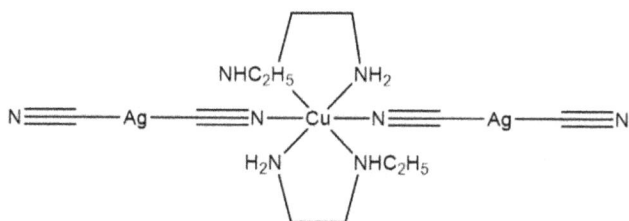

FIGURE 11.22   0D [Cu (N-Eten)$_2$][Ag(CN)$_2$]$_2$. It is a trinuclear bimetallic material consisting of one Cu atom and two Ag atoms bridged by cyanide bridges. The center-symmetric Cu(II) ion forms an elongated coordination structure due to the pseudo-Jahn-Teller effect.

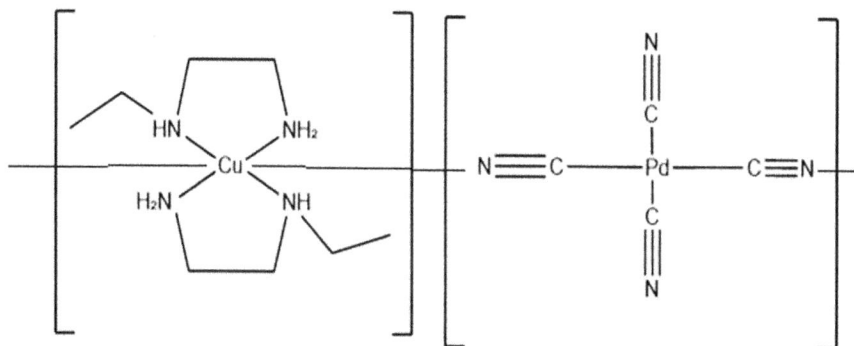

FIGURE 11.23   1D [Cu(N-Eten)$_2$]$_n$[Pd(CN)$_4$]. The Pd complex and the copper precursor are connected by cyanide-bridging, resulting in a zigzag 1D structure.

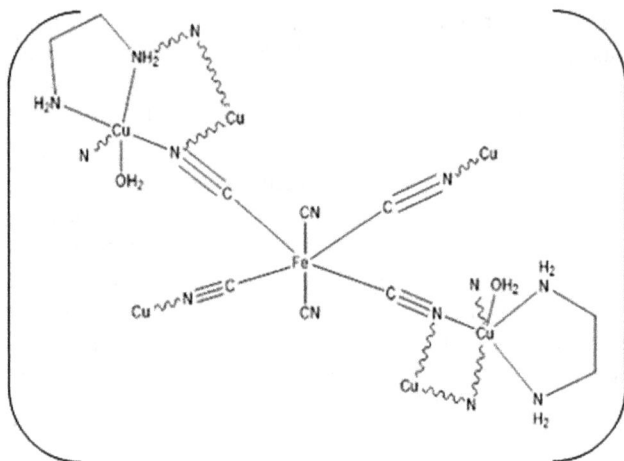

FIGURE 11.24 3D $\{[Cu(en)_2][KFe(CN)_6]\}_n$. In the structure, K interacts with N atoms, causing distortion of the structure. Therefore, the crystal structure shows a distorted octahedral structure. There are two patterns between K and Fe atoms: one bridged by a single cyanide group and the other bridged by two cyanide groups.

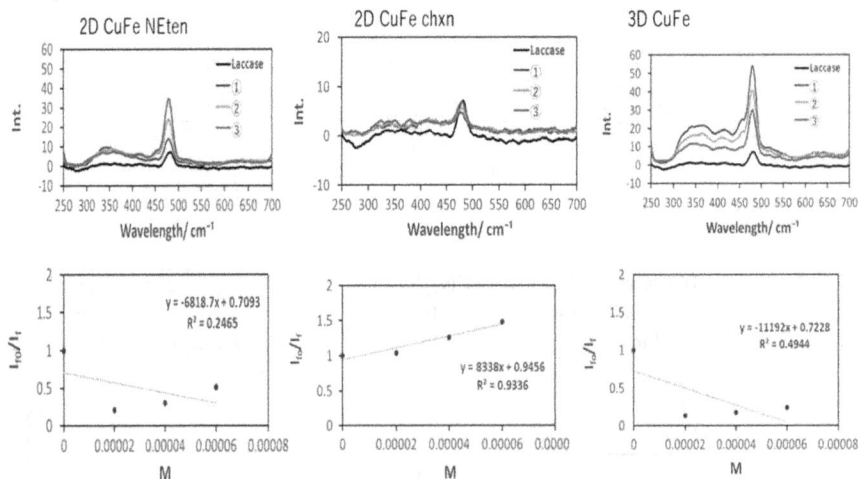

FIGURE 11.25 Fluorescence intensity change and Stern-Volmer plots (except for Mo, instead of docking simulation) for several cyanide-bridged Cu-M complexes.

Spectral and cyclic voltammetric measurements of the complexes were carried out to elucidate the change in enzyme activity due to docking. Stronger fluorescence intensity was observed from each metal complex at around 480 nm. In order to investigate the quenching of the complex materials, the values of each complex were substituted into the Stern-Volmer equation and plotted (Figure 11.25). It was found that the fluorescence intensity increased as the concentration of the metal complex in the solution increased.

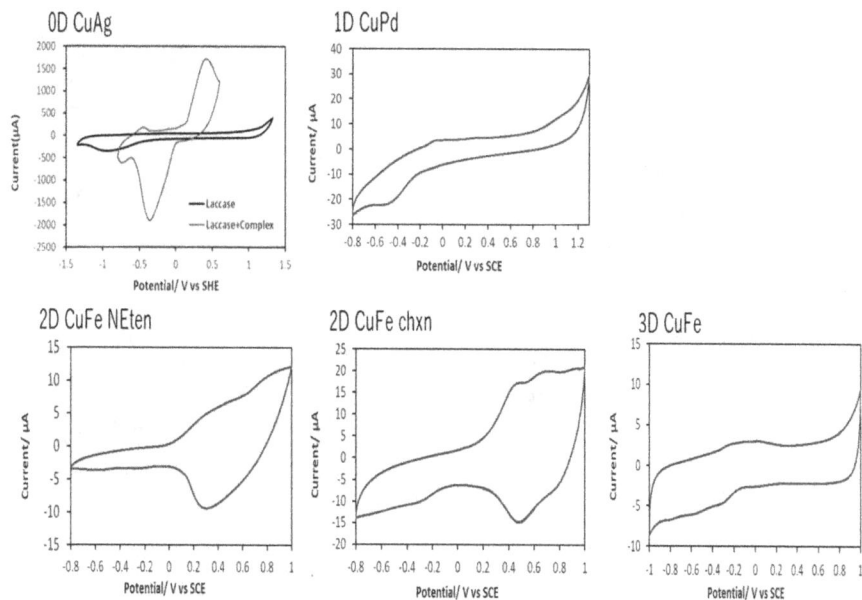

FIGURE 11.26 CV of reduction of oxygen by laccase for cyanide-bridged Cu-M complex-laccase hybrid systems.

Cyclic voltammetry measurements were performed for laccase alone and for the laccase/metal complex complexes for comparison (Figure 11.26). The redox potential of the complexes increased compared to that of laccase alone, confirming that the use of each metal complex as a mediator leads to the improvement of laccase activity.

11.4 CONCLUDING REMARKS

Using MOFs as mediators is difficult because the complex structure derived from MOFs increases the intermolecular distance between MOFs and laccase, making sufficient electron transfer challenging. Here, we considered the mechanism of electron transfer between the electrode and the copper sites inside of the laccase.

Enzymatic fuel cells employ either direct electron transfer (DET) or mediated electron transfer (MET) as the method of electron transfer (Figure 11.27). While DET is a method in which electrons are directly exchanged between the electrode and the enzyme, MET is a method in which the mediator functions as a medium for performing electron relay and sends electrons from the active site of the enzyme to the electrode surface. Therefore, an electrical connection is made between the electrode and the active site of the enzyme. On the other hand, in DET, it is not necessary to consider redoxing the mediator, and each method is characterized in that the mediator does not cause an excess potential. Laccase is an enzyme that can be used with either method. Which method is actually used depends on the distance between the electrode and the enzyme. DET is applied

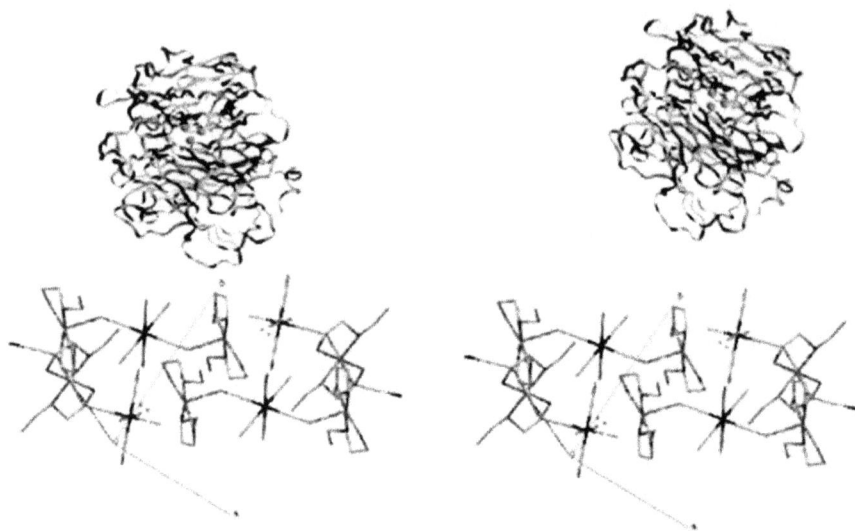

FIGURE 11.27   Contact models of 2D cyanide-bridged MOF and laccase (left) suitable for DET and MET (right) unsuitable for MET (long-distance) but potentially possible for DET.

when the electrode and enzyme are in close proximity. In contrast, if the electrode and the enzyme are not sufficiently close to each other and separated from each other, MET is applied.

Considering these factors and the experimental results again, when the MOF docks with the enzyme to form a complex, a certain distance between the electrode and the enzyme is due to the bulky and complicated structure, which is a feature derived from the MOF. It was speculated that this result may have occurred because it was difficult to directly exchange electrons between the electrode and the enzyme. In addition to using it as a DET [80], it might be possible to act as a mediator that transports electrons during MET, especially for structures with large surface gaps, such as 2D and 3D, among MOFs. Consequently, it seems that a mechanism for donating electrons to laccase, which forms a complex and is far from the electrode, is constructed, and electrochemical properties can be seen.

In summary, MOF can potentially act to improve enzymatic functions to form supramolecules. However, for a specific purpose, for example, metal complex mediators depends on both steric and electrochemical factors, the situation about structural dimensionality of coordination polymers is complicated even if electrochemically proper metals are used.

REFERENCES

[1] Li, H., Eddaoudi, M. & O'Keeffe, M. et al. 1999. "Design and synthesis of an exceptionally stable and highly porous metal-organic framework." *Nature*. 402,276–279. 10.1038/46248

[2] Zhou. H.-C. J. & Kitagawa, S. 2014. "Metal–Organic Frameworks (MOFs)." *Chemical Society Reviews.* 43,5415–5148. 10.1039/C4CS90059F

[3] Wang, Q. & Astruc, D. 2020. "State of the Art and Prospects in Metal–Organic Framework (MOF)-Based and MOF-Derived Nanocatalysis." *Chemical Reviews.* 120,1438–1511. 10.1021/acs.chemrev.9b00223

[4] Liu, J., Chen, L., Cui, H., Zhang, J., Zhang, L. & Su, C.-Y. 2014. "Applications of metal–organic frameworks in heterogeneous supramolecular catalysis." *Chemical Society Reviews.* 43,6011–6061. 10.1039/C4CS00094C

[5] Avery, E., Baumann D. A., Liu, B. B. & Thoi, V. S. 2019. "Metal-organic framework functionalization and design strategies for advanced electrochemical energy storage devices." *Communications Chemistry.* 2,86. 10.1038/s42004-019-0184-6

[6] Liang, W., Wied, P., Carraro, F., Sumby, C. J., Nidetzky, B., Sung, C.-K., Falcaro, T. P. & Doonan, C. J. 2021. "Metal–Organic Framework-Based Enzyme Biocomposites." *Chemical Reviews.* 121,1077–1129. 10.1021/acs.chemrev.0c01029

[7] Lian, X., Fang, Y., Joseph, E., Wang, Q., Li, J., Banerjee, S., Lollar, C., Wang, X. & Zhou, H.-C. 2017. "Applications of metal–organic frameworks in heterogeneous supramolecular catalysis." *Chemical Society Reviews.* 46,3386–3401. 10.1039/C4CS00094C

[8] Sadakiyo, M. 2022. "Support effects of metal-organic frameworks in heterogeneous catalysis." *Nanoscale.* 14,3398-3406. 10.1039/C4CS00094C

[9] Wu, X., Yue, H., Zhang, Y., Gao, X., Li, X., Wang, L., Cao, Y., Hou, M., An, H., Zhang, L., Li, S., Ma, J., Lin, H., Fu, Y., Gu, H., Wei, L. W., Zare, R. N., & Ge, J. 2019. "Packaging and delivering enzymes by amorphous metal-organic frameworks." *Nature Communications.* 10, 5165. 10.1038/s41467-019-13153-x

[10] Feng, D., Liu, T. F. & Su, J. et al. 2015. "Stable metal-organic frameworks containing single-molecule traps for enzyme encapsulation." *Nature Communications.* 6,5979. 10.1038/ncomms6979

[11] Ahmad, R., Shanahan, J., Rizaldo, S., Kissel, D. S. & Stone, K. L. 2020. "Co-immobilization of an Enzyme System on a Metal-Organic Framework to Produce a More Effective Biocatalyst." *Catalysts.* 10,499. 10.3390/catal10050499

[12] Samui, A., & Sahu, S. K. 2018. "One-pot synthesis of microporous nanoscale metal organic frameworks conjugated with laccase as a promising biocatalyst." *New Journal of Chemistry.* 42,4192–4200. 10.1039/c7nj03619a

[13] Patra, S., Sene, S., Mousty, C., Serre, C., Chausse, A., Legrand, L. & Steunou, N. 2016. "Design of Laccase-Metal Organic Framework-Based Bioelectrodes for Biocatalytic Oxygen Reduction Reaction." *ACS Applied Materials & Interfaces.* 8,20012–20022. 10.1021/acsami.6b05289

[14] Liu, S., Kang, L. & Jun, S. C. 2021. "Challenges and Strategies toward Cathode Materials for Rechargeable Potassium-Ion Batteries." *Advanced. Materials.* 33 2004689. 10.1002/adma.202004689

[15] Nihei, M. et al., 2020. "Molecular Prussian Blue Analogues: From Bulk to Molecules and Low-dimensional Aggregates." *Chemistry Letters.* 49, 1206–1215. 10.1246/cl.200428

[16] Ohba, M. & Okawa, H. 2000. "Synthesis and magnetism of multi-dimensional cyanide-bridged bimetallic assemblies." *Coordination Chemistry Reviews.* 198,313–328. 10.1016/S0010-8545(00)00233-2

[17] Ohkoshi, S., Namai, A. & Tokoro, H. 2019. "Humidity sensitivity, organic molecule sensitivity, and superionic conductivity on porous magnets based on cyanide-bridged bimetal assemblies." *Coordination Chemistry Reviews.* 380,572-583. 10.1016/j.ccr.2018.10.004

[18] Nihei, M. 2012. "Multi-Step Phase Transition Based on Intramolecular Electron Transfers in a Cyanide-Bridged Tetranuclear Molecular Square." *Journal of the Crystallographic Society of Japan.* 54,319–324. 10.5940/jcrsj.54.319

[19] Akitsu, T. 2008. "Jahn-Teller Distortion and Electronic Functions of Copper (II) Complexes in Protein and Molecular Crystal Matrix." *Journal of the Crystallographic Society of Japan*. 50,201–205. 10.5940/jcrsj.50.201

[20] Akitsu, T., Einaga, Y. & Yoza, K. 2010. "Thermally-Accessible Lattice Strain and Local Pseudo Jahn-Teller Distortion in Various Dimensional Cu^{II}-M^{III} Bimetallic Cyanide-Bridged Assemblies." *The Open Inorganic Chemistry Journal*. 2,1–10. 1 0.2174/1874098700802010001

[21] Taylor, R. & Wood, P. A. 2019. "A Million Crystal Structures: The Whole Is Greater than the Sum of Its Parts." *Chemical Reviews*. 119,9427–9477. 10.1021/acs.chemrev. 9b00155

[22] Bienko, A., Suracka, K., Mrozinski, J., Kruszynski, R., Bienko, D., Wojciechowska, A. & Boca, R. 2010. "A heterobimetallic cyanide-bridged $Cu^{II}Fe^{III}Cu^{II}$ trimer. Synthesis, crystal structure and magnetic properties." *Polyhedron*. 29,2546–2552. 10.1016/j.poly.2010.05.027

[23] Lim, J. H., Yoon, J. H., Choi, S. Y., Ryu, D. W., Koh, E. K. & Hong, C. S. 2010. "Synthesis, Crystal Structures, and Magnetic Properties of Cyano-Bridged Honeycomblike Layers M^{V}–Cu^{II} (M = Mo, W) Chelated by a Macrocyclic Ligand." *Inorganic Chemistry*. 45,7821–7827. 10.1021/ic060868b

[24] Tazaki, D., Akitsu, T., Silva, P. S. P. & Rodrigues, V. H. S. 2015. "Crystal structures and non-linear optics of several types of bimetallic assemblies comprising chiral Cu(II) complexes." *Polyhedron*. 102,297–307. 10.1016/j.poly.2015.09.042

[25] Yuan, A., Shen, X. & Zhou, H. 2008. "Crystal structure and magnetic properties of a three-dimensional complex constructed from $[CuL]^{2+}$ and $[Fe(CN)_6]^{3-}$ precursors." *Transition Metal Chemistry*. 33,133–136. 10.1007/s11243-007-9032-0

[26] Hong, C. S. & You, Y. S. 2004. "Cyano-bridged Fe(II)–Cu(II) bimetallic assemblies: honeycomb-like and pentanuclear structures." *Inorganica Chimica Acta*. 357,2371–3278. 10.1016/j.ica.2004.04.004

[27] Liu, H. & Wang, D. 2008. "Catena-Poly[[[μ-cyanido-1:2κ^2C:N-tricyanido-1κ^3C-bis (ethylenediamine)-2κ^4N,N'-copper(II)iron(II)]-μ-cyanido-κ^2C:N-[bis(ethylenediamine-κ^2N,N')copper(II)]-μ-cyanido-κ^2N:C] 4.5-hydrate]." *Acta Crystallographica E*. 64, m1148–m114. 10.1107/S1600536808023830

[28] Coronado, E., Gimenez-Saiz, C., Nuez, A., Sanchez, V. & Romero, F. M. 2003. "Stoichiometric Control of the Magnetic Properties in Copper(II) Cyano-Bridged Bimetallic Complexes." *European Journal of Inorganic Chemistry*. 2003,4289–4293 10.1002/ejic.200300641

[29] Cha, M. J., Shin, J. W., Lee, Y. H., Kim, Y., Kim, B. G. & Min, K. S. 2009. "Structure and magnetic properties of a novel branch-like one-dimensional cyano-bridged assembly $[Cu(L)]_3[Fe(CN)_6]_2 \cdot 8H_2O$ (L = 6,13-dimethyl-6-nitro-1,4,8,11 tetraazabicyclo [11.1.1]pentadecane)." *Inorganic Chemistry Communications*. 12,520–522. 10.1016/j.inoche.2009.04.018

[30] Kongchoo, S., Chainok, K., Kantacha, A. & Wongnawa. S. 2017. "Copper(II) complex as a precursor for formation of cyano-bridged pentanuclear Fe^{III}-Cu^{II} bimetallic assembly: Synthesis, characterization, crystal structure and antibacterial activity." *Journal of Chemical Science*. 129,431–440. 10.1007/s12039-017-1255-9

[31] Lim, J. H., Yoon, J. H., Kim, H. C., Hong, C. S. 2006. "Surface Modification of a Six-Capped Body-Centered Cube Ni9W6 Cluster: Structure and Single-Molecule Magnetism" Angewandte Chemie International Edition. 45,7424–7426 10.1002/ anie.200601759

[32] Shen, X.-P., Xu, Y., Zhou, H., Shu, H.-Q. & Yuan, A.-H. 2008. "Crystal structure and magnetic properties of a cyano-bridged bimetallic assembly $[CuL_4]_3[Fe(CN)_6]_2 \cdot 2H_2O$ (L 4 = 3,10-dibutyl-1,3,5,8,10,12-hexaazacyclotetradecane)." *Journal of Molecular Structure*. 892,58–62 . 10.1016/j.molstruc.2008.04.055

[33] Rodriquez-Dieguez, A., Kivekas, R., Sillanpaa, R., Cano, J., Lloret, F., Mckee, V., Stoeckli-Evans, H. & Colacio, E. 2006. "Structural and Magnetic Diversity in Cyano-Bridged Bi- and Trimetallic Complexes Assembled from Cyanometalates and $[M(rac\text{-}CTH)]^{n+}$ Building Blocks (CTH) d,l-5,5,7,12,12,14-Hexamethyl-1,4,8,11 tetraazacyclotetradecane)." *Inorganic Chemistry*. 45,10537-10551. 10.1021/ic061187j

[34] Solomon, E. I., Szilagyi, R. K., DeBeer, G. S. & Basumallick, L. 2004. "Electronic structures of metal sites in proteins and models: Contributions to function in blue copper proteins." *Chemical Reviews*. 104,419–458 10.1021/cr0206317

[35] Quintanar, L., Stoj, C., Taylor, A. B., Hart, P. J., Kosman, D. J. & Solomon, E. I. 2007. "Shall We Dance? How A Multicopper Oxidase Chooses Its Electron Transfer Partner."*Accounts of Chemical Researches*. 40,445–452 10.1021/ar600051a

[36] Mate, D. M. & Alcalde, M. 2017. "Laccase: A multi-purpose biocatalyst at the forefront of biotechnology." *Microbial Biotechnology*. 10,1457-1467 10.1111/1751–7915.12422

[37] Mano, N. & de Poulpiquet A. 2018. "O_2 Reduction in Enzymatic Biofuel Cells." *Chemical Reviews*. 118, 2392–2468. 10.1021/acs.chemrev.7b00220

[38] Goff, A., Holzinger, M. & Cosnier, S. 2015. "Recent progress in oxygen-reducing laccase biocathodes for enzymatic biofuel cells." *Cellular and Molecular Life Sciences*. 72,941–952 10.1007/s00018-014-1828-4

[39] Moehlenbrock, M. J. & Minteer, S. D. 2008. "Extended lifetime biofuel cells." *Chemical Society Reviews*. 37,1188–1196 10.1039/B708013C

[40] Xiao, X., Xia, H.-Q., Wu, R., Bai, L., Yan, L., Magner, E., Cosnier, S., Lojou, E., Zhu, Z. & Liu, A. 2019. "Tackling the Challenges of Enzymatic (Bio)Fuel Cells." *Chemical Reviews*. 119,9509–9558 10.1021/acs.chemrev.9b00115

[41] Mehra, R., Muschiol, J., Meyer, A. S. & Kepp, K. P. 2018. "A structural-chemical explanation of fungal laccase activity." *Scientific Reports*. 8,17285 10.1038/s41598-018-35633-8

[42] Morozova, O. V., Shumakovich, G. P., Shleev, S. V. & Yaropolov, Y. I. 2007. "Laccase-Mediator Systems and Their Applications: A Review." *Applied Biochemistry and Microbiology*. 43,523–535 10.1134/S0003683807050055

[43] Sakai, H., Mita, H., Sugiyama, T., Tokita, Y., Shirai, S. & Kano, K. 2014. "Construction of a Multi-stacked Sheet-type Enzymatic Biofuel Cell." *Electrochemistry*. 82,156–161 10.5796/electrochemistry.82.156

[44] Hitaishi, V. P., Clément, R., Quattrocchi, L., Parent, P., Duché, D., Zuily, L., Ilbert, M., Lojou, E. & Mazurenko, I. 2020. "Interplay between Orientation at Electrodes and Copper Activation of Thermus thermophilus Laccase for O_2 Reduction." *Journal of the American Chemical Society*. 142,1394–1405 10.1021/jacs.9b11147

[45] Agbo, P., Heath. J. R. & Gray, H. B. 2014. "Modeling Dioxigen Reduction at Multicopper Oxidase Cathodes." *Journal of the American Chemical Society*. 136,13882-13887 10.1021/ja5077519

[46] Cheng, C.-Y., Liao, C.-I. & Lin, S.-F. 2015. "Borate-fructose complex: A novel mediator for laccase and its new function for fructose determination." *FEBS Letters*. 589,3107–3112 10.1016/j.febslet.2015.08.032

[47] Tsujimura, S., Kawaharada, M., Nagkagawa, T., Kano, K., Ikeda, T. 2003. "Mediated bioelectrocatalytic O_2 reduction to water at high positive electrode potentials near neutral PH." *Electrochemistry Communications*. 5,138–141 10.1016/S1388-2481(03)00003-1

[48] Kajiwara, K., Yamane, S., Haraguchi, T., Pradhan, S., Sinha, C., Parida, R., Giri, S., Roymahaptra, G., Moon, D. & Akitsu, T. 2019. "Computational Design of Azoanthraquinone Schiff Base Mn Complexes as Mediators for Biofuel Cell Cathode." *Journal of Chemistry and Chemical Engineering*. 13,23–33 10.17265/1934-7375/2019.01.003

[49] Sano, A., Yagi, S., Haraguchi, T. &, Akitsu, T. 2018. "Synthesis of Mn(II) and Cu (II) complexes including azobenzene and its application to mediators of laccase for biofuel cells." *Journal of the Indian Chemical Society*.95,487–494. No doi.

[50] Kunitake, F., Kim, S., Yagi, S., Yamazaki, S., Haraguchi, T. & Akitsu, T. 2019. "Chiral recognition of azo-Schiff base ligands, their Cu(II) complexes and their docking to laccase as mediators." *Symmetry*. 11,666. 10.3390/sym11050666

[51] Mitsumoto, Y., Sunaga, N. & Akitsu, T. 2017. "Polarized light induced molecular orientation in laccase for chiral azosalen Mn(II), Co(II), Ni(II), Cu(II), Zn(II) mediators toward application for biofuel cell." *SciFed Journal of Chemical Research*. 1, 1. No doi.

[52] Takeuchi, Y. & Akitsu, T. 2016. "Anthraquinone Derivative Chiral Schiff Base Copper(II) Complexes for Enzyme Type Bio-Fuel Cell Mediators." *Journal of Electrical Engineering*. 4,189–196 10.17265/2328-2223/2016.04.005

[53] Takeuchi, Y., Sunaga, N. & Akitsu, T. 2017. "Anthraquinone and L-amino Acid Derivatives Schiff Base Cu(II) Complexes as a Mediator between Cathode of Biofuel Cell and Oxygen-reducing Laccase." *Trends in Green Chemistry* 3,1–8. No doi.

[54] Kurosawa, Y., Tsuda, E., Takase, M., Yoshida, N., Takeuchi, Y., Mitsumoto, Y. & Akitsu, T. 2015. "Spectroscopic and Electrochemical Studies on Metalloprotein (Laccase) and Cu(II) Complex Mediators As Model Systems for Biofuel Cell Cathodes." *Threonine: Food Sources, Functions and Health Benefits*. 73–86. No doi

[55] Kashiwagi, K., Tassinari, T., Haraguchi, T., Banerjee-Gosh, K., Akitsu, T. & Naaman, R. 2020. "Electron Transfer via Helical Oligopeptide to Laccase Including Chiral Schiff Base Copper Mediators." *Symmetry*. 12,808 10.3390/sym12050808

[56] Kajiwara, K., Pradhan, S., Haraguchi, T., Sinha, C., Parida, R., Giri, S., Roymahaptra, G. & Akitsu, T. 2020. "Photo-Tunable Azobenzene-Anthraquinone Schiff Base Copper Complexes as Mediators for Laccase in Biofuel Cell Cathode." *Symmetry*. 7,797. 10.3390/sym12050797

[57] Sato, K., Ohsaka, T., Matsuda, H. & Oyama, N. 1983. "Electrocatalytic Reaction of $Fe^{2+/3+}$ Aqua Ions by $[Mo(CN)_8]^{4-/3-}$ Complexes." *Bulletin of the Chemical Society of Japan*. 56,1863–1864 10.1246/bcsj.56.1863

[58] Tomimura, S., Yasukouch, K. & Taniguchi, I. 1989. "Electrochemical analysis of enzyme reactions for sarcosine oxidases." *Bunseki Kagaku*. 38,601–607 10.2116/bunsekikagaku.38.11_601

[59] Couto, S. R. & Toca, J. L. 2006. "Inhibitors of Laccase: A Review." *Current Enzyme Inhibition*. 2,343–352 10.2174/157340806778699262

[60] Ruy, F., Vercesi, A. E. & Kowaltowski, A. J. 2006. "Inhibition of specific electron transport pathways leads to oxidative stress and decreased Candida albicans proliferation." *Journal of Bioenergetics and Biomembranes*. 38,129 135 10.1007/s10863-006-9012-7

[61] Pamidipati, S. & Ahmed, A. 2020. "A first report on competitive inhibition of laccase enzyme by lignin degradation intermediates." *Folia Microbiologica*. 65,431–437 10.1007/s12223-019-00765-5

[62] Kelley, L., Dillard, B., Tempel, W., Chen, L., Shaw, N., Lee, D., Newton, M., Sugar, F., Jenney Jr, F., Lee, H., Shah, C., Poole III. F., Adams, M., Richardson, J., Richardson, D., Liu, Z., Wang, B. & Rose, J. 2007. "Structure of the hypothetical protein PF0899 from Pyrococcus furiosus at 1.85 Å resolution." *Acta Crystallographica*. F.63 549–552.10. 1107/S1744309107024049

[63] Piontek, K., Antorini, M. & Choinowski, T. 2002. "Crystal Structure of a Laccase from the FungusTrametes versicolor at 1.90 Å Resolution Containing a Full Complement of Coppers." *Journal of Biological Chemistry*. 277,37663–37669 10. 1074/jbc.M204571200

[64] Frisch, M. J., Trucks, G. W., Schlegel, H. B., Scuseria, G. E., Robb, M. A., Cheeseman, J. R., Scalmani, G., Barone, V., Mennucci, B. & Petersson, G. A. et al. 2009. "Gaussian 09, Revision D.01." Gaussian Inc Wallingford CT USA.

[65] CCDC GOLD has Proven Success in Virtual Screening, Lead Optimisation, and Identifying the Correct Binding Mode of Active Molecules. Available online: https://www.ccdc.cam.ac.uk/solutions/csd-discovery/Components/Gold/ (accessed on 8 December 2020).

[66] Ahern, J. C., Shilabin, A., Menline, K. M. & Picke, R. D., 2014. "Patterson, H. Photophysical properties of $\{[Ag(CN)_2]^-\}^2$ complexes trapped in a supramolecular electron-acceptor organic framework." *Dalton Transactions.* 43,12044–12049 10.1039/C4DT01110D

[67] Szklarzewicz, J., Matoga, D., Klys, A. & Lasocha, W. 2008. "Ligand-Field Photolysis of $[Mo(CN)_8]^{4-}$ in Aqueous Hydrazine: Trapped Mo(II) Intermediate and Catalytic Disproportination of Hydrazine by Cyano-Ligand Mo(III,IV) Complexes." *Inorganic Chemistry.* 47,5464–5472 10.1021/ic800053e

[68] Qi, Z., Pillet, S., de Graaf, C., Magott, M., Bendeif, E., Guionneau, P., Rouzieres, M., Marvaud, V., Stefanczyk, O., Pinkowicz, D. & Mathoniere, C. 2000. "Photoinduced Mo-CN Bond Breakage in Octacyanomolybdate Leading to Spin Triplet Trapping." *Angewandte Chemie International Edition.* 59,3117–3121 10.1002/anie.201914527

[69] Rawashdeh-Omary, M. A., Omary, M. A. & Patterson, H. H. 2000. "Oligomerization of $Au(CN)_2^-$ and $Ag(CN)_2^-$ Ions in Solution via Ground-State Aurophilic and Argentophilic Bonding." *Journal of the American Chemical Society.* 122,10371–10380 10.1021/ja001545w

[70] Rawashdeh-Omary, M. A., Omary, M. A., Patterson, H. H. & Fackler Jr, J. P. 2001. "Excited-State Interactions for $[Au(CN)_2^-]_n$ and $[Ag(CN)_2^-]_n$ Oligomers in Solution. Formation of Luminescent Gold-Gold Bonded Excimers and Exciplexes." *Journal of the American Chemical Society.* 122,10371–10380 10.1021/ja011176j

[71] Hayashi, T. & Akitsu T. 2015. "Fluorescence, UV-vis, and CD spectroscopic study on docking of chiral salen-type Zn(II) complexes and lysozyme and HAS proteins." *Threonine, Food Sources, Functions and Health Benefits.* 49-72 . No doi.

[72] Patel, E. N., Arthr, R. B., Nicholas, A. D., Reinheimer, E. W., Omary, N. M. A., Brichacek, M. & Patterson, H. H. 2019. "Synthesis, structure and photophysical proerties of a 2D network with gold dicyanide donors coordinated to aza[5] helicence viologen acceptors." *Dalton Transactions.* 48,10288–10297 10.1039/C9DT01823A

[73] Yamaguchi, T. & Sasaki, Y. 1990. "Cation effect on the quenching of the photoexcited state of tetrakis(.mu.-pyrophosphito-P,P')diplatinate(II) by octacyanomolybdate(IV) in aqueous solution." *Inorganic Chemistry.* 29,493–495 10.1021/ic00328a031

[74] Sato, Y. & Akitsu, T. 2021. "Improvement and enhancement of enzyme activity by metal-organic framework (MOF) adsorption." *Journal of Material Sciences & Engineering.* 10, 4. https://www.hilarispublisher.com/open-access/improvement-and-enhancement-of-enzyme-activity-by-metalorganic-framework-mof-adsorption.pdf

[75] Huang, W., Zhang, W., Gan, Y., Yang, J. & Zhang, S. 2020. "Laccase immobilization with metal-organic frameworks: Current status, remaining challenges and future perspectives." *Critical Reviews in Environmental Science and Technology.* 52,1282–1324 10.1080/10643389.2020.1854565

[76] Ogikubo, Y. & Akitsu, T. 2016. "Enhancing Medical or Biological Functions of Laccase by Cyanide-Bridged Cu(II)-Fe(III) Bimetallic Complexes Madiators." *Drug Designing: Open Access.* 5,130–131 10.4172/2169-0138.1000130

[77] Akitsu, T., Sano, K. & Kimoto, Y. 2010. "Structure of Bis(N-ethylethylenediamine) Cu(II) and Dicyano Ag(I) Bimetallic Assembly." *International Journal of Current Chemistry*.14,1–20 . No doi.

[78] Akitsu, T. & Endo, Y. 2009. "catena-Poly[[bis-(N-ethyl-ethylene-di-amine-κN,N') copper(II)]-μ-cyanido-κN:C-[dicyanido-κC-palladium(II)]-μ-cyanido-κC:N]." *Acta Crystallographica E*. 65,m406–m407 10.1107/s160053680900885x

[79] Luo, J., Hong, M., Chen, C., Wu, M. & Gao, D. 2002 "Synthesis, magnetic properties and crystal structures of two compounds: [Cu(en)(H₂O)]₂[Fe(CN)₆]·4H₂O and [Cu (en)₂][KFe(CN)₆] (en=ethylenediamine)." *Inorganica Chimica Acta*. 328,185–190 10.1016/S0020-1693(01)00732-0

[80] Mano, N. 2020 "A bio on bioelectrochemical cells: a conversation with Nicolas Mano." *Nature Communications*. 11,6412 10.1038/s41467-020-19776-9

12 Metal-Organic Framework–Based Electrochemical Immunosensors for Virus Detection

Stephen Rathinaraj Benjamin,
Eli José Miranda Ribeiro Júnior,
Geanne Matos de Andrade, and
Reinaldo Barreto Oriá

CONTENTS

DOI: 10.1201/9781003252061-12

12.1　INTRODUCTION

Viruses have been near humans for a long time. In the past centuries, there have been many yearly epidemics and pandemics. It is a phenomenon when a zoonotic virus (originating from an animal source) leaps or evolves into a human virus that new viruses infect humans and propagate. Beyond the simple transmission of certain emerging viruses, the development of the global transportation network makes it feasible for infections and their vectors to spread quicker and more quickly than they were conceivable, mainly in developing nations. The combination of these variables has a significant impact on the pandemic of infectious diseases [1]. Approximately one-quarter to one-third of the world's deaths are attributed to infectious diseases. Despite significant advances in the pharmaceutical industry, infectious disease transmission is growing due to globalization, expanded travel and trade, urbanization, congested cities, human behavioral changes, pathogen resurgence, and antibiotic abuse [2]. Multiple viruses have emerged in the past two decades, each with a distinct effect on humanity. The worldwide health hazards posed by viral diseases such as influenza, COVID-19, HIV, Zika, Chikungunya, Ebola, and more strongly emphasize the need for quick, sensitive, and selective detection of viruses and post-infection antibodies.

Currently, the gold-standard methods employed for diagnosing viral diseases include ELISA (enzyme-linked immunosorbent assay) and PCR (polymerase chain reaction). Despite having an adequate upper limit of detection (LOD) and accuracy, they need specific reagents and equipment, resulting in restricted accessibility and long reaction times. Immunoassays or immunosensors are now widely used owing to their excellent specificity and sensitivity [3]. Furthermore, more improved analytical instruments for immunoassay, including quartz crystal microbalance (QCM), surface plasmon resonance (SPR), optical detection techniques, such as electrochemical, chemiluminescence, and fluorescence techniques have been investigated based on numerous signal production principles resulting from dynamic interplay between antigens and antibodies [4,5].

Electrochemical (EC) immunosensors have gained popularity due to their rapid assay time, minimal sample consumption, ease of use, and cheap cost. Many electrochemical immunosensors are presently being developed, including label-free and sandwich immunosensors. Label-free immunosensors work by directly detecting antibody-antigen contact without a secondary antibody; however, they lack sensitivity and repeatability. Sandwich immunosensors work by the quantity of a labeled analyte (often the antibody-Ab) measured in the sample is exactly equivalent to the level of antigen present in the sample, the detectable signal rises in direct proportion to the amount of target analyte measured. The rapidly developing scientific area of nanoparticle labels, including the procedures used to synthesize, modify, and employ nanomaterials, opens up new opportunities for the development of improved analytical instruments and equipment. Nanoparticle labels, such as metallic nanoparticles and hybrid nanostructures, have been employed in the construction of sandwich immunosensors and immunoassay systems.

A MOF consists of a coordinated metal cation or cluster network of three dimensions and an organic ligand. They exhibit excellent crystalline structure, porous nature, high surface area, chemical characteristics, thermal and chemical stability. Recently, significant emphasis has been focused on metal-organic frames that are formed by the self-assembly of metal or nanostructured materials and flexible organic ligands [6,7]. For this reason, MOFs have the potential to be used in a wide range of applications, including catalysis, sensors, and gas separation, due to their high pore sizes and enormous specific surface areas [8,9], heterogeneous catalysis [10,11], drug delivery systems [12–14] and material adsorption and storage [15,16]. MOFs have recently employed fluorescence and electrochemical sensors for virus detection.

This chapter explores the recent advances of EC immunosensors based on MOFs and associated nanostructured materials in detecting various targets for the virus diseases diagnosis. The nanostructures/components of MOF-based materials and immunosensor operational processes are addressed. The difficulties of EC biosensors for the early detection of viruses are also addressed. This review will give understanding and insight into the application of MOF structures in electrochemical immunosensing.

12.2 VIRUSES

Viruses can cause deadly diseases such as rabies, dengue fever, hepatitis, Ebola, AIDS, avian influenza, SARS, and MERS. The viral load has been shown to significantly impact the severity of disorders [17]. Some viruses, including hepatitis B and C, can cause chronic diseases or malignancy in addition to acute disorders [18]. Transmission of viruses occurs through direct contact with infected individuals, contact with contaminated surfaces, and inhalation of virus particles in the air [19]. Viral infections account for one-third of deaths worldwide each year [20]. Severe acute respiratory syndrome coronavirus 2 (SARS-CoV-2) recently produced a pandemic that resulted in high mortality rates and an enormous economic disaster worldwide [21].

Over 1,500 animal and human viruses have been identified, with varying structure, content, and genome [22]. In addition to single-stranded DNA (ssDNA) or double-stranded RNA (dsRNA) variants, viruses have an external protein termed the capsid that include one or more structural proteins that are present in a viral genome [23]. Certain virus families have a nucleocapsid envelope that contains the genome and proteins linked with the nucleic acid. The envelope is composed of knobs of glycosylated membrane proteins and glycoproteins. The replication strategies of a virus's genome can be utilized to categorize the virus. The shape of the virus can also be used to categorize it as icosahedral, filamentous, enveloped, or having a head with tail. Up to this point, enveloping viruses like SARS coronavirus (SARS-CoV), avian influenza, Ebola and the Zika virus have been responsible for epidemics [24]. SARS-CoV-2, a recent global outbreak, expanded quickly and infected hundreds of millions of people, prompting the World Health Organization to declare it a pandemic [25]. Rapid and sensitive virus identification has become more important in equally advanced and emerging nations in order to prevent pandemics with viral diseases. Recent research into the construction of electrochemical immunosensors for viral detection is real viruses mentioned above and discussed in depth in the applications section.

12.3 ELECTROCHEMICAL IMMUNOSENSORS

Electrochemical immunosensors are becoming the main bioanalytical instruments that can distinguish various biomarkers for different diseases. They play an increasingly crucial role in the construction of point-of-care (POC) devices. The immunosensors are particularly sensitive due to their accurate antibody-antigen interaction and electrochemical transductions. Notably, the electrochemical sensing properties of an immunosensor rely on the well-designed immobilization of bio-recognizing components on electrode surfaces.

Immunosensors, also known as biosensors, detect antibodies because they include biological components linked to antibodies. The sensor's sensitivity and selectivity can be improved by using an antigen. The analyte has a strong affinity for the antibody, which enables it to recognize it even when other compounds are present in the environment. Most of this material is immobilized at the transducer component, which converts changes in the biological, physical, or chemical action into an electrical signal. An electrical component processes this signal and either digitizes or amplifies it. Figure 12.1 depicts a diagrammatic depiction of an immunosensor in action. The immunosensors are classified into three groups based on the transducer method used: electrochemical immunosensors, optical immunosensors, and piezoelectric immunosensors.

EC immunosensors have received significant interest in research and the business domain between all kinds of biosensors. Compared to mechanical and optical techniques, these approaches are preferred because of their excellent detection range, practical efficiency, and low cost of manufacture. In addition, EC immunosensors enable prototype miniaturization with uncompromising sensing performance. Due to their simple design, these systems can be easily integrated into diverse electronic devices for continuous monitoring. The remarkable performance of EC immunosensors in terms of limits of detection and linear ranges has also been shown, including sensor limits measured up to a sub-nanomolar level (picomolar-pM) and a more extended linear range than other forms of sensors. Nevertheless, the tendency towards extensive discovery and development has now outperformed the performance limitations of this classifications. Additionally, other benefits, including miniaturization and the ability of EC prototypes to integrate, distinguish EC biosensors from other kinds of commercial devices in terms of generating high-performance

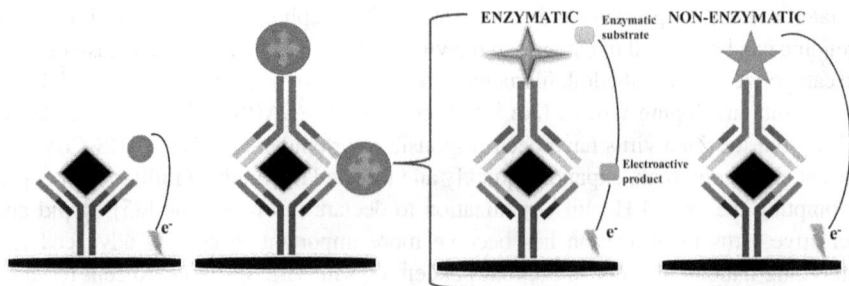

FIGURE 12.1 Schematic representation of an immunosensor construction.

commercial devices. Immunosensors are classified into two categories: labeled and label-free immunosensors. The label-free immunosensors are utilized to directly identify the immunocomplex (e.g., complex antibody structure) by detecting the physical modifications resulting from creating another complex. In contrast, an identifiable sensitivity label is included in the labeled immunosensor, and the immunocomplex is therefore sensitively identified by measuring the label.

Electrochemical immunosensors have effectively integrated immunosensors with electrochemical transducers in the last years. Electrochemical transducers are classified into two categories: amperometric transducers and potentiometric transducers. Amperometric transducers detect a continuous voltage current generated by an enzyme reaction proportional to the analyte concentration on the electrode. However, potentiometric transducers monitor changes in transmembrane potential after an antigen or antibody binding to a particular molecule in a solution [26].

The analyte concentration on surface of the active electrode influences the rate of electron flow. Electrochemical immunosensors typically use direct antibody immobilization followed by electrochemical monitoring analytes. The importance of immobilization varies depending on the target's molecular size. This method is appropriate for analytes of large molecular sizes but not for smaller compounds. Although electrodes' surface immobilization is unstable when antibodies are directly immobilized, this may lead to structural changes in smaller analyte molecules. False findings may thus be produced. Nanomaterial conjugation is typically used to prevent possible loss to the functional group when analytes are directly immobilized to circumvent this restriction.

Electrochemical impedance spectroscopy (EIS) is a very effective label-free technique. It is a useful method for examining biorecognition processes at interfaces between electrodes and electrolytes. This method is non-destructive and can detect sensitive and accurate signals compared to voltammetric biosensors. EIS evaluates electrical interference with less effect, and its reaction is precise, steady, repeatable, and dependable. During an immunoreaction, significant molecules can immobilize on a sensor's surface, changing the charge transfer resistance (R_{ct}). This immobilization hinders the transmission of the redox probe. Electrochemical impedimetric (ECI) immunosensors, which do not need enzyme labeling or a catalyst procedure, are widely employed in clinical diagnostics, food analysis, and environmental monitoring. The immobilization of the capture antibody on the electrode surface is a vital stage in the construction of immunosensors because the efficiency of the antibodies immobilized on the surface of the electrode may have a substantial impact on the analytical performance of immunosensors. The polymer matrix can be used to immobilize antibodies by covalent attachment, physical adsorption, and electrostatic entrapment.

12.4 MOFS UTILIZED IN BIOCOMPOSITES FOR SENSING PURPOSES

MOFs can be designed to have specific characteristics for encapsulating in order to preserve proteins and other biological components from destruction. Adsorption is crucial for entrapping bacteria and viruses. MOFs interact with biomolecules in sensing applications to produce bio-composites with many unique properties, such as nanostructured framework, water durability, and various sensing characteristics.

12.4.1 Structure on a Nanoscale

Most MOF materials are synthesized as micro-sized bulk crystalline at the beginning stages of their development. Because of drawbacks like poor solution dispersibility and insufficient catalytic properties, bulk MOFs have not been widely used in the sensing industry. As a result, nanoscale MOFs have been developed and synthesized, MOFs with a minimum one-dimensional size range of tens or hundreds of nanometers [27]. Compared to their bulk equivalent, nanoscaled MOFs have a larger surface area per unit weight, making them a potential matrix for immobilizing biomolecules. In addition, as the surface area of nanoscaled MOFs increased, so did the number of active sites that were readily accessible, enhancing the MOFs' catalytic activity. Meanwhile, the smaller, nanoscaled MOFs can penetrate the cell via endocytosis and be used to measure a target in vivo using a MOF biocomposite-based sensing platform. The new generation of biocomposite MOF sensors based on nanoscaled MOFs has the potential to substantially enhance MOF selectivity, test time, and application range [28]. Recent studies have shown that two-dimensional (2D) MOF nanosheets, which refer to reducing MOFs from nanometer levels in two dimensions to nanometer levels in one size, may increase the surface-volume ratio and indeed the number of accessible sites, thus improving biomolecule adsorption on the surface [29].

12.4.2 Stability of Water

The water stability of MOFs is critical for sensing applications based on MOF biocomposites. Most biomolecules are known to be water-soluble and hazardous to organic solvents. At this stage in the development process, the MOF structures are easily destroyed in water due to the weak ligand–metal interactions that exist in these MOFs. As a result, much research has focused on developing water-stable MOFs in recent years. In recent years, many MOFs exhibiting the property of water stability have been synthesized. These MOFs can be categorized as either zeolitic imidazolate frameworks (ZIFs), Materials Institute Lavoisiers (MILs), or porous coordination networks (PCNs). Metal carboxylate frameworks with high-valence metal ions, such as Fe^{3+}, Cr^{3+}, and Zr^{4+}, are one form of metal carboxylate framework. Since the coordination number and high charge density of the high-valence metal ions provide a stable coordination connection among the metal ions and ligands and a rigid framework, MOFs are less susceptible to water molecules substituting for organic ligands. Other metal azolate frameworks include tetrazolates, imidazolates, triazolates, and pyrazolates that contain nitrogen-donor molecules. In this class, the most notable example is the zeolitic imidazolate framework. Some ZIF-based biocomposites (such as ZIF-8 and ZIF-70) for detecting diverse targets have been reported recently [30]. Due to the apparent rapid growth of MOFs as a material, a more significant number of MOFs stabilized in water have been successfully made, resulting in improved MOF biocomposite sensor performance.

12.4.3 Sensing Characteristics

Colorimetric, fluorescent, and electrochemical signal reporters based on MOFs can be used in sensing applications because of their various sensing capabilities. It is possible to construct and modify MOF structures with sensing capabilities by rationally selecting organic ligands, metal ions, and synthesis methods. Different MOFs exhibiting various sensing capabilities have been developed and produced to provide biocomposite-based sensor applications. Based on the MOF structure, there are different types of MOFs and composites. Metal ions with integrated sensing capabilities are used as the metal source in the approach to synthesize MOFs. Due to various electrochemical catalytic activity, metal ions, including copper, nickel, cobalt, and cerium ions, were utilized in producing MOFs as an electrochemical signal transducer in bio-composite-based sensors. The addition of a ligand with a specific optical property enhances the optical sensing capabilities of the synthesized MOF. Furthermore, the process of immobilizing or encapsulating different components that function as signal reporters, such as hemin, porphyrins, ferrocene, metal ions, and platinum nanoparticles on the surface of the pores of MOFs, in order to form MOF composites with specific sensing capabilities.

Many synthetic methods have been developed by modifying such principles, including solvothermal, hydrothermal, sonochemical, and microwave-assisted techniques. Hydrothermal: For MOFs, an alternative to the conventional methods (e.g., solvothermal and hydrothermal processes) has been proposed: ultrasonic or microwave-assisted synthesis [31–33]. However, these newer methods are unreliable in terms of repeatability and productivity, impacting particle size [34,35]. Various techniques have been proposed to address these problems, such as direct mixing [36], mechanochemical synthesis [37], and free/wet-drying procedures [38].

12.4.4 Working Principles of MOFs in Electrochemical Immunosensors

The particular functionality of MOFs makes it possible for biomolecules to be attached in both covalent and non-covalent approaches. However, such alteration may not be possible in the case of other materials. MOFs have several adsorption and conjugation sites. These sites include pores, metal centers, hydrophobic ligand rings, and channels in the MOFs. On the other hand, other nanomaterials will find it challenging to create many potential conjugation sites. In addition to pores and metal centers, the MOFs may have several areas for adsorption and conjugation, including hydrophobic rings of the ligand, functional groups ($-NO_2$, -OH), -COOH, $—CONH_2$, $—SO_3H$, –F, and $–CF_3$) of the ligands, and channels in the MOFs [39]. In general, the attachment of biomolecules to metal nanoparticles (such as gold and silver nanoparticles) is facilitated by thiol(-SH) chemistry, adsorption, and electrostatic contact between the particles. In the case of carbonic nanostructures, on the other hand, it is more crucial to take into account hydrophobic, electrostatic, and covalent conjugation than different types of conjugation. Thus, MOFs' unique and remarkable properties were significantly helpful in increasing the chances of successful bioconjugation.

Furthermore, MOF combined with nucleic acid biocomposites are vital in biosensing applications. Nucleic acid-MOF biocomposites can be categorized between those that employ covalent chemical interaction and those that depend on surface adsorption and in situ encapsulation for protein attachment. Since nitrogenous bases and MOF aromatic subunits have extensive electrostatic interactions. On the other hand, MOFs have a low affinity for dsDNA and secondary and tertiary structured ssDNA. Crosslinking agents like glutaraldehyde can also be used to make covalent bonds involving the amino groups of the MOF and the amino groups of the ssDNA. Immunosensors that use MOFs immobilized with antibodies to capture target antigens and provide either an electrochemical or optical signal are often known as antibody-MOF biocomposite sensors. Additionally, the electrostatic interaction coupling antibodies and noble metal nanoparticles can be used to immobilize antibodies.

12.5   MOFS-BASED IMMUNOSENSORS VIRUS DISEASE DETECTION

MOFs have a huge surface area and can be modified structurally to build electrochemical immunosensors. It is an excellent tool for detecting viruses because of the low background signal. Several viruses, including the Avian Leukosis Virus (ALV) [40], HIV [41], Zika [42,43], hepatitis A virus (HAV) [44], Dengue [43], the Japanese Encephalitis Virus (JEV) [45] have been successfully identified using antigen detection techniques using MOF-based detection platforms (Table 12.1).

12.5.1   COVID-19

The World Health Organization (WHO) proclaimed a pandemic of the novel coronavirus disease (COVID-19), which is caused by the SARS-CoV-2 virus, on December 1, 2019. Although 5.94 million COVID-19 deaths were recorded from January 1, 2020, and December 31, 2021, researchers estimate that excess mortality caused 18.2 million deaths globally [48]. The FDA-approved S-protein detection diagnostic techniques identify SARS-CoV-2 proteins in respiratory secretions. They are now available as laboratory-based and near-patient testing.

Electrochemical immunosensors have recently gained popularity as a COVID-19 diagnostic tool owing to their adaptability, sensitivity, selectivity and quick reaction time, and affordable cost. These excellent features make these sensors ideal for accurate biomarkers monitoring and disease early detection. Electrochemical approaches were previously utilized to identify particular antibodies of various deadly viruses. However, surface interference caused by biomolecule adsorption is a significant difficulty with electroanalytical procedures, restricting their use. Designing an electrochemical immunosensor with particular nanomaterials or chemicals is one feasible option. It is critical to develop diagnostic characteristics that quickly identify infected patients and provide treatment options. Therefore, a new MOF-based (SiO2@UiO-66) label-free electrochemical immunosensor for SARS-CoV-2 spike protein was developed [49]. The developed immunosensor low detection value of 100.0 fg mL^{-1} with minimum sample preparation procedures, showing excellent selectivity and reproducibility of SARS-CoV-2 S-protein detection (5.0 min).

TABLE 12.1

Metal-Organic Frameworks (MOFs) for Immunosensors for Different Virus Detection

Virus	Type of Immunosensor/ Transducer	Recognition Matrix	Linear Range	Detection Limit (LOD)	Sample	Ref.
Ebola Virus	Ce-MOF nanoblock, AuNP-GCE	PCSK9	50 fg mL^{-1} to 10 ng mL^{-1}	19.12 $\pm$ 2.69 fg mL^{-1}	Human serum	[46]
Avian Leukosis Virus	rGO-TA-Fe$_3$O$_4$eZIF-Ab$_2$-HRP,DPV	ALV-J antigen^c	152 to 10,000 TCID$_{50}$ mL^{-1}	140 TCID$_{50}$ mL^{-1}	Human serum	[40]
Hepatitis B	Anti-HBsAg /Cu-NH$_2$BDC/GCE CV,DPV,EIS	HBsAg	1 to 500 ng mL^{-1}	0.73 ng mL^{-1}	Human serum	[47]

TCID$_{50}$ mL^{-1} (where TCID50 is the 50% tissue culture infective dose)

AuNP-GCE: Gold nanoparticles – Glassy carbon electrode; rGO-TA-Fe$_3$O$_4$eZIF-Ab$_2$-HRP, DPV: Reduced graphene oxide – tannic acid-ferric oxide nanoparticles zeolitic imidazolate frameworks, HRP-Horseradish peroxidase, DPV- Differential pulse voltammetry; Anti-HBsAg /Cu-NH$_2$BDC-Hepatitis B surface antibody, 2-aminoterephthalic acid.

12.5.2 AVIAN LEUKOSIS

ALV-J (Avian leukosis virus subgroup J) is the frequent natural avian retrovirus linked with neoplastic disorders, immunosuppression, and chicken tumors. Although it is challenging to eliminate, it is essential to rapidly identify ALV-J to separate healthy hens from infected flocks and prevent the transmission of the infection [50]. The use of MOFs in the development of immunosensors has been relatively restricted recently, characterized by the type of the functional MOF group, which usually hinders the covalent bindings of biomolecules, and MOFs' low molecular solubility, which limits electron transmission. The effective modification of the functional surface group and the molecule porosity of MOFs are required for potential widespread use in immunosensors. The zeolitic imidazolate framework (ZIF), amongst MOFs, is probably the most significant because of its adjustable pore sizes, its outstanding chemical stability, and its wide structural variety.

ALV-J is an avian retrovirus that has been identified by Ai and colleagues [40] have developed a sandwich-type electrochemical immunosensor. There were two main parts to this immunosensor: (1) the probe, which was constructed from hollow ZIF-8 structures and functionalized with Ab_2 and HRP (Horseradish peroxidase) on the exterior surface. Glassy carbon electrode (GCE) electrochemical-responsive interface created using tannic acid (TA), Fe_3O_4 nanoparticles, and primary antibodies functionalized with TA (Ab_1). During the synthesis, graphene oxide was reduced to conductive graphene using tannic acid. Tannic acid, which contains reactive cis-diols, was utilized to graft Ab_1 onto the GCE. Due to the Fe_3O_4 nanoparticles simulating HRP's oxidation characteristics, the immunosensor's LOD was improved. The usage of hollow ZIF-8, according to the authors, increased the material's electron transport characteristics, which in turn improved the immunosensor platform's responsiveness. According to the results, the detection range was between 152 to 10,000 times the tissue culture infective dose ($TCID_{50}$), with a detection value of 140 $TCID_{50}$ mL^{-1}. The chemically etched hollow ZIF-8 improved the MOF's electron transport characteristics, and indeed the research demonstrated its applicability in avian serum samples.

12.5.3 HUMAN DEFICIENCY VIRUS (HIV)

For HIV diagnosis, treatment, and AIDS prevention, early identification of HIV biomarkers or genes is critical [51]. Since it takes a long time for HIV antibodies to be produced, current AIDS diagnostic techniques like detecting HIV antibodies and HIV antigens have difficulty keeping up with the growing demand. There will be insufficient HIV antibodies present during this period to provide a conclusive diagnosis [52]. Further nucleic acid-based immunoassays have shown the efficacy of early HIV-1 detection techniques. MOFs' organic linkers have a conjugated-electron system and can form hydrogen bonds with single-stranded DNA. This allows the MOFs and DNA to interact appropriately. MOFs can identify DNA and other biomolecules based on changes in the electrochemical signal. These features expand the scope of MOFs' potential in early detecting malignancies and other disorders.

However, because of their minimal electrochemical reactivity with poor stability, the majority of MOFs possess inadequate practical use in electrochemical bio-sensors [53]. Fluorescence emission spectroscopy is now the most used MOF-based (ZIF-8) biosensor for HIV-1 gene detection [54]. MOFs are good precursors for in-situ pyrolysis-based synthesis of well-dispersed metallic nanoparticles encapsulated in porous graphitic carbon structures. In situ graphitic carbon exhibits strong electron mobility and a long conjugation structure [55]. Metal oxide/carbon nano-composites can be synthesized using MOFs as templates, which improves the electrochemical performance of MOFs [56].

Silver, gold nanoparticles, and carbon nanomaterials have often been employed in treating viral infection or detection. The HIV-1 DNA probe was immobilized on a bimetallic NiCo-based MOF, and the virus was detected electrochemically. Various MOF-on-MOF designs resulted in varied surface and electrochemical characteristics. EIS and DPV (differential pulse voltammetry) revealed the LOD as low as 0.84 and 0.66 pg/mL for Zn-MOF-on-Zr-MOF, Zr-MOF-on-Zn-MOF, respectively. In a recent study, Jia and colleagues found that a different bimetallic MOF, NiCo-based MOF can be utilized to test for HIV-1 DNA by being employed as a sensor scaffold [57]. It is essential to point out that the impacts of the pyrolysis environment on the compositional and sensing characteristics of the generated materials were studied. The findings revealed that various pyrolysis atmospheres (N_2 and H_2) produced pyrolysis products with distinct nanostructures and chemical characteristics, which influenced the fixing of probe DNA and the detection of the target molecule. According to the research results, the pyrolysis result of NiCo-MOF under N_2 was comprised of $NiCo_2O_4$ spinel and CoO. In contrast, the pyrolysis product under H_2 was devoid of CNTs owing to organic ligand degra-dation (i.e., $NiCo_2O_4$/CoO/CoO@CNTs). The proposed immunosensor demon-strated ultra-low detection with a LOD of 16.7 fM by using the intrinsic properties of $NiCo_2O_4$/CoO@CNTs, which include excellent electrical activity, high bio-compatibility, and a significant binding for the probe DNA.

12.5.4 ZIKA VIRUS

A single-stranded RNA virus with high neurotoxic and teratogenic potential, the Zika virus (ZIKV) has been found in the Flaviviridae family [58]. The majority of infected individuals, on the other hand, are asymptomatic or show symptoms that are com-parable to other forms of hyperthermia [59]. General serological methods have dif-ficulties detecting ZIKV because of the cross-reactivity between ZIKV and other members of the same virus family, such as the virus that causes Japanese encephalitis, dengue fever, and yellow fever [60]. Traditional RT-PCR is limited by complicated operational procedures and the time-consuming nature of the technique [61].

MOFs were also increasingly interested in an electrochemiluminescence (ECL), where the energy relaxation of excited species produces an electrochemically prompted optical light process since they may provide a very selective surface to load a considerable amount of ECL luminophores. In contrast to traditional metal com-plexes such as ECL luminophores, luminol, and quantum dots, MOFs built of por-phyrin metal as an organic ligand drew considerable attention to ECL luminophores due to their high effectiveness outstanding quality stabilities. Zhang et al. [42] have developed the RNA-detecting platform based on the ECL technology. They used metal-organic gel (MOG) as an electrode matrix and metal-organic fiber (MOF) as a nanotag, respectively. A Zr-based MOG was used to manufacture MOG, through post-synthesis Au nanoparticles and in situ graphitic carbon nitride loading in Zr-based MOG.

It utilizes Zika virus (ZIKV) RNA as a template analyte and an interchangeable ultra-sensitive solid-state ECL sensor. The ECL platform demonstrated a wide linear range from 0.3 nM to 3 mM, with a low detection value of 0.1 nM. Dengue and Zika virus RNA was detected using an RNA biosensor based on Cu-MOF technology. It's also capable of quenching fluorescently tagged particular probes and restoring fluorescence after the presence of the target. Using the single detection technique, this assay was able to identify DENV and ZIKV at concentrations as low as 332 pM and up to as high as 192 pM using the synchronous fluorescence detection method [62].

12.5.5 HEPATITIS B VIRUS

Hepatitis B virus (HBV) is a Hepanaviridae family DNA viral infection that causes hepatitis B infection [63]. This virus is the primary cause of hepatitis, including acute and chronic hepatitis, cirrhosis, and HCC. Despite the availability of a hepatitis B vaccination since 1982, new cases of infection are reported each year. Hepatitis B virus infection affects about 820 million individuals worldwide, leading to an increased risk for cirrhosis, liver failure, and hepatocellular cancer [64]. Hepatitis B may be prevented and controlled via developing an efficient and fast technique for diagnosing the disorder. Rezki et al. (2021) developed an additional HbsAg electrochemical immunosensor with a lower detection limit than the previously described work. This research used the solvothermal approach to produce amine-functionalized MOF nanospheres. As a result of their electroactive features, MOF nanomaterials enable sensitive detection due to open micro-and mesopores that allow for improved functionalization and the production of electrochemical signals. A covalent bond was formed between the antibodies and the aminated Cu-MOF nanospheres through EDC/NHS coupling. Electrochemical studies were carried out using CV (Cyclic voltammetry), EIS, and DPV, and the linear range was determined to be 1–500 ng mL^{-1}, with a limit of detection of 730 pg mL^{-1} HBsAg.

Copper-based MOFs have recently attracted a lot of attention in biological applications. Lin et al. [65] developed electroactive Cu-MOF for the ultra-sensitive electrochemical HBV DNA biosensor detection of HBV DNA. In addition, Cu-MOFs with strong -conjugate systems can interact with ErGO to enhance sensitivity, and this interaction adds to the overall stability of the sensor. The novel electrochemical biosensor demonstrated good performance in detecting HBV DNA in human serum samples with a dynamic range of 50.0–10.0 nM and a detection limit of 5.2 fM.

Electrochemiluminescence is electrochemical-initiated chemiluminescence, specifically the electrical initiation and relaxation of excited state chemicals' energies to generate optical radiation. Researchers have recently increased interest in the ECL sensor because of its benefits, such as excellent response controllability, broad linear range, and high sensitivity. ECL is a novel kind of immunosensor that combines ECL with immunological reactions. The benefits of this technique are also ECL technologies, often employed in drug analysis, DNA detection and protein detection, plant residues, or heavy metals. Electrochemiluminescence or electrogenerated chemiluminescence is the light activity in which the light produced by the electron transport process is measured when a potential applies to an electrode. The intensity of light of this analysis is dependent on the concentration of the total number of analytes involved. Their excellent sensitivity and selectivity make their biosensor application a viable option [66].

12.5.6 C-REACTIVE PROTEIN (CRP)

C-reactive protein (CRP) is a widely used diagnostic test for detecting cardiovascular disease and inflammation, and it is an acute-phase protein mainly produced by the liver [67]. Average plasma CRP concentrations are less than 1.0 mg/L, and the clinical diagnostic threshold in a healthy human is between 1–3 mg/L [68]. Plasma CRP levels are a significant factor in many disorders, including cardiovascular and inflammatory conditions. The early identification of CRP and pharmacological research, a precise and susceptible analytical methods system is highly beneficial. In addition to their usage as electrode elements for biomolecule immobilization and signaling molecules to enhance the efficacy of CRP immunosensors, a variety of nanomaterials have been tested for use in a variety of applications. ZnO nanoparticles are widely employed in the electrochemical sector owing to their unique properties, which include solid electron transport capabilities and exceptional chemical stability.

Dong's team developed [69] the carbon paste electrode (CPE) and immobilized anti-CRP using IL-dispersed MOFs (Zr-tdc) synthesized from Zr(IV) and 2,5-thiophenedicarboxylate ligand (H$_2$tdc). These researchers also fabricated a new ZnO/porous carbon matrix by thermolyzing a mixed-ligand MOF to detect CRP electrochemically in actual samples. Upon optimization, the EC immunosensor showed a LOD of 5.0 pgmL^{-1} (S/N = 3) in the linear range of 0.01–1,000 ng.mL^{-1}, as well as high level of stability, accuracy, and reproducibility. MOF frameworks have a higher concentration of Cu^{2+} than nanoparticle-based signal tags, which results in stronger electrochemical signals. Later, electroactive MOFs with large surface areas, such as Cu-MOFs and Fe-porphyrin-based MOFs, were utilized as nanocarriers. The redox mediators were employed directly in electrochemical biosensors, without the need for any post-processing or the incorporation of external redox signals. For example, Yang et al. [70] investigated using Cu-MOF as a signal tag for electrochemical immunosensor detection of CRP. The detecting antibodies were labeled with Au/Cu-MOF composites (Ab$_2$). Cu-MOFs were placed on the electrode surface to increase electrochemical signal outputs following the sandwich immunoreaction between CRP and Ab$_2$. The MOFs signal of the generated Cu^{2+} ions could be recognized instantly without the need for acid dissolution or preconcentration, resulting in a substantial reduction in detection stages and an overall increase in inefficiency. Since the 3D infinite extended MOF is loaded with Cu^{2+}, there are a lot of electrochemical signals generated from it. Metal ions function as active redox centers; thus, altering their valence or oxidation state may deteriorate MOF. The electrochemical immunosensor was constructed using the Fc-Zn-MOF signal tag for Ab detection. This sensor has enhanced analytical performance, with an operational range of 1 to 100 ng/mL^{-1} and a limit detection value of 0.03 pg/mL^{-1}.

A novel sandwich-based immunosensing test for the identification of CRP has been designed, which utilizes toluidine blue (Tb) as the signal-generating element and N-doped 3D carbon microstructures as the electrode platform. The proposed biosensor's sensitivity may be improved using N-doped bamboo-like carbon nanotubes. C-reactive protein concentrations were raised to enhance the electrochemical signal, and the LOD value was as high as 166.7 pg mL^{-1} [71].

12.6   OTHER DISEASES

In recent years, the Japanese encephalitis virus (JEV) has emerged as a serious mosquito-borne viral encephalitis in several Asian countries, with a high mortality rate and a serious potential for re-infection. JEV poses a threat to 25 countries, and about 3 billion people, including 700 million children under 15, reside in such high-risk regions [72]. Yang et al. (2020) developed a fluorescent molecularly imprinted (MIP) sensing material that is based on MOF identification of the JEV. During the tests, it was discovered that it has excellent sensitivity and selectivity and that its highest detection limit is 13.0 pmol/L.

The hollow nano box-MOF/AuPt alloy was used to develop the electrochemical immunosensor to detect the protein LAG-3 [73]. The first step was to synthesize ZIF-67 nanocubes in water using surfactants $Co(NO_3)_2$ and 2MI in the presence of CTAB. Hollow nano boxes were generated after applying ultrasonic/thermal treatment to a solution containing the prior MOF. AuPt alloy was formed by reacting H_2PtCl_6 and $HAuCl_4$ with the previously developed hollow nano boxes and a reducing agent, $NaBH_4$, to form the alloy layer. rGO-SnO_2 nanosheets were used to immobilize this material, then coated on a modified GCE electrode. Streptavidin preferentially binds the alloy, and biotin-modified antibodies are subsequently incubated with the electrode. The sensitivity of the modified electrode was increased by incubating it with LAG-3 and antibody-modified silica nanoparticles, which enhanced electrochemical signals by lowering H_2O_2. The biosensor was utilized to identify the analyte in specimens, and it demonstrated no cross-reactivity with other proteins at a detection limit of 1.1 pg/mL, indicating that it was effective.

The biomarker galectin-3 (Gal-3) was detected utilizing an Au@MIL-88(Fe)-NH$_2$ functionalized GCE electrode as an electrochemical sandwich immunosensor [74]. The MOF was synthesized by reacting $FeCl_3$ and 2ATPA with DMF and acetic acid to form a solution. The AuNPs@MOF particles were formed by incubating the material with $HAuCl_4$ and reducing it with $NaBH_4$. This hybrid material was developed by combining the surface-functionalized electrodes with N-doped graphene nanoribbons, which have been demonstrated to enhance conductivity performance. Prior to incubation with the analyte, the surface of the functionalized electrode was coated with antibody Gal-3. The electrochemical signal was obtained by performing a sandwich-type experiment using methylene blue crystals coated with AuPt NPs and then functionalized with Gal-3-Ab$_2$. This study showed a 100 fg/mL–50 nmol^{-1} linear response, with a LOD of 33.33 fg/mL. Potential interfering species generated a signal fluctuation of no more than 1.55% at high concentrations, and the biosensor demonstrated excellent repeatability (RSD 2.75%) across various electrodes. The recovery rate for spiked serum samples was more than 97.99%.

12.7 FUTURE CHALLENGES AND OPPORTUNITIES IN THE FIELD OF METAL-ORGANIC FRAMEWORKS

MOFs and their derived materials in sensing have been extensively discussed, with particular attention paid to their use in electrochemical sensing. MOF sensors have proven to be quite effective at identifying various organic ions and compounds. Different MOF-based composites used in the sensing process were examined as part of this review. Nanocomposite components such as metal/metal oxide NPs, and graphene derivatives should work together synergistically to improve electrochemical detection of different biomarkers in viral diseases. The use of various functionalized MOFs in future research may help composite-based sensors enhance their recognition ability. Additionally, combining several NPs with various benefits may substantially improve the sensing area's efficiency.

Despite the fact that MOF-based materials have shown tremendous promise, more study is needed to enhance their analytical sensitivity, stability, selectivity, and anti-interference characteristics. Furthermore, the critical obstacles with MOF-based viral recognition and diagnostic methodologies are detecting limits, unstable of specific MOFs in aqueous medium and unique physiological conditions, toxic biological effects of particular metal ions in MOFs, and complex in-vivo environments. MOFs have a detection limit in clinical samples that is too low to identify viruses at low doses, reducing sensitivity. Future research should focus on constructing MOF structures with smaller particle sizes to improve fluorescence signal amplification while maintaining high selectivity and quenching efficiency; combining viral RNA into the construction of innovative biosensors using metal nanoparticles such as metal and carbon materials, MOF-derived porous carbon; and developing automatic high – sensitivity MOF-based biosensors with high selectivity. The addition of conductive functional components to MOFs, such as carbon nanomaterials or metal nanoparticles, would significantly enhance high sensitivity by increasing the homogeneous dispersion and the exposition of more active sites.

In the future, researchers should focus on the following to make progress: designing MOF structures with smaller particle sizes to boost fluorescence signals and improve quenching efficiency and selection, including viral RNA integration, implementing innovative immunosensors using nanoparticles like MOF-derived porous carbon, and developing automated ultrasensitive MOF-based immunosensors with rapid detection on a large scale, incorporating electrochemistry technology. Sonochemistry-based green synthesis pathways should be investigated as an additional advantage of MOFs over graphene oxide-based biosensor platforms in detecting target ligands (aptamers).

12.8 CONCLUSION

In recent years, the amount of research on MOF composite-based sensors has risen significantly. MOF composite sensing applications and synergistic processes that improve sensing performance on viral detections were highlighted in this study. MOF composite sensors have excellent sensitivity, selectivity, and stability due to the synergy of their component MOFs' characteristics. The characteristics of MOFs that synergistically improve sensing performance are specifically examined. This study will be a valuable resource for creating cutting-edge virus detection systems using the MOF's exciting new properties. MOFs-based electrochemical immunosensors have interesting applications, and electrochemical-related research is expected to increase rapidly in the next several years as a result of nanoscience advancements.

ACKNOWLEDGMENTS

The authors grateful to the Coordination for the Improvement of Higher Education Personnel (CAPES)-Brazil, PROJECT DATA: CORPORATE NAME: Reinaldo Barreto Oriá, Project Title: Risk factors, outcomes and impact of COVID-19 in patients with chronic diseases in Ceará, Process number: 88881.505364/202001 CAPESEPIDEMICS Program for financial support.

REFERENCES

[1] Tatem AJ, Rogers DJ, Hay SI. Global transport networks and infectious disease spread. *Adv Parasitol* 2006;62. 10.1016/S0065-308X(05)62009-X

[2] Verikios G. The dynamic effects of infectious disease outbreaks: The case of pandemic influenza and human coronavirus. *Socioecon Plann Sci* 2020;71:100898. 10.1016/j.seps.2020.100898

[3] Pei X, Zhang B, Tang J, Liu B, Lai W, Tang D. Sandwich-type immunosensors and immunoassays exploiting nanostructure labels: A review. *Anal Chim Acta* 2013; 758:1–18. 10.1016/j.aca.2012.10.060

[4] Tang D, Niessner R, Knopp D. Flow-injection electrochemical immunosensor for the detection of human IgG based on glucose oxidase-derivated biomimetic interface. *Biosens Bioelectron* 2009;24:2125–2130. 10.1016/j.bios.2008.11.008

[5] Tang D, Yuan R, Chai Y. Ultrasensitive electrochemical immunosensor for clinical immunoassay using thionine-doped magnetic gold nanospheres as labels and horseradish peroxidase as enhancer. *Anal Chem* 2008;80:1582–1588. 10.1021/ac702217m

[6] Wu LL, Wang Z, Zhao SN, Meng X, Song XZ, Feng J, et al. A metal-organic framework/DNA hybrid system as a novel fluorescent biosensor for mercury(II) Ion detection. *Chem – A Eur J* 2016;22. 10.1002/chem.201503335

[7] Peng Z, Jiang Z, Huang X, Li Y. A novel electrochemical sensor of tryptophan based on silver nanoparticles/metal-organic framework composite modified glassy carbon electrode. *RSC Adv* 2016;6. 10.1039/c5ra25251b

[8] Xue DX, Wang Q, Bai J. Amide-functionalized metal-organic frameworks: Syntheses, structures and improved gas storage and separation properties. *Coord Chem Rev* 2019;378. 10.1016/j.ccr.2017.10.026

[9] Kazemi S, Safarifard V. Carbon dioxide capture on metal-organic frameworks with amide-decorated pores. *Nanochem Res* 2018;3. 10.22036/ncr.2018.01.007

[10] Beheshti S, Safarifard V, Morsali A. Isoreticular interpenetrated pillared-layer microporous metal-organic framework as a highly effective catalyst for three-component synthesis of pyrano[2,3-d]pyrimidines. *Inorg Chem Commun* 2018;94:80–84. 10.1016/J.INOCHE.2018.06.002

[11] Hu ML, Safarifard V, Doustkhah E, Rostamnia S, Morsali A, Nouruzi N, et al. Taking organic reactions over metal-organic frameworks as heterogeneous catalysis. *Microporous Mesoporous Mater* 2018;256:111–127. 10.1016/J.MICROMESO.2017.07.057

[12] Luo Z, Jiang L, Yang S, Li Z, Soh WMW, Zheng L, et al. Light-induced redox-responsive smart drug delivery system by using selenium-containing polymer@MOF shell/core nanocomposite. *Adv Healthc Mater* 2019;8. 10.1002/adhm.201900406

[13] Wang D, Zhao C, Gao G, Xu L, Wang G, Zhu P. Multifunctional NaLnF4@MOF-ln nanocomposites with dual-mode luminescence for drug delivery and cell imaging. *Nanomaterials* 2019;9. 10.3390/nano9091274

[14] Suresh K, Matzger AJ. Enhanced drug delivery by dissolution of amorphous drug encapsulated in a water unstable metal–organic framework (MOF). *Angew Chemie – Int Ed* 2019;58. 10.1002/anie.201907652

[15] Sohouli E, Sadeghpour Karimi M, Marzi Khosrowshahi E, Rahimi-Nasrabadi M, Ahmadi F. Fabrication of an electrochemical mesalazine sensor based on ZIF-67. *Meas J Int Meas Confed* 2020;165. 10.1016/j.measurement.2020.108140

[16] Seyfi Hasankola Z, Rahimi R, Shayegan H, Moradi E, Safarifard V. Removal of Hg2+ heavy metal ion using a highly stable mesoporous porphyrinic zirconium metal-organic framework. *Inorganica Chim Acta* 2020;501. 10.1016/j.ica.2019.119264

[17] Vermisoglou E, Panáček D, Jayaramulu K, Pykal M, Frébort I, Kolář M, et al. Human virus detection with graphene-based materials. *Biosens Bioelectron* 2020;166:112436. 10.1016/J.BIOS.2020.112436

[18] Liu Z, Shi O, Zhang T, Jin L, Chen X. Disease burden of viral hepatitis A, B, C and E: A systematic analysis. *J Viral Hepat* 2020;27:1284–1296. 10.1111/jvh.13371

[19] Castaño N, Cordts SC, Kurosu Jalil M, Zhang KS, Koppaka S, Bick AD, et al. Fomite transmission, physicochemical origin of virus–surface interactions, and disinfection strategies for enveloped viruses with applications to SARS-CoV-2. *ACS Omega* 2021;6:6509–6527. 10.1021/acsomega.0c06335

[20] Moriyama M, Hugentobler WJ, Iwasaki A. Seasonality of respiratory viral infections. *Annu Rev Virol* 2020;7:83–101. 10.1146/annurev-virology-012420-022445

[21] Rasmi SAB, Türkay M. Introduction to aggregate planning and strategies, 2021, pp. 1–15. 10.1007/978-3-030-58118-3_1

[22] Greening GE, Cannon JL. Human and animal viruses in food (Including Taxonomy of Enteric Viruses). *Viruses in Foods*, Cham: Springer International Publishing; 2016, pp. 5–57. 10.1007/978-3-319-30723-7_2

[23] Pashchenko O, Shelby T, Banerjee T, Santra S. A comparison of optical, electrochemical, magnetic, and colorimetric point-of-care biosensors for infectious disease diagnosis. *ACS Infect Dis* 2018;4:1162–1178. 10.1021/acsinfecdis.8b00023

[24] Siddharta A, Pfaender S, Vielle NJ, Dijkman R, Friesland M, Becker B, et al. Virucidal activity of world health organization–recommended formulations against enveloped viruses, including zika, ebola, and emerging coronaviruses. *J Infect Dis* 2017;215:902–906. 10.1093/infdis/jix046

[25] Spinelli A, Pellino G. COVID-19 pandemic: Perspectives on an unfolding crisis. *Br J Surg* 2020;107:785–787. 10.1002/bjs.11627

[26] Balahura L-R, Stefan-Van Staden R-I, Van Staden JF, Aboul-Enein HY. Advances in immunosensors for clinical applications. *J Immunoass Immunochem* 2019;40:40–51. 10.1080/15321819.2018.1543704

[27] He C, Liu D, Lin W. Nanomedicine applications of hybrid nanomaterials built from metal–ligand coordination bonds: Nanoscale metal–organic frameworks and nanoscale coordination polymers. *Chem Rev* 2015;115:11079–11108. 10.1021/acs.chemrev.5b00125

[28] Wang Z, Dong P, Sun Z, Sun C, Bu H, Han J, et al. NH 2 -Ni-MOF electrocatalysts with tunable size/morphology for ultrasensitive C-reactive protein detection via an aptamer binding induced DNA walker–antibody sandwich assay. *J Mater Chem B* 2018;6:2426–2431. 10.1039/C8TB00373D

[29] Zhao M, Wang Y, Ma Q, Huang Y, Zhang X, Ping J, et al. Ultrathin 2D metal-organic framework nanosheets. *Adv Mater* 2015;27:7372–7378. 10.1002/adma.201503648

[30] Lyu F, Zhang Y, Zare RN, Ge J, Liu Z. One-pot synthesis of protein-embedded metal–organic frameworks with enhanced biological activities. *Nano Lett* 2014; 14:5761–5765. 10.1021/nl5026419

[31] Xiang Z, Cao D, Shao X, Wang W, Zhang J, Wu W. Facile preparation of high-capacity hydrogen storage metal-organic frameworks: A combination of microwave-assisted solvothermal synthesis and supercritical activation. *Chem Eng Sci* 2010;65. 10.1016/j.ces.2010.02.005

[32] Khan NA, Jhung SH. Facile syntheses of metal-organic framework Cu3(BTC)2(H2O)3 under ultrasound. *Bull Korean Chem Soc* 2009;30. 10.5012/bkcs.2009.30.12.2921

[33] Ni Z, Masel RI. Rapid production of metal-organic frameworks via microwave-assisted solvothermal synthesis. *J Am Chem Soc* 2006;128. 10.1021/ja0635231

[34] Klinowski J, Almeida Paz FA, Silva P, Rocha J. Microwave-assisted synthesis of metal–organic frameworks. *Dalt Trans* 2011;40:321–330. 10.1039/C0DT00708K

[35] Schlesinger M, Schulze S, Hietschold M, Mehring M. Evaluation of synthetic methods for microporous metal-organic frameworks exemplified by the competitive formation of [Cu2(btc)3(H2O)3] and [Cu2(btc)(OH)(H2O)]. *Microporous Mesoporous Mater* 2010;132. 10.1016/j.micromeso.2010.02.008

[36] Zhuang JL, Ceglarek D, Pethuraj S, Terfort A. Rapid room-temperature synthesis of metal-organic framework HKUST-1 crystals in bulk and as oriented and patterned thin films. *Adv Funct Mater* 2011;21. 10.1002/adfm.201002529

[37] Yang H, Orefuwa S, Goudy A. Study of mechanochemical synthesis in the formation of the metal-organic framework Cu3(BTC)2 for hydrogen storage. *Microporous Mesoporous Mater* 2011;143. 10.1016/j.micromeso.2011.02.003

[38] Wee LH, Lohe MR, Janssens N, Kaskel S, Martens JA. Fine tuning of the metal–organic framework Cu3(BTC)2 HKUST-1 crystal size in the 100 nm to 5 micron range. *J Mater Chem* 2012;22:13742. 10.1039/c2jm31536j

[39] Bai K-P, Zhou L-J, Yang G-P, Cao M-X, Wang Y-Y. Four new metal-organic frameworks based on diverse metal clusters: Syntheses, structures, luminescent sensing and dye adsorption properties. *J Solid State Chem* 2020;287:121336. 10.101 6/j.jssc.2020.121336

[40] Liu C, Dong J, Ning S, Hou J, Waterhouse GIN, Cheng Z, et al. An electrochemical immunosensor based on an etched zeolitic imidazolate framework for detection of avian leukosis virus subgroup J. *Microchim Acta* 2018;185:423. 10.1007/s00604-018-2930-3

[41] Du M, Li N, Mao G, Liu Y, Wang X, Tian S, et al. Self-assembled fluorescent Ce(III) coordination polymer as ratiometric probe for HIV antigen detection. *Anal Chim Acta* 2019;1084:116–122. 10.1016/j.aca.2019.08.010

[42] Zhang YW, Liu WS, Chen JS, Niu HL, Mao CJ, Jin BK. Metal-organic gel and metal-organic framework based switchable electrochemiluminescence RNA sensing platform for Zika virus. *Sensors Actuators B Chem* 2020;321:128456. 10.1016/ J.SNB.2020.128456

[43] Xie BP, Qiu GH, Hu PP, Liang Z, Liang YM, Sun B, et al. Simultaneous detection of Dengue and Zika virus RNA sequences with a three-dimensional Cu-based zwitterionic metal–organic framework, comparison of single and synchronous fluorescence analysis. *Sensors Actuators, B Chem* 2018;254:1133–1140. 10.1016/j.snb.2017.06.085

[44] Luo L, Zhang F, Chen C, Cai C. Molecular imprinting resonance light scattering nanoprobes based on pH-responsive metal-organic framework for determination of hepatitis A virus. *Microchim Acta* 2020;187:140. 10.1007/s00604-020-4122-1

[45] Yang J, Feng W, Liang K, Chen C, Cai C. A novel fluorescence molecularly imprinted sensor for Japanese encephalitis virus detection based on metal organic frameworks and passivation-enhanced selectivity. *Talanta* 2020;212:120744. 10.1016/j.talanta.2020.120744

[46] Zhou Y, He J, Zhang C, Li J, Fu X, Mao W, et al. Novel Ce(III) – Metal organic framework with a luminescent property to fabricate an electrochemiluminescence immunosensor. *ACS Appl Mater Interfaces* 2020;12:338–346. 10.1021/acsami.9b19246

[47] Rezki M, Septiani NLW, Iqbal M, Harimurti S, Sambegoro P, Adhika DR, et al. Amine-functionalized Cu-MOF nanospheres towards label-free hepatitis B surface antigen electrochemical immunosensors. *J Mater Chem B* 2021;9:5711–5721. 10.1039/D1TB00222H

[48] Larkin HD. Global COVID-19 death toll may be triple the reported deaths. *JAMA* 2022;327:1438. 10.1001/jama.2022.4767

[49] Mehmandoust M, Gumus ZP, Soylak M, Erk N. Electrochemical immunosensor for rapid and highly sensitive detection of SARS-CoV-2 antigen in the nasal sample. *Talanta* 2022;240:123211. 10.1016/j.talanta.2022.123211

[50] Zhou G, Cai W, Liu X, Niu C, Gao C, Si C, et al. A duplex real-time reverse transcription polymerase chain reaction for the detection and quantitation of avian leukosis virus subgroups A and B. *J Virol Methods* 2011;173. 10.1016/j.jviromet.2011.02.017

[51] Ambinder RF, Bhatia K, Martinez-Maza O, Mitsuyasu R. Cancer biomarkers in HIV patients. *Curr Opin HIV AIDS* 2010;5:531–537. 10.1097/COH.0b013e32833f327e

[52] Zhao W-W, Han Y-M, Zhu Y-C, Zhang N, Xu J-J, Chen H-Y. DNA labeling generates a unique amplification probe for sensitive photoelectrochemical immunoassay of HIV-1 p24 antigen. *Anal Chem* 2015;87:5496–5499. 10.1021/acs.analchem.5b01360

[53] Du M, Chen M, Yang X-G, Wen J, Wang X, Fang S-M, et al. A channel-type mesoporous In(<scp>iii</scp>)–carboxylate coordination framework with high physicochemical stability for use as an electrode material in supercapacitors. *J Mater Chem A* 2014;2:9828–9834. 10.1039/C4TA00963K

[54] Pan Y, Zhan S, Xia F. Zeolitic imidazolate framework-based biosensor for detection of HIV-1 DNA. *Anal Biochem* 2018;546:5–9. 10.1016/j.ab.2018.01.017

[55] Wang Y, Shi R, Lin J, Zhu Y. Enhancement of photocurrent and photocatalytic activity of ZnO hybridized with graphite-like C3N4. *Energy Environ Sci* 2011;4:2922. 10.1039/c0ee00825g

[56] Liu J, Hou M, Yi J, Guo S, Wang C, Xia Y. Improving the electrochemical performance of layered lithium-rich transition-metal oxides by controlling the structural defects. *Energy Environ Sci* 2014;7:705–714. 10.1039/C3EE41664J

[57] Jia Z, Ma Y, Yang L, Guo C, Zhou N, Wang M, et al. NiCo2O4 spinel embedded with carbon nanotubes derived from bimetallic NiCo metal-organic framework for the ultrasensitive detection of human immune deficiency virus-1 gene. *Biosens Bioelectron* 2019;133:55–63. 10.1016/j.bios.2019.03.030

[58] Adegoke O, Morita M, Kato T, Ito M, Suzuki T, Park EY. Localized surface plasmon resonance-mediated fluorescence signals in plasmonic nanoparticle-quantum dot hybrids for ultrasensitive Zika virus RNA detection via hairpin hybridization assays. *Biosens Bioelectron* 2017;94:513–522.

[59] Faria HAM, Zucolotto V. Label-free electrochemical DNA biosensor for zika virus identification. *Biosens Bioelectron* 2019;131:149–155. 10.1016/j.bios.2019.02.018

[60] Cabral-Miranda G, Cardoso AR, Ferreira LCS, Sales MGF, Bachmann MF. Biosensor-based selective detection of Zika virus specific antibodies in infected individuals. *Biosens Bioelectron* 2018;113:101–107. 10.1016/j.bios.2018.04.058

[61] Liu S-Q, Li X, Deng C-L, Yuan Z-M, Zhang B. Development and evaluation of one-step multiplex real-time RT-PCR assay for simultaneous detection of Zika virus and Chikungunya virus. *J Med Virol* 2018;90:389–396. 10.1002/jmv.24970

[62] Xie B-PP, Qiu G-HH, Hu P-PP, Liang Z, Liang Y-MM, Sun B, et al. Simultaneous detection of Dengue and Zika virus RNA sequences with a three-dimensional Cu-based zwitterionic metal–organic framework, comparison of single and synchronous fluorescence analysis. *Sensors Actuators B Chem* 2018;254:1133–1140. 10.1016/j.snb.2017.06.085

[63] Liang TJ. Hepatitis B: The virus and disease. *Hepatology* 2009;49:S13–S21. 10.1002/hep.22881

[64] Chen J, Chen Q, Gao C, Zhang M, Qin B, Qiu H. A SiO 2 NP–DNA/silver nanocluster sandwich structure-enhanced fluorescence polarization biosensor for amplified detection of hepatitis B virus DNA. *J Mater Chem B* 2015;3:964–967. 10.1039/C4TB01875C

[65] Lin X, Lian X, Luo B, Huang X-C. A highly sensitive and stable electrochemical HBV DNA biosensor based on ErGO-supported Cu-MOF. *Inorg Chem Commun* 2020;119:108095. 10.1016/j.inoche.2020.108095

[66] Munge BS, Stracensky T, Gamez K, DiBiase D, Rusling JF. Multiplex immunosensor arrays for electrochemical detection of cancer biomarker proteins. *Electroanalysis* 2016;28:2644–2658. 10.1002/elan.201600183

[67] Kuo YC, Lee CK, Lin CT. Improving sensitivity of a miniaturized label-free electrochemical biosensor using zigzag electrodes. *Biosens Bioelectron* 2018;103:130–137. 10.1016/J.BIOS.2017.11.065

[68] Kowalczyk A, Sęk JP, Kasprzak A, Poplawska M, Grudzinski IP, Nowicka AM. Occlusion phenomenon of redox probe by protein as a way of voltammetric detection of non-electroactive C-reactive protein. *Biosens Bioelectron* 2018;117:232–239. 10.1016/j.bios.2018.06.019

[69] Dong S, Zhang D, Cui H, Huang T. ZnO/porous carbon composite from a mixed-ligand MOF for ultrasensitive electrochemical immunosensing of C-reactive protein. *Sensors Actuators, B Chem* 2019;284. 10.1016/j.snb.2018.12.150

[70] Liu T-ZZ, Hu R, Zhang X-BB, Zhang K-LL, Liu Y, Zhang X-BB, et al. Metal–organic framework nanomaterials as novel signal probes for electron transfer mediated ultrasensitive electrochemical immunoassay. *Anal Chem* 2016;88:12516–12523. 10.1021/acs.analchem.6b04191

[71] Li M, Xia X, Meng S, Ma Y, Yang T, Yang Y, et al. An electrochemical immunosensor coupling a bamboo-like carbon nanostructure substrate with toluidine blue-functionalized Cu(ii)-MOFs as signal probes for a C-reactive protein assay. *RSC Adv* 2021;11:6699–6708. 10.1039/D0RA09496J

[72] Coker RJ, Hunter BM, Rudge JW, Liverani M, Hanvoravongchai P. Emerging infectious diseases in southeast Asia: Regional challenges to control. *Lancet* 2011;377. 10.1016/S0140-6736(10)62004-1

[73] Xu W, Qin Z, Hao Y, He Q, Chen S, Zhang Z, et al. A signal-decreased electrochemical immunosensor for the sensitive detection of LAG-3 protein based on a hollow nanobox-MOFs/AuPt alloy. *Biosens Bioelectron* 2018;113:148–156. 10.1016/j.bios.2018.05.010

[74] Janegitz BC, Silva TA, Wong A, Ribovski L, Vicentini FC, Taboada Sotomayor M del P, et al. The application of graphene for in vitro and in vivo electrochemical biosensing. *Biosens Bioelectron* 2017;89:224–233. 10.1016/j.bios.2016.03.026

13 Future Challenges and Opportunities in the Field of Metal-Organic Frameworks

Naresh Kumar and Nidhi Goel

CONTENTS

13.1 INTRODUCTION

Porous materials or metal organic frameworks, particularly chemically and thermally stable, contain regions of empty space and have wide domestic as well as industrial applications such as energy storage, catalyst, adsorbent, gas separation, super-capacitors, fuel cells, sensor, ion-exchange and various other life-saving things [1]. The properties of MOFs can be controlled by both metal and linkers [2]. In 1830, Zeise's reported the first inorganic/organic hybrid organometallic platinum species. The synthesized complex is well known as Zeise's salt, which triggers the field of organometallic chemistry as well as the curiosity towards the metal–organic interface. Later, these kinds of materials were explored for their physical properties by expanding the chemical connectivity towards the higher one/two/three dimensions in crystalline as well as amorphous structures [3]. Inspite of entirely inorganic, silicious

DOI: 10.1201/9781003252061-13

zeolites provided a landmark in the field and revealed the process of obtaining the novel chemical properties with desired porosity as well as the potential of designed scaffold bearing heterogeneous catalytic sites [4]. Expansion from aluminosilicates to bulky chalcogenides, superior organic anions as well as the metals ahead of IVth group helped to design and synthesizes the novel materials with isostructure and narrative topology which provide the extraordinary connectivity [5]. Introduction of organic bridging ligands in the porous coordination complexes invented a new family of materials that are popularly called metal-organic frameworks (MOFs). The 21st century has brought a massive growth in the synthetic strategies, characterization and industrial applications of MOFs [6]. They are made up of clusters/metal cations linked through coordination bonding with polytopic organic linkers. The metal ion serves as a base/arm for linker molecules in endless combinations to make the repeating cage like hollow structures, which contain a surprisingly huge internal surface area for various potential applications. Scientists have synthesized MOFs with a surface area of over 7,800 m^2/g [7]. Li and co-researchers provided the first breakthroughs to synthesize MOFs with permanent microporosity restraining Langmuir surface area of 310 m^2/g [8]. Later on, scientists succeeded to design and synthesized the various MOFs with enhanced surface area upto 3,800 and 5,200 m^2/g in 2005 and 2009, respectively [9,10]. The combination of above-mentioned properties along with the control offered by MOFs in the pore space offers multifunctional compounds which is hard to get with other class of materials [11]. Thus, MOFs render the exclusive atomic-level structural diversity as compared to other porous materials that leads towards the uniform and tunable porosity, flexible network topology, particular geometry, and chemical functionality. Although the syntheses of these materials are mainly focused on the properties related to porosity but they have the potential far beyond this, which is still needed for exploration. This chapter explores the challenges with MOFs and their plausible solutions for the betterment of human life.

13.2   CHALLENGES IN THE SYNTHESIS OF MOFS

The various synthetic approaches significantly influence the electrochemical properties of synthesized MOFs. Synthesis of MOFs template and their transformation process into targeted porous material with functionalized groups are the two main points which should be considered prior to design the desired materials [12]. The amendment of metal source and organic linkers/ligands in appropriate solvent system results in the synthesis of desired novel porous MOFS with various structures and porosity [13]. Thus, to create the novel future porous materials (MOFs) with enormous possible topologies, the choice of synthetic route is critically important. The following paragraphs provide an overview of some basic challenges encountered during the synthesis of MOFs.

13.2.1   Approaches Towards the Templates/Precursors Synthesis of MOFs

The fabrication of well-controlled functionalized MOFs chiefly depends on its components, such as template/precursor and metal ions. The various methods used for the synthesis of templates/precursors includes electrospinning, ligand competition growth, epitaxial growth, self-templated solvothermal approach, self-sacrificing template method, interfacial engineering, directional engraving, spray-drying techniques, microwave-assisted synthesis and emulsion-based interfacial assembly. The MOFs provide the high porosity, copious unsaturated metal sites and large surface area but they display low electronic conductivity [14]. Making hierarchical MOFs templates/precursors with complex constructions is feasible by using the directional engraving process. The method mainly depends on the selection of an appropriate oxidative linker. Unique etching routes (inward or outward) can be preferred by adjusting the growth of oxidative species for the preparation of a hollow or yolk-shell like MOFs. In spite of the bond shearing via engraving with oxidizing agents like nitric acid, peroxymonosulfate etc. under diverse solvent conditions such as acidic, basic and neutral, the etched MOF crystals remain inclusive microporosity and crystallinity. Likewise, the epitaxial growth can also lead to the formation of hollow MOFs structures. As the outer shell grows epitaxially on the internal core, the central yolk-shell particle collapses by etching leaving a cavity, which then spreads out gradually to form the hollow MOFs. In addition to serving as a protective layer, the thin outer shell could be functionalized to increase active sites and increase skeleton stability [15–17]. The simple emulsion-based approach can also use to synthesize the hollow MOFs spheres with adjustable shell thickness [18]. This technology uses the micro- and nano-sized emulsion droplets which serve both as micro-reactor and transient spherical templates/precursors and leads towards the construction of different nanomaterials@MOFs with the perfect maintenance of their distinctive morphologies. The thickness of shell may be attuned by the alteration of precursor concentration and time of crystallization [19]. Self-sacrificing template method is also an effective method to synthesize MOFs. In the growth process, raw materials can be used as templates and sacrificed to produce MOFs with good confinement. In addition to the excellent conductivity and increased active sites that the obtained MOFs exhibit, they also retain a number of advantages that lead to fast mass/charge transfer and enhanced electrochemical performance [20,21]. Spray-drying technique is a low cost and fast method extensively used to fabricate the self-assembly nano-MOFs. In this atomizing process, the solution contain templates/precursors are injecting simultaneously with the compressing gas such as inert gas, air at different speeds to get the micro-droplets spray. A heated gas stream treats the suspended precursor droplet, making the solvent rapidly evaporate and the precursors disperse toward the surface radially. Slow increase in the precursor(s) concentration on these compressive droplets with evaporation results in the fabrication of MOFs crystal as the value reached the critical concentration. The resulted MOF crystals possess limited mobility and accumulate on the surface thus subsequently turns into a well-defined shell [22,23]. So this technique is suitable for the synthesis of MOFs possessing the hollow or micro/nano-scale spheres [24]. To construct the different MOFs hybrids with great porosity, electrospinning technology is a method of choice which receives the extensive worldwide attention. The technique engages two main conditions namely direct electrospinning and surface decoration; whereas concentration of solution, injection rate, viscosity, operating voltage and distance, etc. are the major factors affecting the structure of synthesized MOFs [25]. In the direct electrospinning method, a solution of polymer and MOFs particles are mixed consistently followed by the electrospun to construct the MOF/nanofiber composites while the surface decoration technology initially produce the nanofiber of polymer using electrospinning and finally the growth of MOFs particle on its surface. Thus, the fabricated MOFs are found to have pleasing structural flexibility, high volume density, controlled size with great porosity. Many research groups have constructed the MOFs using this method [26–29].

13.2.2 APPROACHES FOR MOFs-DERIVED POROUS MATERIALS

The chemical and thermal instability of MOFs limits their applications in various fields especially in the storage of energy. Thus, the uses of appropriate amendment play decisive role in the conversion of MOFs templates/precursors to functionalized derivatives. As compared to the traditional methods for the synthesis of templates/ precursors, strategies based on MOFs-derivatives can construct the controlled porous materials [13]. These modification methods may be categorized into self-pyrolysis (in air/inert atmosphere), chemical reaction with solid/liquid and relevant gases/vapors ligands based on their reaction mechanisms. In the in situ air or inert atmosphere, functionalized materials via carbonization can be constructed by the self-pyrolysis of MOFs templates/precursors [30]. MOFs also offer the building blocks after calcination. The final products like metal oxides/carbides and nano-sized metal particles embedded in carbon frameworks to fabricate the heterojunction structure which not even maintain the original structure of MOFs but exhibit great porosity [31,32]. The direct polymerization of MOFs templates/ precursors under high temperature results the formation of core-shell/hollow metal oxides [33–36]. Here, the ion exchange and carbonization occurs simultaneously for the formation of new ionic bonds and finally derivatives of MOFs. The discussed method also saves the raw materials and time for fabrication [37]. Metal ions present in the templates/precursors of MOFs coordinate with liquid organic ligands and transformed into the uniform nucleation and the crystals of MOFs are grown on pre-build hybrids surface that result in the formation of homogeneous hetero-materials [38–40]. The reaction of MOFs templates/precursors with solid organic ligands in the absence of any solvent(s) is an efficient for the fabrication of MOF derived functional materials. Metal ions, liberated out from MOFs templates/precursors, along with solid organic ligands construct the homogeneous mixtures through coordination reaction [41]. The reaction followed the similar kind of mechanism including the breaking of metal-based template bonds and formation of fresh bonds in the mixture; however, the reaction kinetics is relatively slow compared to the above-mentioned other methods. The use of external forces such as ball milling, mortar gridding, heating and high pressure encourages the reaction as the surface contact reactions does not occur automatically [42].

13.3 APPLICATIONS AND CHALLENGES IN MOFS

MOFs have various applications in the field of material science. Their wide variety of applications has the basis of modulation in their chemical, physical, electrical and optical properties by various means. Herein, we have discussed some of their applications and challenges related to the real commercial use of these porous coordination polymers.

13.3.1 APPLICATION OF MOFs IN ELECTRIC CONDUCTIVITY

The expansion of MOF-derived materials towards the electrical conductivity opens a new horizon for novel technologies used in electrolysis [43], energy storage [44], thermoelectric [45], sensors [46], and photovoltaics [11]. In order to function as an electrically conductive MOF, the band dispersion needs to be sufficient for charge carriers to be mobile through the band conduction [47]. However, it is noticed that the charge hoping (depends on the spatial separation and density of states between hopping sites) is the main mechanism of MOF's electrical conductivity in most of the cases [48,49]. The use of mixed redox states such as Fe^{3+} imperfections to Fe^{2+}-based MOFs has been proven to encourage the higher electrical conductivity, attributed to hole delocalization, is a productive approach in this regard [50]. Likewise, MOFs that exhibit electrical conductivity can be formed by intentionally forming the organic holes through air oxidation viz. the materials fabricated from hexa-iminotriphenylenesemiquinonate or tetrathiafulvalene-tetrabenzoate [51,52].

Amendments in the organic ligand are usual and certainly more synthetically tractable towards the conductive MOFs, given the variety of probable targets. Though, the localization of charge plays supportive role in various applications which depends on the discrete states including catalysis, photonics etc. and the low bulk conductivity materials may be valuable in several electro-chromic devices but mobile charge with disperse band carriers along with low effective mass are advantageous in most of the electronic devices [53]. The charge locality is noticeable in flat bands (<0.1 eV dispersion bands value) in the molecular orbitals of crystalline materials that primarily are as a result of poor matching between energy levels at the metal-organic interface [54]. The literature towards this matching of energy levels is available primarily for thiolated ligands where the band type conductivity opens the magnificent challenges in MOF chemistry [55]. Various contemplation have been used in disperse bands for ideal energy level contacts. In this regard, theoretical tools are available to compute the eigenvalues and occupation to visualize an estimate of energetic contacts of organic as well as inorganic components of MOF [56]. Complexity of molecular orbitals along with the inadequate electronic mixing at the interface of metal and organic constituents of MOF makes them less feasible for these kinds of computations [57]. Thus, the advancement of computational tools will play a decisive role for the novel designing and better understanding of this subfield.

The understanding of semiconductors in the field of transport opens up new applications of MOFs in defect chemistry. This is a relatively new area and does not possess much information even though no virtual study is available that may address the influence of defects on electrical conductivity of MOFs [58]. Filling this void of information is the necessity of time as MOFs permit the admirable control upon the defect concentration (concentration of charge carrier) via conventional redox reactions. Some other noticeable defect control, which destabilizes the MOF, includes vacancies, schottky and interstitials. Though some MOFs such as linker vacancies in zirconium carboxylates exhibit resilience against the high defect concentrations while post-synthetic exchange of metals as well as ligands has also been noticed in many MOFs [58]. So the redox control along with compositional defect chemistry plays a leader role for the future development of MOFs with great potential as electrically conductive materials.

13.3.2 MOFs for the Remediation of Water Pollutants

About 2/3 of Earth's surface is covered by water and its pollution is the chief environmental issue of current era. The removal of nuclear as well as industrial aqueous wastes including radionuclides (^{235}U, ^{137}Cs, ^{89}Sr, ^{65}Zn, ^{59}Fe, ^{57}Co, etc.), anion species (CrO_4^{2-}, AsO_4^{3-}, and CN^-) and heavy metals (Hg^{2+}, Pb^{2+}, Cd^{2+}, and Tl^+) should be the future priority [59]. A variety of physical, chemical as well as biological methods have been investigated to eliminate the above-mentioned water pollutants, but the level of success has not achieved yet. Some economical methods like precipitation are in use to remove the metals from contaminated water but they failed to achieve the safety guidelines level of minimum concentration along with the production of huge sludge [60]. In this regard, photocatalysis has also showed some potential, but its major drawback is the removal of unwanted by-products such as subtraction of Cr^{3+} from Cr(VI) through another treatment method so consuming both time and money [61]. Likewise, the cost-effective biological methods also exhibit the same kind of disadvantages [62]. The limitations of these mentioned approaches boost the development of novel, effective MOF-based techniques *viz.* ion exchange and sorption for wastewater treatment [59]. Inorganic materials such as layered double hydroxides and zeolites clays as well as organic resins have been used in traditional methods of ion-exchange; however, their negative aspects in slow sorption, poor regeneration, reusability, low chemical and thermal stability limits their broad use [63–65].

The discussed problems can be overcome by the use of MOFs as sorbent materials and these materials have been explored for their potential towards the treatment of wastewater as to their unique chemical and thermal properties [66–77]. MOFs containing sorption properties are believed to be next-generation materials for ion-exchange [78,79]. Therefore, the MOF-based ion exchangers are competent towards the quick sorption, superior selectivity, chemical (resistance in acid/base) and thermal (400–500°C) stability along with high yield and effective cost over conventional ion exchangers [80–84]. In addition, extra research efforts are needed to further explore the mechanism of ion-exchange of MOFs by using the advanced technologies. The sorption properties of MOFs towards variety of ions including radionuclides (possess immense environmental potential) are yet to be explored for the betterment of Earth's atmosphere and life.

13.3.3 Uses of MOFs in Gas Adsorption and Separation

Gases play a vital role to accomplish the day-to-day need of energy in various industries as well as in households. Thus, the exploration of low cost and eco-friendly materials for the separation and storage of gases is the scientific priority. In this regard, MOFs provide a ray of hope as these materials are well equipped with the exceptional porosity, superb thermal and chemical stability along with high surface area, which makes them an appropriate transportation vehicle as well as unique tool for separation and purification to solve the purpose as herein the density of stored gas is enhanced via a host-guest interaction mechanism [7,85,86].

The utilization of MOFs as adsorbents has begun in the late 1990s when understanding was developed towards their permanent porosity [8,87]. Initially, the MOFs were used for the storage of methane (a fuel gas) in 1997 [87]. Afterward, these materials were also applied for the storage of hydrogen and acetylene in 2003 and 2005, respectively [88,89]. The advancement of framework interpenetration and isoreticular principle facilitate the precise curb of porosity of MOFs [90–94]. In addition, the use of advanced techniques including gas chromatography, fixed-bed adsorption and assessment of gas sorption site using crystallography spectacularly enhanced the use of MOFs in commercial gas separation [95–99]. Yagi and co-researchers first fabricated a MOF for the storage of hydrogen gas [88] with the continuous improvement in the uptake capacity [100]. Later on, various research groups have developed a variety of MOFs for hydrogen storage [101–105]. Research also concludes that at high pressure and 77 K temperature, the gravimetric hydrogen storage potential of MOFs is generally related to their pore volume and surface area [7,106–108]. The storage of hydrogen is essential to function at near ambient temperature in the hydrogen fuel cells and this job is extremely challenging as the gas interacts with MOFs mainly through the weak van der Waals interactions. As per our understanding, there are no MOFs that have been tested that can conjoin the targets settled by the U.S. Department of Energy for gravimetric capacity [86]. The low CO_2 emission and high octane number make the methane as an attractive fuel for transportation and ultimately inspire the scientific community towards the various materials for their storage opportunities [109,110]. Methane interacts with MOFs in a moderate manner that make its storage as real at high pressure and room temperature. After the independent pioneer work by Kitagawa and Yaghi [87,90], significant advancements have been made in this field [111–114]. Simultaneous elongation of organic linkers along with the incorporation of functional sites into the fabricated MOFs has been found to enhance the methane storage capacity efficiently [115,116]. Acetylene is another important gas extensively used in the manufacturing of various chemical and electrical products. Though, its harmless storage is still a tedious challenge due to the explosive nature under compression pressure of above 0.2 MPa. In industry, there are particular cylinders filled with porous materials to store the acetylene, but it causes the massive expense as well as acetone pollutants. Thus, the exploration to search the novel safe methods for the storage of acetylene is required. In this context, MOFs have grown up as a significant material as adsorbent of acetylene storage [117]. Matsuda and colleagues in 2005 reported the first MOF to store acetylene [89]. Subsequently, a number of MOFs have been developed for this job with great potential [118–121]. Various approaches such as porosity optimization, functionalization of organic linkers and induction of open metal sites for the specific binding of acetylene molecules were found suitable for the improvement of integrations of acetylene and MOFs for the augmentation of storage capacity [122].

Besides the storage of vital useful gases, various MOFs have exhibited immense possibilities to use them as crucial adsorbent on the basis of their unique porosity. This MOF-based separation has seen the fast escalation during this century [122–127]. Bloch and co-workers fabricated MOF-based material for the speedy separation of acetylene/ethane. Similarly, Yang and colleagues developed unique MOFs based on hydroxyl–functionalization, which showed the great selectivity toward the separation of acetylene/ethane [128]. Specific tuning of MOFs pore size can achieve the great level separation of the same [129]. Likewise, many MOFs with great separation potential towards the variety of gases and other compounds have been designed and synthesized [129–132].

13.3.4   APPLICATIONS OF MOFs IN THE FOOD INDUSTRY

MOFs possess the potential to tailor their porosity, stability and surface-to-volume ratio under specific/desired conditions along with poor water solubility [35,133]. They are found to exhibit the versatile applications in the food industry, including adsorbents for food contamination of heavy metals, toxic dyes, etc., sensing of volatile organic compounds, carrier for bioactive molecules, role in food packaging, elimination of unwanted materials [134–138]. The γ-cyclodextrin–based MOFs synthesized through the green chemistry approach and potassium ions as linkers showed the specific interest in the food industry compared to the conventional one that contains heavy metals and organic linkers [139]. Thus, these cyclodextrin-based MOFs are fabricated by using the edible and non-toxic materials like starch that also serve as an economical approach. Chopra and colleagues fabricated three MOFs and evaluated them as ethylene scavengers. The results showed that these sachet-type ethylene absorbers MOFs lengthen the postharvest life of bananas and they may be explored as plant growth regulators [140]. Further research is needed in this field as it is in its early stage. Recently, MOFs have also been explored as carriers for antimicrobial agents to reduce/inhibit the pathogenic growth that is the chief cause of foodborne disease [141,142]. Duan and co-workers developed a HKUST-1 MOF using a silver nanoparticle that showed a strong antimicrobial activity against *Staphylococcus aureus* [143]. On the other hand, Wu and team synthesized (thymol)-loaded Zn@MOFs that was found to exhibit potential inhibition against *Escherichia coli* O157:H7 [144]. Controlled release of active molecules via carriers into the small intestine is a key approach in the food industry. In this context, Lashkari and colleagues fabricated a HKUST-1 MOF and corroborated the organized release of active molecules [145]. Recently UiO-66-based carriers were developed for curcumin loading and results confirmed the controlled release of curcumin for 3 hours in the gastrointestinal conditions [146]. Besides these key roles in the food industry, MOFs have also been extensively used to detect and remove the various food contaminants including heavy metals, dyes, herbicides, antibiotics, etc. [147–150].

13.3.5   MOFs AND VACCINES

Vaccination is an effective protective approach against numerous diseases and saved many lives in recent times [151]. Recently, MOFs have been explored for various biological functions including bio-sensing, bio-imaging, carrier system for various bioactive molecules, drug delivery, etc. for the betterment of human beings [152–156]. Some proteins are dissolved upon entering into the human skin. These proteins can be freezed by using the MOF-based biocompatible polymers that ultimately turns into the vaccines. This is a unique innovation in the healthcare sector for the proper supply of affected vaccines in the remote areas. Design and development of protein encapsulated MOFs have gained wide attention by the scientific community [156–158]. In this framework, mesoporous MOFs were synthesized by Ma and colleagues for the immobilization of microperoxidase-11 and results concluded that the immobilized enzyme exhibited the superior catalysis activity as compared to the non-immobilized [159]. Ge and research team also developed a method for enzyme embedding by using MOFs [160]. Likewise, Liang and co-workers applied a bio-mimetic mineralization process for the successful encapsulation of bio-macromolecules such as proteins within the MOFs [161]. Zhang and co-workers also designed a novel approach for the development of an effective subunit vaccine based on MOFs [162]. Thus, the field needs more and more research for the proper understanding of these biological applications of MOFs.

13.3.6 MOF-BASED MATERIALS USED IN ELECTROCATALYSIS AND BATTERIES

With the advancement of synthetic strategies, MOF-derived porous materials have been synthesized in numerous geometries including nanosheets/cubes, shuttle, hollow sphere/octahedron, core-double shell polyhedron and many more. These structural variations showed an excellent potential in electrocatalysis as well as rechargeable batteries (needs of future). The porous materials derived from MOFs and carbon supported by metal oxides, metal particles, selenides, sulfides, and phosphides have shown a great potential towards the electrochemical effects and offers a unique class of materials with huge applications [163]. These kinds of porous materials provide exclusive functional architectures along with consistent distribution of active sites and possess the potential to swap the expensive metal catalyst used in electrolysis. Hollow carbon cavity holds the minute metal particles that can avoid the particle agglomeration and augmenting the mechanical strength so that the structural stability of electrode material can be advanced. Specific metal/ multi-metals and heteroatoms may be incorporated by altering the metal centers and organic ligands in these MOFs [13]. Chen and co-workers fabricated a MOF-derived yolk–shell Co@C–N with increased catalytic activity. The fabricated hollow material exhibits a high surface area in which C-N nanosheets were conductive for the stabilization of small particles of Co, whereas the C-N shell facilitates the transfer of mass [164]. Zhang and colleagues synthesized a 3D MOF-based Co@N-C bifunctional catalyst with hierarchical rod-like configuration where the Co atoms were embedded uniformly in N-doped carbon rings. This MOF was found to exhibit the electrocatalytic activities in an exceptionally positive manner [31]. Wang and research group applied the molten salt-assisted solid state assembly strategy and synthesized Fe-Co alloy/N-doped mesoporous composites and Zn-based MOFs that showed the excellent bifunctional electrocatalytic activities [165]. The same kind of materials (CoNi@NC) were synthesized by using cobalt-nickel allow nanoparticles. The MOF performed the outstanding oxygen electrocatalytic activities [166,167]. Jin and colleagues have developed a NiCo MOF composite that offers prosperous redox active sites for the high energy storage in lithium batteries [168]. MOF-derived metal oxides supported by carbon are the hot topic of research for anode materials used in rechargeable batteries as metal oxides offer numerous benefits, including chemical stability, plentiful reserve capability, non-toxicity, and large theoretical capacity. Incorporation of carbon materials in these MOFs advances the total conductivity and also helps to maintain the buffer system of volume variation in metal oxides at the time of cyclic process. MOF-based hollow spheres of CuO were found to carry the great potential for sodium storage [169] and iron-based MOFs showed the regulation of redox reaction occurs within the Li-S batteries [170]. Likewise, the TiO_2- and CoO-based MOFs also exhibited the vital storage potential in lithium and sodium ion batteries [171,172]. Additionally, the multicomponent systems of metal oxides provide the combined benefits of single component system via the uninterrupted contraction and expansion, which leads towards the stability of skeleton along with enhanced electrochemical potential of materials. This approach has been successfully tested to increase the storage capacity of lithium and sodium ion batteries [173–177]. In addition to the above discussion, potassium ion batteries present a novel kind of secondary energy storage devices with a high value of energy density as a substitute to lithium ion batteries. Yi and co-workers fabricated the NC@CoP/NC polyhedron as the anode material used in the potassium ion batteries [16]. They provide a cheap and abundant source along with comparable redox potential to lithium [178–181]. Furthermore, these materials are derived from MOFs through environment-friendly and low-cost approaches, but some severe concerns including low-voltage and high-voltage polarization along with poor specific capacity need more exploration prior to their commercial scale use in the future [36].

13.3.7 MOFs Toxicity

This is a weakly explored area which needs more attention to authenticate the safety of these promising porous coordination polymers/MOFs towards the human health as well as environmental issues [182]. MOFs toxicity is among the major challenges in front of the scientific community that should be taken seriously prior to their large-scale commercial uses in the real world. Toxicity effects related to MOFs (if present) are plausibly due to the various metal ions and functional groups present in the organic ligands [183,184]. The solvent used during the synthetic approach and the size of MOF crystals also play a key role in their toxicity [185]. Due to the invasive nature and the nanoscale particle size of MOFs, their pores are engaged by various solvents including water, dimethyl sulfoxide, dimethylformamide, acetone, diethylene formamide and many others used during the synthesis [186,187]. A little amount of these solvents may be retained in these nanoscale pores of activated MOFs that can be substituted through the heat/vacuum or other mechanism. Besides the pore size and metal ions, toxicity of MOFs also depends on some other factors including their structural topology and kind of organic linkers [183,184].

13.4 CONCLUSIONS

The malleable chemical composition and high surface area of MOFs make a special place for them in multidisciplinary fields with plentiful advantages for the betterment of life. Research on the various properties of MOFs in a variety of fields including optics, chemical engineering, electronics, pharmaceuticals and many more provide them a hot place in the current scientific era. We have discussed the diverse, exciting viewpoints necessary for the advancement in the synthesis of MOFs. Further, we have explored the prime applications of MOFs in the field of catalysis, gas separation, electrical conductivity, electrocatalysis, rechargeable batteries, etc. The plausible opportunities along with the current challenges in described applications have also been illustrated for continuous expansion of the scientific community towards the metal-organic frameworks chemistry that shall fulfill their bright future.

ACKNOWLEDGMENTS

The authors are thankful to BHU and IIT Indore for research environment. N.K. is thankful to the University Grants Commission (UGC), India for postdoctoral fellowship. N.G. gratefully acknowledges the financial support from UGC, New Delhi (Letter No. F.30–431/2018(BSR), M-14–55) and IoE, BHU (Letter No. R/Dev/D/IoE/Seed Grant/2020-21).

REFERENCES

[1] Bennett TD, Coudert FX, James SL, Cooper AI (2021) The changing state of porous materials. *Nat Mater.* (9):1179–1187. 10.1038/s41563-021-00957-w

[2] Allendorf MD, Bauer CA, Bhakta RK, Houk RJT (2009) Luminescent metal-organic frameworks. *Chem Soc Rev.* 38(5):1330–1352. 10.1007/978-1-4614-7388-6

[3] Ozin GA (1992) Nanochemistry: Synthesis in diminishing dimensions. *Adv Mater.* 4(10):612–649. 10.1016/S1359-0294(96)80084-7

[4] Davis ME (1993) New vistas in zeolite and molecular sieve catalysis. *Acc Chem Res.* 26(3):111–115. 10.1038/nature02529

[5] Hendon CH, Butler KT, Ganose AM, Román-Leshkov Y, Scanlon DO, Ozin GA, Walsh A (2017) Electroactive nanoporous metal oxides and chalcogenides by chemical design. *Chem Mater.* 29(8):3663–3670. 10.1021/acs.chemmater.7b00464

[6] Furukawa H, Cordova KE, O'Keeffe M, Yaghi OM (2013) The chemistry and applications of metal-organic frameworks. *Science* 341(6149):1230444. 10.7150/thno.31918

[7] Farha OK, Eryazici I, Jeong NC, Hauser BG, Wilmer CE, Sarjeant AA, Snurr RQ, Nguyen ST, Yazaydın AÖ, Hupp JT (2012) Metal-organic framework materials with ultrahigh surface areas: Is the sky the limit? *J Am Chem Soc.* 134(36):15016–15021. 10.1007/s40010-014-0125-9

[8] Li H, Eddaoudi M, Groy TL, Yaghi OM (1998) Establishing microporosity in open metal–organic frameworks: Gas sorption isotherms for Zn(BDC) (BDC = 1,4-Benzenedicarboxylate). *J Am Chem Soc.* 120(33):8571–8572. 10.1126/science.12304

[9] Férey G, Mellot-Draznieks C, Serre C, Millange F, Dutour J, Surblé S, Margiolaki I. (2005) A chromium terephthalate-based solid with unusually large pore volumes and surface area. *Science* 309(5743):2040–2042. 10.1126/science.1116275

[10] Koh K, Wong-Foy AG, Matzger AJ (2009) MOF@MOF: Microporous core-shell architectures. *Chem Commun (Camb).* 41:6162–6164. 10.1039/b904526k

[11] Stavila V, Talin AA, Allendorf MD (2014) MOF-based electronic and opto-electronic devices. *Chem Soc Rev.* 43(16):5994–6010. 10.1039/c4cs00096j

[12] Yang J, Zhang Y, Liu Q, Trickett CA, Monge MA, Cong H, Aldossary A, Deng H, Yaghi OM, Gutiérrez-Puebla E (2017) Principles of designing extra-large pore openings and cages in zeolitic imidazolate frameworks. *J Am Chem Soc.* 139(18):6448–6455. 10.1007/s10311-019-00933-6

[13] Wang L, Zhu Y, Du C, Ma X, Cao C (2020) Advances and challenges in metal-organic framework derived porous materials for batteries and electrocatalysis. *J Mater Chem A* 8:24895–24919. 10.1039/D0TA08311A

[14] Masoomi MY, Morsali A, Dhakshinamoorthy A, Garcia H (2019) Mixed-metal MOFs: Unique opportunities in metal-organic framework (MOF) functionality and design. *Angew Chem Int Ed Engl.* 58(43):15188–15205. 10.1007/s10562-021-03865-5

[15] Xu T, Sun K, Gao D, Li C, Hu X, Chen G (2019) Atomic-layer-deposition-formed sacrificial template for the construction of an MIL-53 shell to increase selectivity of hydrogenation reactions. *Chem Commun.* 55(53):7651–7654. 10.1039/C9CC02727K

[16] Yi Y, Zhao W, Zeng Z, Wei C, Lu C, Shao Y, Guo W, Dou S, Sun J (2020) ZIF-8@ ZIF-67-derived nitrogen-doped porous carbon confined CoP polyhedron targeting superior potassium-ion storage. *Small* 16(7):1906566. 10.1002/smll.201906566

[17] Wang Z, Shen K, Chen L, Li Y (2022) Scalable synthesis of multi-shelled hollow N-doped carbon nanosheet arrays with confined Co/CoP heterostructures from MOFs for pH-universal hydrogen evolution reaction. *Sci. China Chem.* 65:619–629. 10.1007/s11426-021-1175-2

[18] Yang Y, Wang F, Yang Q, Hu Y, Yan H, Chen YZ, Liu H, Zhang G, Lu J, Jiang HL, Xu H (2014) Hollow metal-organic framework nanospheres via emulsion-based interfacial synthesis and their application in size-selective catalysis. *ACS Appl Mater Interfaces* 6(20):18163–18171. 10.1002/adma.201803291

[19] Huo J, Marcello M, Garai A, Bradshaw D (2013) MOF-polymer composite microcapsules derived from Pickering emulsions. *Adv Mater.* 25(19):2717–2722. 10.1002/adma.201204913

[20] Li J, Lu S, Huang H, Liu D, Zhuang Z, Zhong C (2018) ZIF-67 as continuous self-sacrifice template derived NiCo2O4/Co, N-CNTs nanocages as efficient bifunctional electrocatalysts for rechargeable Zn-air batteries. *ACS Sustainable Chem. Eng.* 6(8):10021–10029. 10.1021/acssuschemeng.8b01332

[21] Liu T, Li P, Yao N, Kong T, Cheng G, Chen S, Luo W (2019) Self-sacrificial template-directed vapor-phase growth of MOF assemblies and surface vulcanization for efficient water splitting. *Adv. Mater.* 31(21):1806672. 10.1002/adma.201806672

[22] Carne-Sanchez A, Imaz I, Cano-Sarabia M, Maspoch D (2013) A spray-drying strategy for synthesis of nanoscale metal-organic frameworks and their assembly into hollow superstructures. *Nat. Chem.* 5(3):203–211. 10.1038/nchem.1569.

[23] Avci-Camur C, Troyano J, Pérez-Carvajal J, Legrand A, Farrusseng D, Imaz I, Maspoch D (2018) Aqueous production of spherical Zr-MOF beads via continuous-flow spray-drying. *Green Chem.* 20:873–878. 10.1039/C7GC03132G

[24] Zhang A, Li XY, Zhang S, Yu Z, Gao X, Wei X, Wu Z, Wu WD, Chen XD (2017) Spray-drying-assisted reassembly of uniform and large micro-sized MIL-101 microparticles with controllable morphologies for benzene adsorption. *J. Colloid Interface Sci.* 506:1–9. 10.1016/j.jcis.2017.07.022

[25] Xue J, Wu T, Dai Y, Xia Y (2019) Electrospinning and electrospun nanofibers: Methods, materials, and applications. *Chem Rev.* 119(8):5298–5415. 10.1021/acs.chemrev.8b00593

[26] Doan TLH, Dao TQ, Tran HN, Tran PH, Le TN (2016) An efficient combination of Zr-MOF and microwave irradiation in catalytic Lewis acid Friedel-Crafts benzoylation. *Dalton Trans.* 45(18):7875–7880. 10.1039/C6DT00827E

[27] Guo W, Sun W, Lv LP, Kong S, Wang Y (2017) Microwave-assisted morphology evolution of Fe-based metal-organic frameworks and their derived Fe_2O_3 nanostructures for Li-ion storage. *ACS Nano* 11(4):4198–4205. 10.1021/acsnano. 7b01152

[28] Lee EJ, Bae J, Choi KM, Jeong NC (2019) Exploiting microwave chemistry for activation of metal-organic frameworks. *ACS Appl. Mater. Interfaces* 11(38):35155–35161. 10.1021/acsami.9b12201

[29] Laha S, Chakraborty A, Maji TK (2020) Synergistic role of microwave and perturbation toward synthesis of hierarchical porous MOFs with tunable porosity. *Inorg. Chem.* 59(6):3775–3782. 10.1021/acs.inorgchem.9b03422

[30] Tian Y, Huang H, Liu G, Bi R, Zhang L (2019) Metal-organic framework derived yolk-shell NiS2/carbon spheres for lithium-sulfur batteries with enhanced polysulfide redox kinetics. *Chem. Commun.* 55(22):3243–3246. 10.1039/c9cc00486f

[31] Zhang M, Dai Q, Zheng H, Chen M, Dai L (2018) Novel MOF-derived Co@N-C bifunctional catalysts for highly efficient Zn-air batteries and water splitting. *Adv Mater.* 30(10):1705431. 10.1002/adma.201705431

[32] Ahmed S, Shim J, Sun HJ, Park G (2020) Transition metals (Co or Ni) encapsulated in carbon nanotubes derived from zeolite imidazolate frameworks (ZIFs) as bifunctional catalysts for the oxygen reduction and evolution reactions. *Phys. Status Solidi A* 217(12):1900969. 10.1002/pssa.201900969

[33] Han Y, Li J, Zhang T, Qi P, Li S, Gao X, Zhou J, Feng X, Wang B (2018) Zinc/Nickel-doped hollow core-shell Co_3O_4 derived from a metal-organic framework with high capacity, stability, and rate performance in lithium/sodium-ion batteries. *Chem. Eur. J.* 24(7):1651–1656. 10.1002/chem.201704416

[34] Li Z, Zhao J, Nie J, Yao S, Wang J, Feng X (2020) Co_3O_4/NiO/C composites derived from zeolitic imidazolate frameworks (ZIFs) as high-performance anode materials for Li-ion batteries. *J Solid State Electrochem.* 24(5):1133–1142. 10. 1007/s10008-020-04595-1

[35] Meng J, Liu X, Niu C, Pang Q, Li J, Liu F, Liu Z, Mai L (2020) Advances in metal-organic framework coatings: Versatile synthesis and broad applications. *Chem. Soc. Rev.* 49(10):3142–3186. 10.1039/C9CS00806C

[36] Wang L, Han Z, Zhao Q, Yao X, Zhu Y, Ma X, Wu S, Cao C (2020) Engineering yolk–shell P-doped NiS2/C spheres via a MOF-template for high-performance sodium-ion batteries. *J Mater Chem A* 8(17):8612–8619. 10.1039/D0TA02568B

[37] Young C, Wang J, Kim J, Sugahara Y, Henzie J, Yamauchi Y (2018) Controlled chemical vapor deposition for synthesis of nanowire arrays of metal-organic frameworks and their thermal conversion to carbon/metal oxide hybrid materials. *Chem. Mater.* 30(10):3379–3386. 10.1021/acs.chemmater.8b00836

[38] Tian T, Huang L, Ai L, Jiang J (2017) Surface anion-rich NiS2 hollow microspheres derived from metal-organic frameworks as a robust electrocatalyst for the hydrogen evolution reaction. *J. Mater. Chem. A* 5(39):20985–20992. 10.1039/C7TA06671F

[39] Bai XJ, Chen D, Li LL, Shao L, He WX, Chen H, Li YN, Zhang XM, Zhang LY, Wang TQ, Fu Y, Qi W (2018) Fabrication of MOF thin films at miscible liquid-liquid interface by spray method. *ACS Appl Mater Interfaces* 10(31):25960–25966. 10.1021/acsami.8b09812

[40] Fan S, Li G, Yang G, Guo X, Niu X (2019) $NiSe_2$ nanooctahedra as anodes for high-performance sodium-ion batteries. *New J. Chem.* 43(32):12858–12864. 10. 1039/C9NJ02631B

[41] Yang J, Feng X, Lu G, Li Y, Mao C, Wen Z, Yuan W (2018) Correction: NaCl as a solid solvent to assist the mechanochemical synthesis and post-synthesis of hierarchical porous MOFs with high I_2 vapour uptake. *Dalton Trans.* 47(17):6250-6250. 10.1039/C8DT90057D

[42] Tanaka S, Kida K, Nagaoka T, Ota T, Miyake Y (2013) Mechanochemical dry conversion of zinc oxide to zeolitic imidazolate framework. *Chem Commun.* 49(72):7884–7886. 10.1039/C3CC43028F

[43] Wang W, Xu X, Zhou W, Shao Z (2017) Recent progress in metal-organic frameworks for applications in electrocatalytic and photocatalytic water splitting. *Adv Sci.* 4(4):1600371. 10.1002/advs.201600371

[44] Sheberla D, Bachman JC, Elias JS, Sun CJ, Shao-Horn Y, Dincă M (2017) Conductive MOF electrodes for stable supercapacitors with high areal capacitance. *Nat. Mater.* 16(2):220–224. 10.1038/nmat4766

[45] Erickson KJ, Léonard F, Stavila V, Foster ME, Spataru CD, Jones RE, Foley BM, Hopkins PE, Allendorf MD, Talin AA (2015) Thin film thermoelectric metal-organic framework with high seebeck coefficient and low thermal conductivity. *Adv Mater.* 27(22):3453–3459. 10.1002/adma.201501078

[46] Campbell MG, Liu SF, Swager TM, Dincă M (2015) Chemiresistive sensor arrays from conductive 2D metal-organic frameworks. *J Am Chem Soc.* 137(43):13780–13783. 10.1021/jacs.5b09600

[47] Allendorf MD, Schwartzberg A, Stavila V, Talin AA (2011) A roadmap to implementing metal-organic frameworks in electronic devices: Challenges and critical directions. *Chem Eur J.* 17(41):11372–11388. 10.1002/chem.201101595

[48] Liu J, Zhou W, Liu J, Howard I, Kilibarda G, Schlabach S, Coupry D, Addicoat M, Yoneda S, Tsutsui Y, Sakurai T, Seki S, Wang Z, Lindemann P, Redel E, Heine T, Wöll C (2015) Photoinduced charge-carrier generation in epitaxial MOF thin films: High efficiency as a result of an indirect electronic band gap? *Angew Chem Int Ed Engl.* 54(25):7441–7445. 10.1002/anie.201501862

[49] Hendon CH, Walsh A, Dincă M (2016) Frontier orbital engineering of metal-organic frameworks with extended inorganic connectivity: Porous alkaline-earth oxides. *Inorg Chem.* 55(15):7265–7269. 10.1021/acs.inorgchem.6b00979

[50] Sun L, Hendon CH, Park SS, Tulchinsky Y, Wan R, Wang F, Walsh A, Dincă M (2017) Is iron unique in promoting electrical conductivity in MOFs? *Chem Sci.* 8(6):4450–4457. 10.1039/C7SC00647K

[51] Narayan TC, Miyakai T, Seki S, Dincă M (2012) High charge mobility in a tetrathiafulvalene-based microporous metal-organic framework. *J. Am. Chem. Soc.* 134(31):12932–12935. 10.1021/ja3059827

[52] Park SS, Hontz ER, Sun L, Hendon CH, Walsh A, Van Voorhis T, Dincă M (2015) Cation-dependent intrinsic electrical conductivity in isostructural tetrathiafulvalene-based microporous metal–organic frameworks. *J Am Chem Soc.* 137(5):1774–1777. 10.1021/ja512437u

[53] Wade CR, Li M, Dincă M (2013) Facile deposition of multicolored electrochromic metal–organic framework thin films. *Angew Chem Int Ed.* 52(50):13377–13381. 10.1002/anie.201306162

[54] Sun L, Hendon CH, Minier MA, Walsh A, Dincă M (2015) Million-fold electrical conductivity enhancement in Fe_2(DEBDC) versus Mn(DEBDC) (E= S, O). *J Am Chem Soc.* 137(19):6164–6167. 10.1021/jacs.5b02897

[55] Herm ZR, Wiers BM, Mason JA, van Baten JM, Hudson MR, Zajdel P, Brown CM, Masciocchi N, Krishna R, Long JR (2013) Separation of hexane isomers in a metal-organic framework with triangular channels. *Science* 340(6135):960–964. 10.1126/science.1234071

[56] Davies DW, Butler KT, Jackson AJ, Morris A, Frost JM, Skelton JM, Walsh A (2016) Computational screening of all stoichiometric inorganic materials. *Chem.* 1(4):617–627. 10.1016/j.chempr.2016.09.010

[57] Wilmer CE, Leaf M, Lee CY, Farha OK, Hauser BG, Hupp JT, Snurr RQ (2011) Large-scale screening of hypothetical metal-organic frameworks. *Nat Chem.* 4(2):83–89. 10.1038/NCHEM.1192

[58] Cliffe MJ, Wan W, Zou X, Chater PA, Kleppe AK, Tucker MG, Wilhelm H, Funnell NP, Coudert FX, Goodwin AL (2014) Correlated defect nanoregions in a metal-organic framework. *Nat Commun.* 5:4176. 10.1038/ncomms5176

[59] Kumar P, Pournara A, Kim K-H, Bansal V, Rapti S, Manos MJ (2017) Metal-organic frameworks: Challenges and opportunities for ion-exchange/sorption applications. *Prog Mater Sci.* 86:25–74. 10.1016/j.pmatsci.2017.01.002

[60] Tonini DR, Gauvin DA, Soffel RW, Freeman WP (2003) Achieving low mercury concentrations in chlor-alkah wastewaters. *Environ Prog.* 22(3):167–173. 10.1002/ep.670220314

[61] Hoch LB, Mack EJ, Hydutsky BW, Hershman JM, Skluzacek JM, Mallouk TE (2008) Carbothermal synthesis of carbon-supported nanoscale zero-valent iron particles for the remediation of hexavalent chromium. *Environ Sci Technol.* 42(7):2600–2665. 10.1021/es702589u

[62] Narayani M, Shetty KV (2013) Chromium-resistant bacteria and their environmental condition for hexavalent chromium removal: A review. *Crit Rev Environ Sci Technol.* 43(9):955–1009. 10.1080/10643389.2011.627022

[63] Dyer A, Pillinger M, Harjula R, Amin S (2000) Sorption characteristics of radio-nuclides on synthetic birnessite-type layered manganese oxides. *J Mater Chem.* 10(8):1867–1874. 10.1039/B002435J

[64] Benhammou A, Yaacoubi A, Nibou L, Tanouti B (2005) Adsorption of metal ions onto Moroccan stevensite: Kinetic and isotherm studies. *J Colloid Interface Sci.* 282(2):320–326. 10.1016/j.jcis.2004.08.168

[65] Fan Q, Li Z, Zhao H, Jia Z, Xu J, Wu W (2009) Adsorption of Pb(II) on paly-gorskite from aqueous solution: Effects of pH, ionic strength and temperature. *Appl Clay Sci.* 45(3):111–116. 10.1016/j.clay.2009.04.009

[66] Eddaoudi M, Moler DB, Li H, Chen B, Reineke TM, O'keeffe M, Yaghi OM (2001) Modular chemistry: Secondary building units as a basis for the design of highly porous and robust metal-organic carboxylate frameworks. *Acc Chem Res.* 34(4):319–330. 10.1021/ar000034b

[67] Ferey G (2008) Hybrid porous solids: Past, present, future. *Chem Soc Rev.* 37(1):191–214. 10.1039/B618320B

[68] Horike S, Shimomura S, Kitagawa S (2009) Soft porous crystals. *Nat Chem.* 1(9):695–704. 10.1038/nchem.444

[69] Farha OK, Hupp JT (2010) Rational design, synthesis, purification, and activation of metal–organic framework materials. *Acc Chem Res.* 43(8):1166–1175. 10.1021/ar1000617

[70] Zheng S, Wu T, Zhang J, Chow M, Nieto RA, Feng P, Bu X (2010) Porous metal carboxylate boron imidazolate frameworks. *Angew Chem Int Ed.* 49(31):5362–5366. 10.1002/ange.201001675

[71] Burtch NC, Jasuja H, Walton KS (2014) Water stability and adsorption in metal-organic frameworks. *Chem Rev.* 114(20):10575–10612. 10.1021/cr5002589

[72] Canivet J, Fateeva A, Guo Y, Coasne B, Farrusseng D (2014) Water adsorption in MOFs: Fundamentals and applications. *Chem Soc Rev.* 43(16):5594–5617. 10.1039/C4CS00078A

[73] Zhao X, Liu D, Huang H, Zhang W, Yang Q, Zhong C (2014) The stability and defluoridation performance of MOFs in fluoride solutions. *Microporous Mesoporous Mater*. 185:72–78. 10.1016/j.micromeso.2013.11.002

[74] Kumar P, Deep A, Kim KH (2015) Metal organic frameworks for sensing applications. *TrAC, Trends Anal Chem*. 73:39–53. 10.1016/j.trac.2015.04.009

[75] Taylor JM, Komatsu T, Dekura S, Otsubo K, Takata M, Kitagawa H (2015) The role of a three dimensionally ordered defect sublattice on the acidity of a sulfonated metal-organic framework. *J Am Chem Soc*. 137(35):11498–14506. 10.1021/jacs.5 b07267

[76] Howarth AJ, Liu Y, Li P, Li Z, Wang TC, Hupp JT, Farha OK (2016) Chemical, thermal and mechanical stabilities of metal-organic frameworks. *Nat Rev Mater*. 1:15018. 10.1038/natrevmats.2015.18

[77] Kaur R, Kim KH, Paul AK, Deep A (2016) Recent advances in the photovoltaic applications of coordination polymers and metal organic frameworks. *J Mater Chem A*. 4(11):3991–4002. 10.1039/C5TA09668E

[78] Howarth AJ, Liu Y, Hupp JT, Farha OK (2015) metal-organic frameworks for applications in remediation of oxyanion/cation-contaminated water. *CrystEngComm*. 17(38):7245–7253. 10.1039/C5CE01428J

[79] Rapti S, Pournara A, Sarma D, Papadas IT, Armatas GS, Tsipis AC, Lazarides T, Kanatzidis MG, Manos MJ (2016) Selective capture of hexavalent chromium from an anion-exchange column of metal organic resin-alginic acid composite. *Chem Sci*. 7(3):2427–2436. 10.1039/C5SC03732H

[80] Kandiah M, Nilsen MH, Usseglio S, Jakobsen S, Olsbye U, Tilset M, Larabi C, Quadrelli EA, Bonino F, Lillerud KP (2010) Synthesis and stability of tagged UiO-66 Zr-MOFs. *Chem Mater*. 22(24):6632–6640. 10.1021/cm102601v

[81] Yee KK, Reimer N, Liu J, Cheng SY, Yiu SM, Weber J, Stock N, Xu Z (2013) Effective mercury sorption by thiol-laced metal-organic frameworks: In strong acid and the vapor phase. *J Am Chem Soc*. 135(21):7795–7798. 10.1021/ja400212k

[82] Morris W, Briley WE, Auyeung E, Cabezas MD, Mirkin CA (2014) Nucleic acid–metal organic framework (MOF) nanoparticle conjugates. *J Am Chem Soc*. 136(20):7261–7264. 10.1021/ja503215w

[83] Hu Z, Zhao D (2015) De facto methodologies toward the synthesis and scale-up production of UiO-66-type metal-organic frameworks and membrane materials. *Dalton Trans*. 44(44):19018–19040. 10.1039/C5DT03359D

[84] Rapti S, Pournara A, Sarma D, Papadas IT, Armatas GS, Hassan YS, Alkordi MH, Kanatzidis MG, Manos MJ (2016) Rapid, green and inexpensive synthesis of high quality UiO-66 amino-functionalized materials with exceptional capability for removal of hexavalent chromium from industrial waste. *Inorg Chem Front*. 3(5):635–644. 10.1039/C5QI00303B

[85] Furukawa H, Ko N, Go YB, Aratani N, Choi SB, Choi E, Yazaydin AO, Snurr RQ, O'Keeffe M, Kim J, Yaghi OM (2010) Ultrahigh porosity in metal-organic frameworks. *Science* 329(5990):424–428. 10.1126/science.1192160

[86] Li H, Li L, Lin RB, Zhou W, Zhang Z, Xiang S, Chen B (2019) Porous metal-organic frameworks for gas storage and separation: Status and challenges. *EnergyChem* 1(1):100006. 10.1016/j.enchem.2019.100006

[87] Kondo M, Yoshitomi T, Matsuzaka H, Kitagawa S, Seki K (1997) Three-dimensional framework with channeling cavities for small molecules: {[M$_2$ (4,4'-bpy)$_3$(NO$_3$)$_4$]•xH$_2$O}$_n$ (M = Co, Ni, Zn). *Angew. Chem. Int. Ed*. 36(16):1725–1727. 10.1002/anie.199717251

[88] Rosi NL, Eckert J, Eddaoudi M, Vodak DT, Kim J, O'Keeffe M, Yaghi OM (2003) Hydrogen storage in microporous metal-organic frameworks. *Science* 300(5622): 1127–1129. 10.1126/science.1083440

[89] Matsuda R, Kitaura R, Kitagawa S, Kubota Y, Belosludov RV, Kobayashi TC, Sakamoto H, Chiba T, Takata M, Kawazoe Y, Mita Y (2005) Highly controlled acetylene accommodation in a metal-organic microporous material. *Nature* 436(7048):238–241. 10.1038/nature03852

[90] Eddaoudi M, Kim J, Rosi N, Vodak D, Wachter J, O'Keeffe M, Yaghi OM (2002) Systematic design of pore size and functionality in isoreticular MOFs and their application in methane storage. *Science* 295(5554):469–472. 10.1126/science.1 067208

[91] Friedrichs OD, Foster MD, O'Keeffe M, Proserpio DM, Treacy MMJ, Yaghi OM (2005) What do we know about three-periodic nets? *J Solid State Chem.* 178(8):2533–2554. 10.1016/j.jssc.2005.06.037

[92] Chen B, Ma S, Hurtado EJ, Lobkovsky EB, Zhou HC (2007) A triply inter-penetrated microporous metal–organic framework for selective sorption of gas molecules. *Inorg Chem.* 46(21):8490–8492. 10.1021/ic7014034

[93] Xue M, Ma S, Jin Z, Schaffino RM, Zhu GS, Lobkovsky EB, Qiu SL, Chen B (2008) Robust metal–organic framework enforced by triple-framework inter-penetration exhibiting high H_2 storage density. *Inorg Chem.* 47(15):6825–6828. 10.1021/ic800854y

[94] Zhai QG, Bu X, Zhao X, Li DS, Feng P (2017) Pore space partition in metal-organic frameworks. *Acc Chem Res.* 50(2):407–417. 10.1021/acs.accounts.6b00526

[95] Rowsell JL, Spencer EC, Eckert J, Howard JA, Yaghi OM (2005) Gas adsorption sites in a large-pore metal-organic framework. *Science* 309(5739):1350–1354. 10.1126/science.1113247

[96] Chen B, Liang C, Yang J, Contreras DS, Clancy YL, Lobkovsky EB, Yaghi OM, Dai S (2006) A microporous metal-organic framework for gas-chromatographic separation of alkanes. *Angew Chem Int Ed.* 45(9):1390–1393. 10.1002/anie.2005 02844

[97] Mueller U, Schubert M, Teich F, Puetter H, Schierle-Arndt K, Pastré J (2006) Metal-organic frameworks—prospective industrial applications. *J Mater Chem.* 16(7):626–636. 10.1039/B511962F

[98] Bárcia PS, Zapata F, Silva JA, Rodrigues AE, Chen B (2007) Kinetic separation of hexane isomers by fixed-bed adsorption with a microporous metal–organic framework. *J Phys Chem. B* 111(22):6101–6103. 10.1021/jp0721898

[99] Vaidhyanathan R, Iremonger SS, Shimizu GK, Boyd PG, Alavi S, Woo TK (2010) Direct observation and quantification of CO_2 binding within an amine-functionalized nanoporous solid. *Science* 330(6004):650–653. 10.1126/science.11 94237

[100] Kaye SS, Dailly A, Yaghi OM, Long JR (2007) Impact of preparation and handling on the hydrogen storage properties of $Zn_4O(1,4$-benzenedicarboxylate$)_3$ (MOF-5). *J Am Chem Soc.* 129(46):14176–14177. 10.1021/ja076877g

[101] Murray LJ, Dinca M, Long JR (2009) Hydrogen storage in metal-organic frame-works. *Chem. Soc. Rev.* 38(5):1294–1314. 10.1039/B802256A

[102] Yan Y, Yang S, Blake AJ, Schröder M (2014) Studies on metal-organic frameworks of Cu(II) with isophthalate linkers for hydrogen storage. *Acc Chem Res.* 47(2):296–307. 10.1021/ar400049h

[103] Gualdrón DAG, Colón YJ, Zhang X, Wang TC, Chen YS, Hupp JT, Yildirim T, Farha OK, Zhang J, Snurr RQ (2016) Evaluating topologically diverse metal-organic frameworks for cryo-adsorbed hydrogen storage. *Energy Environ. Sci.* 9(10):3279–3289. 10.1039/C6EE02104B

[104] Ahmed A, Liu Y, Purewal J, Tran LD, Foy AGW, Veenstra M, Matzger AJ, Siegel DJ (2017) Balancing gravimetric and volumetric hydrogen density in MOFs. *Energy Environ. Sci.* 10(1):2459–2471. 10.1039/C7EE02477K

[105] Kapelewski MT, Runčevski T, Tarver JD, Jiang HZH, Hurst KE, Parilla PA, Ayala A, Gennett T, FitzGerald SA, Brown CM, Long JR (2018) Record high hydrogen storage capacity in the metal-organic framework Ni₂(m-dobdc) at near-ambient temperatures. *Chem Mater.* 30(22):8179–8189. 10.1021/acs.chemmater.8b03276

[106] Farha OK, Yazaydın AÖ, Eryazici I, Malliakas CD, Hauser BG, Kanatzidis MG, Nguyen ST, Snurr RQ, Hupp JT (2010) De novo synthesis of a metal-organic framework material featuring ultrahigh surface area and gas storage capacities. *Nat Chem.* 2(11):944–948. 10.1038/nchem.834

[107] Farha OK, Wilmer CE, Eryazici I, Hauser BG, Parilla PA, O'Neill K, Sarjeant AA, Nguyen ST, Snurr RQ, Hupp JT (2012) Designing higher surface area metal-organic frame- works: are triple bonds better than phenyls? *J Am Chem Soc.* 134(24):9860–9863. 10.1021/ja302623w

[108] Yuan D, Zhao D, Sun D, Zhou HC (2010) An isoreticular series of metal-organic frameworks with dendritic hexacarboxylate ligands and exceptionally high gas-uptake capacity. *Angew Chem Int Ed.* 49(31):5357–5361. 10.1002/anie.201001009

[109] He Y, Zhou W, Qian G, Chen B (2014) Methane storage in metal-organic frameworks. *Chem Soc Rev.* 43(16):5657–5678. 10.1039/C4CS00032C

[110] Li B, Wen HM, Zhou W, Xu JQ, Chen B (2016) Porous metal-organic frameworks: Promising materials for methane storage. *Chem* 1(4):557–580. 10.1016/j.chempr.2016.09.009

[111] Peng Y, Krungleviciute V, Eryazici I, Hupp JT, Farha OK, Yildirim T (2013) Methane storage in metal-organic frameworks: Current records, surprise findings, and challenges. *J Am Chem Soc.* 135(32):11887–11894. 10.1021/ja4045289

[112] Li B, Wen HM, Wang H, Wu H, Tyagi M, Yildirim T, Zhou W, Chen B (2014) A porous metal-organic framework with dynamic pyrimidine groups exhibiting record high methane storage working capacity. *J Am Chem Soc.* 136(17):6207–6210. 10.1021/ja501810r

[113] Zhang M, Zhou W, Pham T, Forrest KA, Liu W, He Y, Wu H, Yildirim T, Chen B, Space B, Pan Y, Zaworotko MJ, Bai J (2017) Fine tuning of MOF-505 analogues to reduce low-pressure methane uptake and enhance methane working capacity. *Angew. Chem. Int. Ed.* 56(38):11426–11430. 10.1002/anie.201704974

[114] Wen HM, Li B, Li L, Lin RB, Zhou W, Qian G, Chen B (2018) A metal-organic framework with optimized porosity and functional sites for high gravimetric and volumetric methane storage working capacities. *Adv Mater.* 30(16):1704792. 10.1002/adma.201704792

[115] Lin JM, He CT, Liu Y, Liao PQ, Zhou DD, Zhang JP, Chen XM (2016) A metal-organic framework with a pore size/shape suitable for strong binding and close packing of methane. *Angew Chem Int Ed.* 55(15):4674–4678. 10.1002/anie.201511006

[116] Yan Y, Kolokolov DI, da Silva I, Stepanov AG, Blake AJ, Dailly A, Manuel P, Tang CC, Yang S, Schröder M (2017) Porous metal-organic polyhedral frameworks with optimal molecular dynamics and pore geometry for methane storage. *J Am Chem Soc.* 139(38):13349–13360. 10.1021/jacs.7b05453

[117] Zhang Z, Xiang S, Chen B (2011) Microporous metal-organic frameworks for acetylene storage and separation. *CrystEngComm* 13(20):5983–5992. 10.1039/C1CE05437F

[118] Xiang S, Zhou W, Gallegos JM, Liu Y, Chen B (2009) Exceptionally high acetylene uptake in a microporous metal–organic framework with open metal sites. *J Am Chem Soc.* 131(34):12415–12419. 10.1021/ja904782h

[119] Cai J, Wang H, Wang H, Duan X, Wang Z, Cui Y, Yang Y, Chen B, Qian G (2015) An amino-decorated NbO-type metal-organic framework for high C₂H₂ storage and selective CO₂ capture. *RSC Adv.* 5(94):77417–77422. 10.1039/C5RA12700A

[120] Wen HM, Wang H, Li B, Cui Y, Wang H, Qian G, Chen B (2016) A microporous metal-organic framework with lewis basic nitrogen sites for high C_2H_2 storage and significantly enhanced C_2H_2/CO_2 separation at ambient conditions. *Inorg Chem.* 55(15):7214–7218. 10.1021/acs.inorgchem.6b00748

[121] Moreau F, da Silva I, Al Smail NH, Easun TL, Savage M, Godfrey HG, Parker SF, Manuel P, Yang S, Schröder M (2017) Unravelling exceptional acetylene and carbon dioxide adsorption within a tetra-amide functionalized metal-organic framework. *Nat Commun.* 8:14085. 10.1038/ncomms14085

[122] Li JR, Kuppler RJ, Zhou HC (2009) Selective gas adsorption and separation in metal-organic frameworks. *Chem Soc Rev.* 38(5):1477–1504. 10.1039/B802426J

[123] Li JR, Sculley J, Zhou HC (2012) Metal-organic frameworks for separations. *Chem Rev.* 112(2): 869–932. 10.1021/cr200190s

[124] Herm ZR, Bloch ED, Long JR (2014) Hydrocarbon separations in metal-organic frameworks. *Chem. Mater.* 26(1):323–338. 10.1021/cm402897c

[125] Bao Z, Chang G, Xing H, Krishna R, Ren Q, Chen B (2016) Potential of microporous metal-organic frameworks for separation of hydrocarbon mixtures. *Energy Environ Sci.* 9(12):3612–3641. 10.1039/C6EE01886F

[126] Yoon JW, Chang H, Lee SJ, Hwang YK, Hong DY, Lee SK, Lee JS, Jang S, Yoon TU, Kwac K, Jung Y, Pillai RS, Faucher F, Vimont A, Daturi M, Férey G, Serre C, Maurin G, Bae YS, Chang JS (2017) Selective nitrogen capture by porous hybrid materials containing accessible transition metal ion sites. *Nat Mater.* 16(5):526–531. 10.1038/nmat4825

[127] Zhao X, Wang Y, Li DS, Bu X, Feng P (2018) Metal-organic frameworks for separation. *Adv Mater.* 30(37):1705189. 10.1002/adma.201705189

[128] Yang S, Ramirez-Cuesta AJ, Newby R, Garcia-Sakai V, Manuel P, Callear SK, Campbell SI, Tang CC, Schröder M (2014) Supramolecular binding and separation of hydrocarbons within a functionalized porous metal-organic framework. *Nat Chem.* 7(2):121–129. 10.1038/nchem.2114

[129] Lin RB, Li L, Zhou HL, Wu H, He C, Li S, Krishna R, Li J, Zhou W, Chen B (2018) Molecular sieving of ethylene from ethane using a rigid metal-organic framework. *Nat Mater.* 17(12):1128–1133. 10.1038/s41563-018-0206-2

[130] Liao PQ, Zhang WX, Zhang JP, Chen XM (2015) Efficient purification of ethene by an ethane-trapping metal-organic framework. *Nat Commun.* 6:8697. 10.1038/ncomms9697.

[131] Cadiau A, Adil K, Bhatt PM, Belmabkhout Y, Eddaoudi M (2016) A metal-organic framework–based splitter for separating propylene from propane. *Science* 353(6295):137–140. 10.1126/science.aaf6323

[132] Hazra A, Jana S, Bonakala S, Balasubramanian S, Maji TK (2017) Separation/ purification of ethylene from an acetylene/ethylene mixture in a pillared-layer porous metal-organic framework. *Chem. Commun.* 53(36):4907–4910. 10.1039/ C7CC00726D

[133] Wang H, Lashkari E, Lim H, Zheng C, Emge TJ, Gong Q, Yam K, Li J (2016) The moisture-triggered controlled release of a natural food preservative from a microporous metal-organic framework. *Chem Comm.* 52(10):2129–2132. 10.1039/C5 CC09634K

[134] Khan NA, Jung BK, Hasan Z, Jhung SH (2015) Adsorption and removal of phthalic acid and diethyl phthalate from water with zeolitic imidazolate and metalorganic frameworks. *J Hazard Mater.* 282:194–200. 10.1016/j.jhazmat.2014.03.047

[135] Reed DA, Xiao DJ, Gonzalez MI, Darago LE, Herm ZR, Grandjean F, Long JR (2016) Reversible CO scavenging *via* adsorbate-dependent spin state transitions in an Iron(II)-triazolate metal-organic framework. *J Am Chem Soc.* 138(17):5594–5602. 10.1021/jacs.6b00248

[136] Manousi N, Zachariadis GA, Deliyanni EA, Samanidou VF (2018) Applications of metal-organic frameworks in food sample preparation. *Molecules* 23(11):2896. 10.3390/molecules23112896

[137] Mashhadzadeh AH, Taghizadeh A, Taghizadeh M, Munir MT, Habibzadeh S, Salmankhani A, Stadler FJ, Saeb MR (2020) Metal-organic framework (MOF) through the lens of molecular dynamics simulation: Current status and future perspective. *J. Compos. Sci.* 4(2):75. 10.3390/jcs4020075

[138] Sharanyakanth PS, Radhakrishnan M (2020) Synthesis of metal-organic frameworks (MOFs) and its application in food packaging: A critical review. *Trends Food Sci Technol.* 104:102–116. 10.1016/j.tifs.2020.08.004

[139] Smaldone RA, Forgan RS, Furukawa H, Gassensmith JJ, Slawin AM, Yaghi OM, Stoddart JF (2010) Metal-organic frameworks from edible natural products. *Angew Chem Int Ed Engl.* 49(46):8630–8634. 10.1002/anie.201002343

[140] Chopra S, Dhumal S, Abeli P, Beaudry R, Almenar E (2017) Metal-organic frameworks have utility in adsorption and release of ethylene and 1-methylcyclopropene in fresh produce packaging. *Postharvest Biol Technol.* 130:48–55. 10.1016/j.postharvbio.2017.04.001

[141] Gutierrez TJ, Ponce AG, Alvarez VA (2017) Nano-clays from natural and modified montmorillonite with and without added blueberry extract for active and intelligent food nanopackaging materials. *Mater Chem Phys.* 194:283–292. 10.1016/j.matchemphys.2017.03.052

[142] Gutierrez TJ (2018) Active and intelligent films made from starchy sources/ blackberry pulp. *J Polym Environ.* 26(6):2374–2391. 10.1007/s10924-017-1134-y

[143] Duan C, Meng J, Wang X, Meng X, Sun X, Xu Y, Zhao W, Ni Y (2018) Synthesis of novel cellulose-based antibacterial composites of Ag nanoparticles@metal-organic frameworks@ carboxymethylated fibers. *Carbohydr Polym.* 193:82–88. 10.1016/j.carbpol.2018.03.089

[144] Wu Y, Luo Y, Zhou B, Mei L, Wang Q, Zhang B (2019) Porous metal-organic framework (MOF) carrier for incorporation of volatile antimicrobial essential oil. *Food Control* 98:174–178. 10.1016/j.foodcont.2018.11.011

[145] Lashkari E, Wang H, Liu L, Li J, Yam K (2017) Innovative application of metal-organic frameworks for encapsulation and controlled release of allyl isothio-cyanate. *Food Chem.* 221:926–935. 10.1016/j.foodchem.2016.11.072

[146] Ma P, Zhang J, Liu P, Wang Q, Zhang Y, Song K, Li R, Shen L (2020) Computer-assisted design for stable and porous metal-organic framework (MOF) as a carrier for curcumin delivery. *LWT–Food Sci Technol.* 120:108949. 10.1016/j.lwt.2019.108949

[147] Hashemi B, Zohrabi P, Raza N, Kim KH (2017) Metal-organic frameworks as advanced sorbents for the extraction and determination of pollutants from environmental, biological, and food media. *Trends Analyt Chem.* 97:65–82. 10.1016/j.trac.2017.08.015

[148] Rocío-Bautista P, Termopoli V (2019) Metal-organic frameworks in solid-phase extraction procedures for environmental and food analyses. *Chromatographia* 82(8):1191–1205. 10.1007/s10337-019-03706-z

[149] Zhang Y, Li G, Wu D, Li X, Yu Y, Luo P, Chen J, Dai C, Wu Y (2019) Recent advances in emerging nanomaterials based food sample pretreatment methods for food safety screening. *Trends Analyt Chem.* 121:115669. 10.1016/j.trac.2019.115669

[150] Villa CC, Sanchez LT, Valencia GA, Ahmed S, Gutierrez TJ (2021) Molecularly imprinted polymers for food applications: A review. *Trends Food Sci Technol.* 111:642–669. 10.1016/j.tifs.2021.03.003

[151] Lizotte PH, Wen AM, Sheen MR, Fields J, Rojanasopondist P, Steinmetz NF, Fiering S (2016) *In situ* vaccination with cowpea mosaic virus nanoparticles suppresses metastatic cancer. *Nat Nanotechnol.* 11(3):295–303.10.1038/nnano.2015.292

[152] Horcajada P, Chalati T, Serre C, Gillet B, Sebrie C, Baati T, Eubank JF, Heurtaux D, Clayette P, Kreuz C, Chang JS, Hwang YK, Marsaud V, Bories PN, Cynober L, Gil S, Férey G, Couvreur P, Gref R (2010) Porous metal-organic-framework nanoscale carriers as a potential platform for drug delivery and imaging. *Nat Mater.* 9(2):172–178. 10.1038/nmat2608

[153] Lu G, Li S, Guo Z, Farha OK, Hauser BG, Qi X, Wang Y, Wang X, Han S, Liu X, DuChene JS, Zhang H, Zhang Q, Chen X, Ma J, Loo SC, Wei WD, Yang Y, Hupp JT, Huo F (2012) Imparting functionality to a metal-organic framework material by controlled nanoparticle encapsulation. *Nat Chem.* 4(4):310–316. 10.1038/nchem.1272

[154] Li Y, Tang J, He L, Liu Y, Liu Y, Chen C, Tang Z (2015) Core-shell upconversion nanoparticle@metal-organic framework nanoprobes for luminescent/magnetic dual-mode targeted imaging. *Adv Mater.* 27(27):4075–4080. 10.1002/adma.201501779

[155] Zheng H, Zhang Y, Liu L, Wan W, Guo P, Nyström AM, Zou X (2016) One-pot synthesis of metal-organic frameworks with encapsulated target molecules and their applications for controlled drug delivery. *J Am Chem Soc.* 138(3):962–968. 10.1021/jacs.5b11720

[156] Wang HS, Wang YH, Ding Y (2020) Development of biological metal-organic frameworks designed for biomedical applications: from bio-sensing/bio-imaging to disease treatment. *Nanoscale Adv.* 2(9):3788–3797. 10.1039/D0NA00557F

[157] Feng D, Liu TF, Su J, Bosch M, Wei Z, Wan W, Yuan D, Chen YP, Wang X, Wang K, Lian X, Gu ZY, Park J, Zou X, Zhou HC (2015) Stable metal-organic frameworks containing single-molecule traps for enzyme encapsulation. *Nat Commun.* 6:5979. 10.1038/ncomms6979

[158] Bilal M, Adeel M, Rasheed T, Iqbal HMN (2019) Multifunctional metal-organic frameworks-based biocatalytic platforms: Recent developments and future prospects. *J Mater Res Technol.* 8(2):2359–2371. 10.1016/J.JMRT.2018.12.001

[159] Lykourinou V, Chen Y, Wang XS, Meng L, Hoang T, Ming LJ, Musselman RL, Ma S (2011) Immobilization of MP-11 into a mesoporous metal-organic framework, MP-11@mesoMOF: A new platform for enzymatic catalysis. *J Am Chem Soc.* 133(27):10382–11385. 10.1021/ja2038003

[160] Lyu F, Zhang Y, Zare RN, Ge J, Liu Z (2014) One-pot synthesis of protein-embedded metal-organic frameworks with enhanced biological activities. *Nano Lett.* 14(10):5761–5765. 10.1021/nl5026419

[161] Liang K, Ricco R, Doherty CM, Styles MJ, Bell S, Kirby N, Mudie S, Haylock D, Hill AJ, Doonan CJ, Falcaro P (2015) Biomimetic mineralization of metal-organic frameworks as protective coatings for biomacromolecules. *Nat Commun.* 6:7240. 10.1038/ncomms8240

[162] Zhang Y, Wang F, Ju E, Liu Z, Chen Z, Ren R, Qu X (2016) Metal-organic-framework-based vaccine platforms for enhanced systemic immune and memory response. *Adv. Funct. Mater.* 26(35): 6454–6461. 10.1002/adfm.201600650

[163] Wu Y, Huang Z, Jiang H, Wang C, Zhou Y, Shen W, Xu H, Deng H (2019) Facile synthesis of uniform metal carbide nanoparticles from metal-organic frameworks by laser metallurgy. *ACS Appl Mater Interfaces* 11(47):44573–44581. 10.1021/acsami.9b13864

[164] Chen H, Shen K, Mao Q, Chen J, Li Y (2018) Nanoreactor of MOF-derived yolk–shell Co@C–N: Precisely controllable structure and enhanced catalytic activity. *ACS Catal.* 8(2):1417–1426. 10.1021/acscatal.7b03270

[165] Wang Y, Pan Y, Zhu L, Yu H, Duan B, Wang R, Zhang Z, Qiu S (2019) Solvent-free assembly of Co/Fe-containing MOFs derived N-doped mesoporous carbon nanosheets for ORR and HER. *Carbon* 146:671–679. 10.1016/j.carbon.2019.02.002

[166] Guan BY, Lu Y, Wang Y, Wu M, Lou XWD (2018) Porous Iron–Cobalt alloy/nitrogen-doped carbon cages synthesized via pyrolysis of complex metal-organic framework hybrids for oxygen reduction. *Adv Funct Mater.* 28(10):1706738. 10.1002/adfm.201706738

[167] Ning H, Li G, Chen Y, Zhang K, Gong Z, Nie R, Hu W, Xia Q (2019) Porous N-doped carbon-encapsulated CoNi alloy nanoparticles derived from MOFs as efficient bifunctional oxygen electrocatalysts. *ACS Appl Mater Interfaces* 11(2):1957–1968. 10.1021/acsami.8b13290

[168] Jin J, Zheng Y, Huang S, Sun P, Srikanth N, Kong LB, Yan Q, Zhou K (2019) Directly anchoring 2D NiCo metal-organic frameworks on few-layer black phosphorus for advanced lithium-ion batteries. *J Mater Chem A* 7(2):783–790. 10.1039/C8TA09327J

[169] Zhang Z, Feng J, Ci L, Tian Y, Xiong S (2016) Mental-organic framework derived CuO hollow spheres as high performance anodes for sodium ion battery. *Mater Technol.* 31(9):497–500. 10.1080/10667857.2016.1189024

[170] Liu G, Feng K, Cui H, Li J, Liu Y, Wang M (2020) MOF derived in-situ carbon-encapsulated Fe$_3$O$_4$@C to mediate polysulfides redox for ultrastable Lithium-sulfur batteries. *Chem Eng J.* 381:122652. 10.1016/j.cej.2019.122652

[171] Yang Y, Fu Q, Zhao H, Mi Y, Li W, Dong Y, Wu M, Lei Y (2018) MOF-assisted three-dimensional TiO$_2$@C core/shell nanobelt arrays as superior sodium ion battery anodes. *J Alloys Compd.* 769:257–263. 10.1016/j.jallcom.2018.07.366

[172] Pang Y, Chen S, Xiao C, Ma S, Ding S (2019) MOF derived CoO-NCNTs two-dimensional networks for durable lithium and sodium storage. *J Mater Chem A* 7(8):4126–4133. 10.1039/C8TA10575H

[173] Zou F, Hu X, Li Z, Qie L, Hu C, Zeng R, Jiang Y, Huang Y (2014) MOF-derived porous ZnO/ZnFe$_2$O$_4$/C octahedra with hollow interiors for high-rate lithium-ion batteries. *Adv Mater.* 26(38):6622–6628. 10.1002/adma.201402322

[174] Zhang X, Qin W, Li D, Yan D, Hu B, Sun Z, Pan L (2015) Metal-organic framework derived porous CuO/Cu$_2$O composite hollow octahedrons as high performance anode materials for sodium ion batteries. *Chem Commun.* 51(91):16413–16416. 10.1039/C5CC06924F

[175] Kim AY, Kim MK, Cho K, Woo JY, Lee Y, Han SH, Byun D, Choi W, Lee JK (2016) One-step catalytic synthesis of CuO/Cu$_2$O in a graphitized porous c matrix derived from the Cu-based metal-organic framework for Li- and Na-ion batteries. *ACS Appl Mater Interfaces* 8(30):19514–19523. 10.1021/acsami.6b05973

[176] Kaneti YV, Zhang J, He YB, Wang Z, Tanaka S, Hossain MSA, Pan ZZ, Xiang B, Yang QH, Yamauchi Y (2017) Fabrication of an MOF-derived heteroatom-doped Co/CoO/carbon hybrid with superior sodium storage performance for sodium-ion batteries. *J Mater Chem A* 5(29):15356–15366. 10.1039/C7TA03939E

[177] Zhang W, Cao P, Zhang Z, Zhao Y, Zhang Y, Li L, Yang K, Li X, Gu L (2019) Nickel/cobalt metal-organic framework derived 1D hierarchical NiCo$_2$O$_4$/NiO/carbon nanofibers for advanced sodium storage. *Chem Eng J.* 364:123–131. 10.1016/j.cej.2019.01.144

[178] Wu X, Leonard DP, Ji X (2017) Emerging non-aqueous potassium-ion batteries: Challenges and opportunities. *Chem Mater.* 29(12):5031–5042. 10.1021/acs.chemmater.7b01764

[179] Yao Q, Zhang J, Shi X, Deng B, Hou K, Zhao Y, Guan L (2019) Rational synthesis of two-dimensional G@porous FeS$_2$@C composite as high-rate anode materials for sodium/potassium ion batteries. *Electrochim Acta* 307:118–128. 10.1016/j.electacta.2019.03.184

[180] Ji B, Yao W, Zheng Y, Kidkhunthod P, Zhou X, Tunmee S, Sattayaporn S, Cheng HM, He H, Tang Y (2020) A fluoroxalate cathode material for potassium-ion batteries with ultra-long cyclability. *Nat Commun.* 11(1):1225. 10.1038/s41467-02 0-15044-y

[181] Lee J, Kim S, Park J, Jo C, Chun J, Sung Y, Lim E, Lee J (2020) A small-strain niobium nitride anode with ordered mesopores for ultra-stable potassium-ion batteries. *J Mater Chem A* 8(6):3119–3127. 10.1039/C9TA11663J

[182] Ettlinger R, Lächelt U, Gref R, Horcajada P, Lammers T, Serre C, Couvreur P, Morris RE, Wuttke S (2022) Toxicity of metal-organic framework nanoparticles: From essential analyses to potential applications. *Chem Soc Rev.* 51(2):464–484. 10.1039/D1CS00918D

[183] Barea E, Montoro C, Navarro JA (2014) Toxic gas removal – Metal-organic frameworks for the capture and degradation of toxic gases and vapours. *Chem Soc Rev.* 43(16):5419–5430. 10.1039/C3CS60475F)

[184] Sajid M (2016) Toxicity of nanoscale metal organic frameworks: A perspective. *Environ Sci Pollut Res Int.* 23(15):14805–14807. 10.1007/s11356-016-7053-y

[185] Kumar P, Anand B, Tsang YF, Kim KH, Khullar S, Wang B (2019) Regeneration, degradation, and toxicity effect of MOFs: Opportunities and challenges. *Environ Res.* 176:108488. 10.1016/j.envres.2019.05.019

[186] Mohanty P, Linn NM, Landskron K (2010) Ultrafast sonochemical synthesis of methane and ethane bridged periodic mesoporous organosilicas. *Langmuir.* 26(2): 1147–1151. 10.1021/la902239m

[187] Dharmarathna S, King'ondu CK, Pedrick W, Pahalagedara L, Suib SL (2012) Direct sonochemical synthesis of manganese octahedral molecular sieve (OMS-2) nanomaterials using cosolvent systems, their characterization, and catalytic applications. *Chem. Mater.* 24(4):705–712. 10.1021/cm203366m

Index

Note: *Italicized* and **bold** page numbers refer to figures and tables.

313

For Product Safety Concerns and Information please contact our EU
representative GPSR@taylorandfrancis.com
Taylor & Francis Verlag GmbH, Kaufingerstraße 24, 80331 München, Germany